EXPLORING
EARTH
SCIENCE

ROBERT E. KILBURN
Science Curriculum Coordinator—Newton, Massachusetts

PETER S. HOWELL
Junior High Science Teacher—Needham, Massachusetts

D1449776

ALLYN AND BACON, INC.

Boston   Rockleigh, N.J.   Atlanta   Dallas   Belmont, Calif.

ROBERT E. KILBURN has taught junior high science for nine years full-time, and for 16 years part-time in his present position as Science Curriculum Coordinator for the Newton, Massachusetts public schools. Prior to teaching, Dr. Kilburn spent four years as a science researcher, the last two as research biophysicist at the General Electric Research Laboratory.

PETER S. HOWELL has taught junior high science for 16 years. Besides teaching, Mr. Howell has been a contributing author and editor of textbooks in science, and has written science materials for educational television.

Cover photo: NASA

Title page photo: WESKEMP: Alain Perceval

Editor: Gene Moulton
Designer: Debby Welling
Photo Researcher: Janice Thalin
Preparation Buyer: Patricia Hart

ISBN: 0-205-06733-6

Printed in the United States of America.

Library of Congress Catalog Card Number 79-54811

1 2 3 4 5 6 7 8 9 88 87 86 85 84 83 82 81 80

PREFACE

The recent changes in science teaching are a response to the mushrooming growth of scientific information. Teachers can no longer depend upon easy generalizations as in the past; ideas become obsolete almost as soon as they are presented. Therefore, modern science programs emphasize the processes by which information is obtained and made meaningful. Such programs help young people "uncover" science rather than "cover" it.

Modern approaches to science teaching demand modern textbooks. This revision of EXPLORING EARTH SCIENCE has the same two goals which made the previous editions successful: (1) to bring young people into contact with their environment in such a way as to stimulate a desire to investigate, and (2) to provide them with an understanding of the methods and philosophies of science so that they can carry out investigations with a minimum of direct guidance from others.

This major revision has incorporated the suggestions of hundreds of teachers who used the previous editions. Their suggestions included organizing the book into fewer topics. This change allows classes more time to focus on each major theme. Each major theme, or unit, has been divided into three parts. The first part consists of several chapters of guided investigations, designed to give students an accurate sense of the nature of science and its limitations. The second part of each unit consists of one chapter devoted to having students work independently on a research project of their own design. There can be no better way to understand the strengths and weaknesses of scientific investigation than to experience it first-hand. The final chapters of each unit utilize readings to extend the learnings of earlier chapters, and deal with important topics not possible to study first-hand. This unit structure allows students to (1) learn the skills of scientific investigation and knowledge-building, (2) practice the skills independently, and (3) utilize their understandings in readings on related topics. Critical attitudes and habits developed through independent thinking will stir continued interest in learning long after the years of formal schooling have ended.

The present authors are indebted to Dr. Walter A. Thurber whose vision and efforts led to the original editions of this series. We hope the present editions maintain the high quality associated with all his work.

CONTENTS

GEOLOGY 1

GEOLOGY

Minerals and Their Uses

The solid part of the earth is made up of rocks. Rocks in turn are made up of minerals, and minerals are composed of chemical elements.

A few minerals are composed of only one element; gold and sulfur are examples. Most minerals, however, are compounds of

two or more elements: red iron ore is a compound of iron and oxygen.

On these four pages are thirteen important minerals. Which ones can you identify from the pictures? Learn to identify all of them.

KEY TO PICTURES

1. Pyrite (fool's gold) 2. Garnet 3. Hematite (red iron ore) 4. Talc 5. Feldspar 6. Mica 7. Quartz 8. Halite (table salt) 9. Gypsum 10. Galena (lead ore) 11. Calcite 12. Magnetite (black iron ore) 13. Graphite

BECOMING ACQUAINTED WITH MINERALS

To begin the study of minerals, collect samples of all the minerals illustrated on the previous two pages. Add specimens of other minerals that are common in your region. Learn to recognize and name each specimen.

It is helpful to have your own collection of minerals. Begin with specimens found locally. Add others by purchase or trade when possible.

General Appearances. Pick up a specimen and examine all sides of it closely. Note its general appearance. Note the way it feels in your hands.

Don't be surprised if some specimens look unlike photographs of the same minerals in this book. Many minerals show various colors. Some have two or more forms.

Compare minerals with each other. Which ones are easy to identify by general appearance? Decide why these specimens are easy to identify.

Studying Differences in Minerals. Minerals are often identified by looking for their differences. What differences have you already noticed? List these.

Choose one mineral for special study. In what ways is this mineral different from other minerals? Study your specimens the same way.

A Suggestion for Record Keeping. Make a chart like the one below. At the tops of the columns list the differences you have noted. You will probably need to add more columns.

Fill in the columns with words describing the minerals. For example, under density tell whether a mineral feels light or heavy when compared with other minerals of the same size. Test hardness by trying to scratch one specimen with another.

Self-Testing. Test yourself in several ways to find out how well you know your specimens. Name the specimens as you pick them up. Find the specimens as someone else calls out the names. Write down the names as someone else holds up the specimens.

Mineral	General Appearance	Density	Hardness	Color
Galena	Like Metal	Heavy	Soft	Bluish Gray
Mica				
Hematite				

Pyrite—Metallic luster

Feldspar—Glassy luster

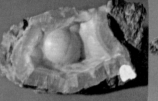
Chalcedony—Waxy luster

Limonite—Earthy luster

IDENTIFICATION TESTS

Some of the differences between minerals can be determined at a glance. Other differences must be discovered by using tests. Several tests used by geologists are simple and can be used anywhere. Some of these are described in the following paragraphs. More complicated tests are described in books about minerals.

Color. A few minerals have only one color; color is useful in their identification. However, most minerals may be found in several colors because of impurities. Colors should be noted as a help in recognizing the different minerals.

Luster. The way in which light is reflected from the surface of a mineral is called its *luster*. A mineral that reflects light like a piece of polished metal is said to have a *metallic luster*. Other lusters are *glassy, waxy, satiny, pearly, greasy,* and *earthy.*

Describe the luster of each of the minerals you have studied.

Hematite

Streak. The picture here shows the streak made when a piece of hematite is scratched on a rough white surface. Note the difference in color between the specimen and its streak.

The color of the specimen is due to the light that reflects from its surface. The color of the streak is due to light passing through the tiny particles that make up the streak. The light passes through the particles, reflects from the white surface, and passes back through the particles.

Several minerals can be identified by their streak. Colors of specimens of the same mineral may be different but the color of the streak is always the same.

For streak tests, you can use the backs of ceramic tiles like those that line bathroom walls. Get them from building contractors or suppliers.

6

Fingernail 2.5

Penny 3

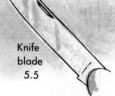

Window glass 5.5

Knife blade 5.5

Steel file 6.5

HARDNESS SCALE

Mineral	Simple Test
1. Talc	1. Fingernail scratches it easily.
2. Gypsum	2. Fingernail barely scratches it.
3. Calcite	3. Copper penny just scratches it.
4. Fluorite	4. Steel knife scratches it easily.
5. Apatite	5. Steel knife scratches it.
6. Feldspar	6. Steel knife does not scratch it; it scratches window glass easily.
7. Quartz	7. Hardest common mineral; it scratches steel and hard glass easily.
8. Topaz	8. Harder than any common mineral.
9. Corundum	9. It scratches topaz.
10. Diamond	10. Hardest of all minerals.

Higher numbered minerals scratch lower numbered ones.

Hardness. Minerals usually differ in their hardness. Therefore, hardness tests are important in the identification of minerals.

Hardness is usually determined by trying to scratch one specimen with another. Geologists have devised a scale ranging from one to ten. They have based this scale on ten minerals beginning with talc as number one and ending with diamond as number ten.

You do not need a complete set of the ten minerals to test hardness. You can use the common materials shown above to find the hardness of most specimens.

When making a test, draw a sharp corner of one material across the flat face of another. Then try to rub off any dust to make sure whether or not a scratch was made.

Magnet Test. Magnetite, a black iron ore, is the only common mineral that is attracted to a magnet. To test a mineral for this property, break off small pieces and touch them with a strong magnet. If they cling to the magnet, they may be pieces of magnetite.

Feel. Rub your fingers over the surface of a mineral and note any unusual feeling. Most minerals are either rough or smooth, but some feel gritty, some powdery, and some greasy. Graphite, shown at the right, feels greasy.

Magnetite

Graphite

7

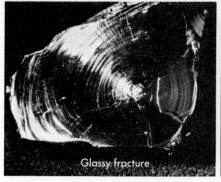
Glassy fracture

Grainy fracture

Earthy fracture

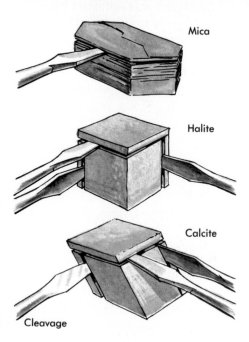
Mica

Halite

Calcite

Cleavage

Halite

Breaking Tests. The way a mineral breaks apart is often an important clue to its identification. The kind of break is determined by the forces that hold the atoms together.

Some minerals break irregularly, leaving rough surfaces. This is called *fracture*. Other minerals split apart cleanly leaving smooth, flat surfaces. This is called *cleavage*. A mineral may cleave in one direction and fracture in all other directions.

A few minerals can be broken apart with the fingers. These are called weak minerals. Stronger minerals can only be broken with a hammer.

Types of Fracture. Many minerals break with rough or grainy fracture. A few break along curved surfaces as glass breaks. A few of the types of fracture are shown above.

Cleavage. Cleavage may be in one, two, three, or more directions. The angles between the cleavage surfaces may also differ.

Mica cleaves in only one direction and fractures in all others. Therefore, a block of mica splits up into thin sheets. Each sheet fractures when bent or twisted.

Halite (table salt) cleaves in three directions, each at right angles to the others. Therefore, halite breaks up into cubes or rectangular blocks.

Calcite also cleaves in three directions, but the surfaces are not at right angles. Therefore, calcite breaks up into blocks that lean to one side.

Specific Gravity. Equal-sized pieces of two different minerals may not weigh the same. One mineral is said to have a greater specific gravity than the other.

The *specific gravity* of a substance is the weight of a sample compared with the weight of an equal volume of water. For example, a mineral with a specific gravity of 3.0 is three times as heavy as the same volume of water.

TESTING FOR CARBONATE (such as calcite). Add 2 or 3 drops of hydrochloric acid. If bubbles are given off, a carbonate is present.

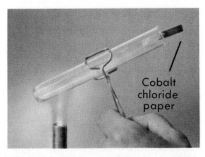

TESTING FOR WATER IN A MINERAL. Heat some small pieces in a test tube. If clear drops of liquid condense in the cool part of the tube, test them with blue cobalt chloride paper. If the paper turns pink, the drops contain water.

MAKING COBALT CHLORIDE PAPER. Make a concentrated solution of cobalt chloride in water. Dip strips of white paper towel into the red solution. When the strips dry (turn blue) they can be used to test for water.

To find the specific gravity of a mineral, first weigh a specimen. Then weigh an equal volume of water. Compare the two weights.

$$\text{Specific gravity} = \frac{\text{Weight of specimen}}{\text{Weight of an equal volume of water}}$$

The picture above shows a way to get a volume of water equal to that of a specimen being tested. Fill a plastic cup with water up to a hole in the side. Place the specimen in the cup. The water which is forced out through the hole has the same volume as the specimen. Collect and weigh the water. (Be sure to subtract the weight of the container.)

Chemical Tests. For a simple and useful test, put two or three drops of dilute hydrochloric acid on a specimen. (**CAUTION:** Do not get hydrochloric acid on your hands or clothes.) If bubbles are given off, the specimen contains some kind of carbonate, probably calcium carbonate. Calcite is crystalline calcium carbonate.

Other tests include heating a specimen to discover what products are given off by the chemical breakdown of the mineral. Cobalt chloride paper can be used to find out whether water is among the products.

A mineral might dissolve in water. Litmus paper can be used to find out if the solution is acidic or basic. Other chemical tests will be described later in this chapter.

SUMMARY QUESTIONS

1. If you found four minerals, how would you decide which is the hardest?
2. Why is a streak test more useful than knowing a mineral's color?

Surface of weathered flint

Flint

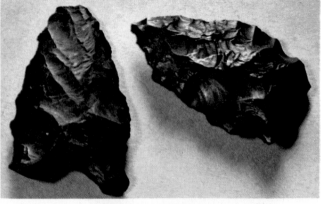

Arrowheads of flint

USEFUL PROPERTIES OF MINERALS

The usefulness of a mineral depends upon its properties, either physical or chemical or both. Physical properties include hardness, color, and cleavage. Chemical properties include solubility and reactions when heated.

Early humans used large quantities of flint because flint is hard and fractures into sharp-edged chips. It made the best cutting tools available at the time. Clay also came into early use because it fuses into a glassy mass when heated to a very high temperature, thus serving for bricks and pottery. Through the centuries, we have steadily discovered new uses for the special properties of minerals.

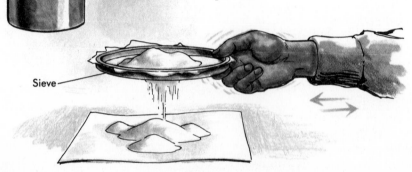

Pipe fittings

Sieve

Talc. Test the hardness of talc with a fingernail. Scrape off a little powder and note the feel of the powder between your fingers.

Make some talcum powder. Grind up a piece of talc in a mortar as shown at the left. Sift the powder through a nylon stocking. Stretch the cloth across embroidery hoops as shown at the left.

Pink talcum powder can be made by grinding a small piece of hematite with the talc. For scented talcum powder, add a drop of perfume to the powder.

Halite. This important mineral is commonly known as table salt. Its chemical name is *sodium chloride*. This tells us that the mineral is a compound of sodium and chlorine. The chemical formula is NaCl.

Electrolysis of Sodium Chloride. Connect two pieces of pencil lead to two dry cells in series. Attach these to a plastic bottle cap or small glass dish as shown below.

Fill the container with table salt. Add water drop by drop until the salt is moist and has a thin layer of water on top.

Watch the reaction. Note the odor given off. Test the liquid by each pencil lead (or pole) with red and blue litmus paper. Describe the changes in the litmus paper.

Products of Electrolysis. Sodium chloride (NaCl) and water (H_2O) are both broken up as an electric current passes through them. Some important chemicals are produced.

1. Chlorine gas is given off at one pole. This can be identified by its odor. At which pole is chlorine gas forming? Where have you noted this odor before? From which compound did the chlorine come?

2. Bubbles of hydrogen are given off at the other pole. From which compound did the hydrogen come?

3. A solution of sodium hydroxide (NaOH) is produced at one pole. Sodium hydroxide is a base. It turns red litmus paper blue. At which pole is sodium hydroxide produced? Find out how this chemical is used.

4. Some of the chlorine unites with water to form a strong bleach called hypochlorous acid (HOCl). At which pole is it formed? What does it do to the litmus paper? How are bleaches used?

Interior of a salt mine.

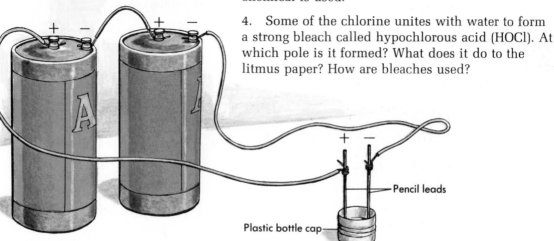
Blue litmus
Red litmus
Pencil leads
Plastic bottle cap

11

A scene in the White Sands National Monument where the sand is made up of almost pure gypsum.

Gypsum. Gypsum is usually found as a solid, dull-looking rock of white or gray color. Sometimes it is found as large, clear crystals called *selenite*.

The chemical name for gypsum is *hydrated calcium sulfate*. Its chemical formula is $CaSO_4 \cdot 2H_2O$. The name and formula tell us that gypsum is a compound of calcium, sulfur, and oxygen combined with two molecules of water. The two molecules of water are called *water of crystallization*.

Dehydrating Gypsum. Break up some gypsum into small pieces. Heat them in a test tube over a flame.

Note any changes that take place in the appearance of the gypsum. Look for drops of liquid in the top of the test tube. Test these drops with cobalt chloride paper. What is the liquid? What is happening to the gypsum?

Dump the contents of the test tube on a metal or glass surface. When they have cooled, try to crush some of the pieces between your fingers. Compare them with pieces of the original gypsum.

Why is the powder called *dehydrated* gypsum?

Uses of Gypsum. Gypsum has been called the second most used building material, following wood in importance. Large quantities of gypsum are used in the manufacture of plaster and plasterboard. How are these used in buildings?

Gypsum is cheap as well as useful. It is plentiful and easily mined. Much of it is near the surface where it can be scooped up at small expense.

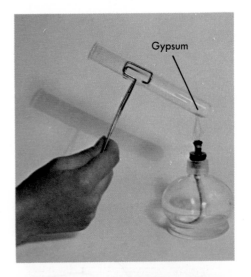

Gypsum

Gypsum

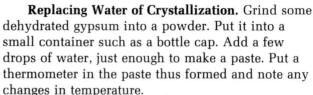

Replacing Water of Crystallization. Grind some dehydrated gypsum into a powder. Put it into a small container such as a bottle cap. Add a few drops of water, just enough to make a paste. Put a thermometer in the paste thus formed and note any changes in temperature.

Half an hour later examine the paste. What has happened to it? Explain the changes in the paste in terms of water of crystallization.

Plaster of Paris. Dehydrated gypsum is sold as plaster of Paris. It is sometimes used for plastering the inside walls of buildings. It is also used by doctors for making the casts that keep broken bones from moving. Plaster of Paris can be purchased in hardware stores and drug stores.

Hydrating Plaster of Paris. Add plaster of Paris to water, stirring it constantly, until the mixture is about like pancake batter. Then pour the mixture into a paper cup.

Observe the plaster over a period of several hours. What change takes place during the first fifteen minutes? Note any temperature changes in the plaster. When is the plaster first hard enough to keep from flowing if the dish is tipped slightly? When is it hard enough to keep from crumbling when the edges are pinched?

Casts and Molds. The diagrams below show how to make a cast of a leaf. The leaf is pressed into soft clay and then removed. The impression of the leaf in the clay is called a *mold*.

Plaster of Paris is then poured into the mold. When the plaster has hardened, and has been removed from the mold, the plaster forms a *cast* of the leaf.

Plaster of Paris is used to make casts of animal footprints in mud. Detectives also use plaster of Paris to make casts of human footprints and auto tire prints.

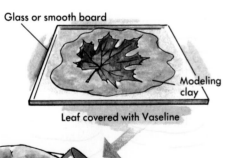

Glass or smooth board

Modeling clay

Leaf covered with Vaseline

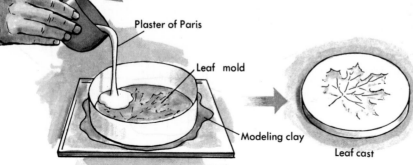

Plaster of Paris

Leaf mold

Modeling clay

Leaf cast

Pyrite

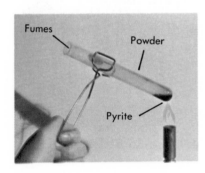

Fumes

Powder

Pyrite

Pyrite. This golden, metallic-looking mineral has fooled so many people that it has been named "fool's gold." It is not a metal, however, as can be determined by pounding it. What does pounding do to a metal? What does pounding do to pyrite?

Pyrite gives off large sparks when it is struck by a hammer. These sparks are responsible for the scientific name of the mineral. *Pyrite* is taken from the old Greek word meaning "fire."

Roasting Pyrite. Crush some pyrite to a powder. Note the sparks when the pyrite is struck. Note the odor.

Put a teaspoonful of the powder in a test tube. Heat it red hot. Note any changes inside the test tube.

Moisture may collect first in the top of the test tube. Then a solid may collect. What is the color of this solid?

Note the odor coming from the test tube. (**CAUTION:** Do not hold the test tube directly under your nose. Instead, fan some of the fumes toward your face with your hand.) Do you know what the odor is?

Dump the contents of the test tube on a flat rock. After the powder is cool, test it with a magnet. What happens?

What two elements have you found in pyrite? Chemists tell us that these are the only two elements in the mineral. Look up the chemical name of the mineral in a dictionary or chemistry book. Do your findings agree with the chemical name and formula?

Vermiculite. Vermiculite is an interesting and useful relative of mica. It looks much like mica but contains water.

Break some vermiculite into small pieces and put them into a test tube. Heat them over a flame. Note what happens. Look for moisture near the top of the test tube.

What does heating do to the moisture in the vermiculite? Why does the vermiculite expand? In what ways does vermiculite resemble popcorn and puffed wheat?

Examine the puffed vermiculite. Why do you think it is sometimes used as a heat and sound insulator in buildings? What other uses does it have?

Making Lime. Crush some calcite (calcium carbonate) into a powder. Put a quarter teaspoonful

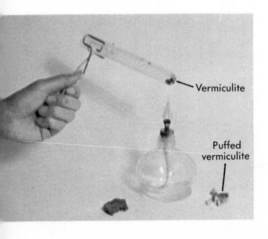

Vermiculite

Puffed vermiculite

of the powder into a test tube and heat it with a gas flame until it glows red for five or ten minutes. While it is heating, thrust a lighted splint of wood into the test tube. Note the result.

Let the test tube cool. Put a thermometer in the powder, add two or three drops of water, and note any changes in temperature. Test the liquid in the test tube with litmus paper.

Calcium Oxide. Calcium carbonate breaks up chemically when heated to a very high temperature. Carbon dioxide is given off and calcium oxide remains. Discuss this chemical equation for the change:

$$CaCO_3 \rightarrow CO_2 + CaO$$

Calcium oxide unites readily with water to form calcium hydroxide, $Ca(OH)_2$. Write a chemical equation for this change.

Calcium hydroxide dissolved in water is called limewater. It is used to test for carbon dioxide. When carbon dioxide is bubbled through limewater, it forms a milky mixture of calcium carbonate. Write a chemical equation for this change.

Lime Mortar. Before portland cement was invented, bricks and building stones were held together by a mortar of sand and calcium hydroxide. The mortar hardened slowly, uniting with carbon dioxide from the air to produce a kind of limestone. Portland cement is used today because it hardens faster.

Powdered calcium carbonate

Rotating kiln for making calcium oxide. Crushed marble (calcium carbonate) tumbles down the tube. Flames heat the marble and drive off the carbon dioxide.

Mining iron ore from
an open pit mine.

Iron Ores. There are three common iron ores—magnetite, hematite, and limonite. All three are compounds of iron and oxygen; they are iron oxides.

To obtain iron from these ores, you have to remove the oxygen. The process of removing oxygen from a compound is called *reduction*. Reduction yields a metal such as iron uncombined with any other element.

Iron ores are reduced with some form of carbon, either charcoal or coke. At high temperatures, carbon unites with oxygen in the ore leaving the iron as a metal.

Hematite. Hematite varies in color from earthy red to metallic black, but its streak is always reddish brown. Red rocks and soils usually owe their color to hematite.

The chemical formula for hematite is Fe_2O_3. Explain what this formula means in terms of atoms.

Reducing Hematite. Grind some hematite to a powder. Test the powder with a magnet. Is it magnetic?

Grind some charcoal to a powder. Mix equal quantities of charcoal and hematite powders. Put a teaspoonful of the mixture into a test tube.

Heat the mixture with a gas flame until it glows red hot. Then put a glowing splint of wood into the test tube. What happens? What conclusion can you draw?

Dump the hot mixture on a pottery plate or flat

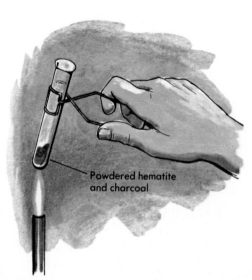

Powdered hematite
and charcoal

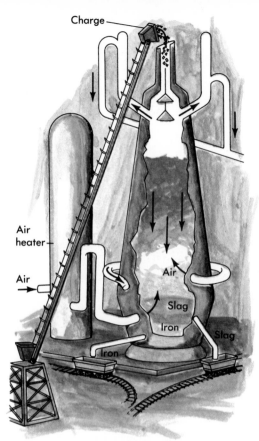

Ore (for iron), coke (for carbon), and limestone (to unite with the impurities) are dumped in at the top of the blast furnace. Burning gas and hot air start the chemical reactions to reduce the iron ore to iron.

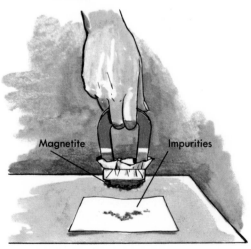

rock and let it cool. Test the powder with a magnet again. What has happened to the hematite?

Limonite. This iron ore varies in color from earthy brown to metallic black. Its streak is always brown. Many rocks and soils owe their brown color to limonite. Iron rust and limonite are much alike chemically.

The chemical formula for limonite is written $Fe_2O_3 \cdot n(H_2O)$. Compare this formula with that of hematite. In what ways are the two minerals alike? In what ways are they different? The n in the formula indicates that the amount of water of crystallization varies.

Dehydrating Limonite. Grind some limonite to a powder. Put a teaspoonful of the powder into a test tube and heat it over a gas flame. Look for drops of liquid in the top of the tube. Test this liquid with cobalt chloride paper. What is the liquid? Where did it come from?

After the powder has cooled, examine it closely. What changes can be seen?

Rub some of the larger pieces of dehydrated limonite on a streak plate. Compare the streak with that of unheated limonite and with that of hematite. What conclusions can you draw?

Magnetite. Magnetite is another iron ore. Its chemical formula is Fe_3O_4. Compare this formula with the formula for hematite.

How many atoms are indicated in each formula? Which mineral contains the higher percentage of iron? Which contains the higher percentage of oxygen? If two blocks of these minerals are of equal size, which should weigh more?

Magnetite that has been exposed to air and water for a long time often becomes reddish brown like limonite. Explain this change in chemical terms.

Concentrating a Magnetite Ore. Iron ores usually contain quartz and other impurities. Magnetite ores can be concentrated with strong magnets.

Grind some magnetite to a fine powder. Weigh the powder. Wrap a piece of paper around a strong magnet and separate the magnetic ore from its impurities. Weigh the concentrated ore and weigh the impurities. Calculate the percentage of ore and impurities in the original sample.

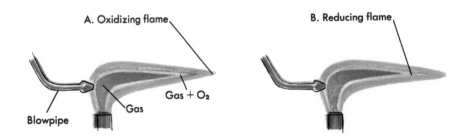

A. Oxidizing flame

B. Reducing flame

Gas + O₂

Gas

Blowpipe

Galena. Galena is a compound of sulfur and lead. Its chemical formula is PbS. (Pb is for "plumbum," which is Latin for lead.)

Galena looks much like a metal. Tap a small piece of galena with a hammer. What happens? How does a bit of metal behave?

Heat breaks the chemical union of lead and sulfur. However, if galena is heated in air, the lead and sulfur oxidize immediately.

Reducing Galena with a Blowpipe. A gas flame has two regions, an inner cone of gas without oxygen, and an outer envelope where the gas mixes with oxygen and burns. A blowpipe concentrates the flame upon a small area and causes intense heating. Different effects can be obtained by using different parts of the flame.

Study the two types of blowpipe flames shown above. A specimen held at *A* is oxidized. Why? A specimen at *B* is heated in the absence of air; elements which unite with hydrogen or carbon of the gas are removed, and the specimen is reduced.

Make a small hollow in a block of charcoal and place in it some bits of galena. Direct a reducing flame on the galena until a molten drop forms. Cool the drop and examine it. Try to cut it with a knife. Tap it with a hammer. Rub it on white paper. What has been produced? (**CAUTION:** Do not breathe in through the tube.)

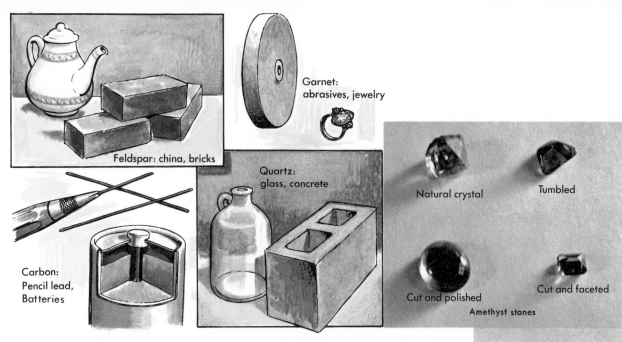

Garnet: abrasives, jewelry

Feldspar: china, bricks

Quartz: glass, concrete

Carbon: Pencil lead, Batteries

Natural crystal

Tumbled

Cut and polished

Cut and faceted

Amethyst stones

Other Useful Properties of Your Minerals. The table lists some of the properties and uses of minerals in your collection. Study each use listed. Which properties of the minerals are necessary for each use?

Mineral	Useful Properties	Uses
Feldspar	Weathers to form clay.	Pottery, china, bricks.
Garnet	Hard, breaks along sharp edges.	Garnet paper, jewelry.
Graphite	Rubs off easily, resists high temperature, greasy feel, conducts electricity.	Pencil lead, crucibles and other heating containers, clutch throw-out bearings, lubricant, electrodes.
Mica	Electrical insulator, heat insulator.	To support heating wires in toaster, iron; as insulating material in electronic capacitors.
Quartz	Weathers to form sand, transparent when pure, hard, shiny on flat surfaces.	Major ingredient in glass, molds for castings, cement; exists in a variety of forms, such as flint, agate, amethyst, jasper, opal, aventurine.

Aventurine

Tiger Eye

Agate

Fluorite

Amethyst   Sapphire

REVIEW QUESTIONS

1. What test can be used to tell whether the crystals shown at *A* are quartz or calcite?
2. What is the mineral shown at *B*?
3. What type of cleavage is shown at *C*?
4. How is oxygen removed from iron ores?
5. What elements make up pyrite?
6. What is meant by the statement that galena has a specific gravity of 7.5?
7. Why is quartz such a common mineral?
8. What happens to gypsum when it is heated to a high temperature?
9. What properties made flint so important during the Stone Age?
10. What similar chemical properties do galena and pyrite share?
11. What similar physical properties do galena and pyrite share?
12. What similar property makes talc and graphite useful?
13. Add more headings and minerals to a table like the one below to summarize how each mineral discussed in this chapter was changed.

THOUGHT QUESTIONS

1. Judging from the formulas, which contains the greater amount of oxygen, magnetite or hematite? What are the proportions?
2. How many directions of cleavage are shown by the fragment of fluorite?
3. Under what conditions would beds of salt and gypsum be deposited from sea water?
4. Why is a sapphire classed as a precious stone while an amethyst is classed as semiprecious?
5. Why does cutting facets on a gem such as a sapphire increase its brilliance?

WAYS OF CHANGING MINERALS

By Grinding	By Electricity		
talc	halite		

2

The Nature of Rocks

Sometimes, large pieces of pure mineral are found in nature. Usually, however, two or more minerals are found together. Such material is called rock. Two rocks are said to be the same type if they have about the same kinds of minerals in the same amounts. Rocks that contain different minerals or quite different amounts of the same minerals are given different names.

New rocks are constantly being formed on the earth's surface. Volcanoes bring molten minerals to the surface where they harden. Sediments are carried into lakes and oceans where they become compressed and hardened.

Other rocks, deep within the earth, are sometimes changed by heat and pressure. New minerals may form, or a rock may melt completely and become an entirely different rock when it hardens.

SEDIMENTARY ROCKS

At one time, the earth's surface consisted of bare rock. Wind and water wore away small bits of minerals and moved them to other locations. These small bits are called sediments. *Sediments can be readily seen in the mineral particles which make up sand and other soils. Other sediments can be seen by examining muddy water with a hand lens.*

Sometimes sediments settle in one place for thousands of years. Under these conditions, each particle may become cemented to other particles next to it. The hardened mass is called sedimentary rock. *Sedimentary rocks may also form when a solution containing dissolved minerals dries up. For example, when oceans dry up, sedimentary rock rich in halite and gypsum may form. The remains of plant sediments may form a sedimentary rock called* coal.

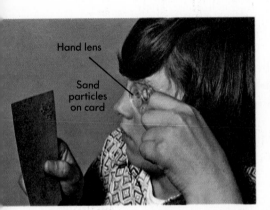

Hand lens

Sand particles on card

Making Sediments. The two most common minerals on the earth's surface are quartz and feldspar. Thus, the two most common sediments carried by streams and deposited in rivers are quartz and feldspar sediments. Since these minerals are plentiful and therefore inexpensive, most people do not consider them important. However, they do have many important uses. Quartz is the major ingredient of glass. (Recall that quartz and glass both cleave with sharp edges.) Feldspar is a major ingredient of china, pottery, and bricks.

Put some pieces of broken glass tubing or test tubes into one mortar and about the same amount of small pieces of brick into another. As a safety precaution, wear safety glasses and gloves, and make a cardboard shield for the pestle as shown above. Have two students grind each sample equally for about five minutes. Then cautiously feel each sample. Examine each sample with a hand lens as shown here. Compare the quartz and

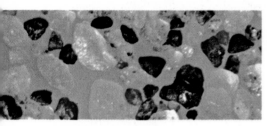

Quartz sand (magnified)

feldspar sediments. Compare the sediments with the original glass and brick.

Analysis of Sand Sediments. Bring to class a sample of sand. Get it from sandy soil, from along a creek or shore, from a construction site, or from a child's sandbox. Place a small sample of the sand on a sheet of paper. Examine the particles with a hand lens. Use your mineral identification skill to identify quartz and other minerals. Examine 100 particles in one region of the sample. Make a table as shown below to record your observations. Examine other samples of sand in the same way.

MINERAL COMPOSITION OF 100 SAND PARTICLES

Sample	No. Grains of Quartz	No. Grains of Feldspar	No. Grains of _____	No. Grains of _____	No. Grains of Unknown Mineral
Bouton's Beach	96	2			
Bank of Io Creek	82				

What mineral or minerals are commonly found in sands? What is the most common mineral in most sands in your region? How does the sand found in nature differ from the quartz sand made by grinding up glass?

Analyzing Sand for Magnetite. Move a magnet through a small mound of sand. Use a hand lens to observe any particles which cling to the magnet. What sand has the highest percentage of magnetite?

Identifying Feldspar Sediments. When quartz in the earth's rocks wears away, or *erodes*, the small particles of quartz are very similar to the original quartz minerals. When feldspar erodes, however, the particles are not only smaller, they also change chemically. Thus, it is much more difficult to recognize feldspar sediments than quartz sediments. Feldspar sediments are the chief ingredients of *clays*.

Rub dry powdered clay between your fingers and note the feel. Clay is made up of extremely small, flat flakes that feel smooth. Fine sand feels gritty.

Wet some clay and note its odor. Note also that it is plastic; the flat flakes cling together when wet, but slide past each other because the water reduces the friction between them. Ordinary mud is not sticky.

Bedded clay

Producing Bedded Sediments. Mix three tablespoonsful of coarse sand with one tablespoonful of powdered clay in 250 ml of water. Stir the mixture rapidly, and at the same time pour it into a liter jar half full of water. Note what happens immediately. Examine the jar again half an hour later. Compare the rates at which sand and clay settle out.

Remove a cupful of water. Add a mixture of sand and clay to the cup and pour this back into the jar. Repeat the process at intervals until several layers have been formed.

Causes of Bedding. Study the photograph of bedded clay above. The thin layers are made up of fine particles. The thick layers are made up chiefly of larger particles.

Geologists believe that bedding of this type is caused by seasonal changes in the flow of streams. If they are correct, at what season might the thick layers of large particles have been deposited? When might the thin layers of small particles have been deposited? How many layers were probably deposited each year?

Sorting of Sediments. The diagram below shows the mouth of a river emptying into a bay. Explain why the sediments are sorted as shown here. How might seasonal changes affect the deposits in each place? How might a flood or a year without rain affect them? Make a diagram to show the type of deposits if five years of dry weather were followed by five years of heavy rains.

Making Artificial Rock. Natural rocks form too slowly for classroom observation. The following experiment deals with making an artificial rock.

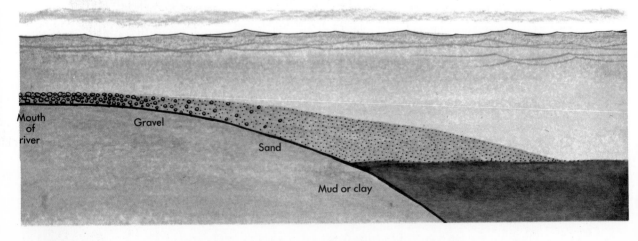

Mouth of river

Gravel

Sand

Mud or clay

There are certain resemblances between this experiment and a real situation. There are also many differences. Conclusions will be drawn in terms of the resemblances; the differences will be ignored but not forgotten.

Add coarse sand in a paper cup to a depth of 2 cm. Add enough saturated solution of Epsom salt to cover the sand. Mix them well.

Let the mixture stand until dry (two or three days). Cut off the paper cup and study the newly made rock with a hand lens. Where is the Epsom salt? How does it hold the sand together?

MAKING A SATURATED SOLUTION OF EPSOM SALT. Add 250 ml of water to 125 ml of Epsom salt. Stir until most of the salt dissolves. The liquid is then saturated Epsom salt solution.

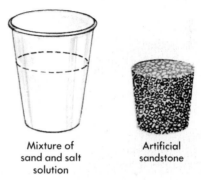

Mixture of sand and salt solution

Artificial sandstone

Formation of Natural Rocks. The artificial rock just formed is held together by the Epsom salt which crystallized around the grains of sand. Natural sediments are sometimes cemented in the same way.

Other processes also help form sedimentary rocks. The weight of overlying sediments squeezes together the sediments beneath. Compression may make the particles soften and flow together. They become interlocked when they harden again.

These three processes, of which little is known, are slow, and usually take place under the surface of the earth.

Natural Cements. The most common cements in sedimentary rocks are calcium carbonate, silicon dioxide, and iron oxide. All are deposited from solutions made up of small amounts of minerals dissolved in ground water. Calcium carbonate may crystallize as calcite. Silicon dioxide may crystallize as quartz.

These three chemicals may also be deposited as a moist, jellylike substance called a *gel*. Such a gel slowly hardens, loses its water, and forms microscopic crystals.

Breccia

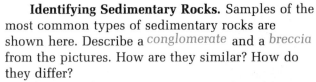
Breccia

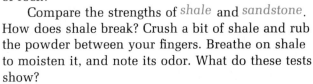
Conglomerate

Conglomerate

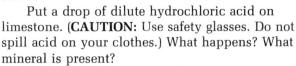
Conglomerate

Identifying Sedimentary Rocks. Samples of the most common types of sedimentary rocks are shown here. Describe a conglomerate and a breccia from the pictures. How are they similar? How do they differ?

Examine a freshly broken piece of sandstone with a hand lens. How would you identify this type of rock?

Compare the strengths of shale and sandstone. How does shale break? Crush a bit of shale and rub the powder between your fingers. Breathe on shale to moisten it, and note its odor. What do these tests show?

Put a drop of dilute hydrochloric acid on limestone. (**CAUTION:** Use safety glasses. Do not spill acid on your clothes.) What happens? What mineral is present?

Impure Sedimentary Rocks. Sandstone sometimes contains a little clay. Such a rock is called a shaley-sandstone. A sandstone that contains some calcium carbonate is called a limey-sandstone.

What might you call a shale that contains sand? What might you call a limestone that contains clay?

Colors of Sedimentary Rocks. Pure quartz sandstone and pure calcite limestone are white. Small amounts of impurities may give these rocks other colors.

Carbon in a rock darkens it, making it gray or black, depending upon the amount present. Carbon can be identified by heating a rock enough to oxidize the carbon. Crush a piece of black shale to a

Brown Shale

Gray Sandstone

Red Sandstone

Black Shale

Shell Limestone

Layered Limestone

powder, and heat the powder red hot in a test tube. If the powder becomes white or gray, the color is due to carbon.

Red iron oxide (hematite) colors rocks bright red. Hydrated iron oxide (limonite) colors rocks yellow or brown. Crush a yellow or brown rock to a powder, and heat the powder to drive off the water of hydration. If the rock becomes red, the color was due to hydrated iron oxide.

Make a collection of sedimentary rocks found in your region. How are these rocks used?

OVERSIMPLIFICATION THROUGH CLASSIFICATION. In order to simplify what they observe, scientists devise classification systems. One such classification system divides sedimentary rocks into groups such as limestone, shale, sandstone, and conglomerate. However, nature does not restrict itself to these cut-and-dried classifications. Close examination of sedimentary layers usually shows sediments which gradually change from lime to limey-clay, to clay, to sandy clay, to sand, to sandy gravel, to gravel. As these sediments change to rocks, some regions of rocks fit the general description of a shale or a sandstone. Many other rock regions, however, best fit into a "crack" in the classification system, resulting in names like sandy-shale or limey-shale. Do not expect every rock you examine to fit exactly into a single place on the classification system.

Cements and Crystals. Sedimentary rocks consist of more than sediments which are cemented together. Sedimentary rocks frequently contain crystals. The processes of cementing sediments and growing crystals are very similar.

Both processes depend on the fact that most minerals dissolve to some extent in water. Even minerals that are almost insoluble in cold water dissolve in the extremely hot water found in regions of volcanic activity.

Under certain conditions a dissolved mineral leaves the solution. If the process is slow, the atoms collect in a well-oriented structure and become a crystal.

Growing Crystals. Slowly add 60 grams of alum (sold in grocery stores) to 125 ml of water. Heat the mixture while stirring it until all of the alum has dissolved. When the hot water holds all the alum it can in solution, the solution is said to be *saturated*.

Pour the liquid into a small jar. Hang a small paper clip by a thread in the jar as shown at the right. Cover the jar.

As the solution cools, it can no longer hold all of the dissolved alum. It tends to become *supersaturated*. Some of the alum will leave the solution as crystals.

After a few hours, examine the crystals which have formed. Which ones seem to be the most perfect? What happens to crystals that grow close together?

Try growing crystals of other chemicals, such as copper sulfate, Rochelle salt, and Epsom salt.

Sudden and Slow Cooling. Heat two tablespoonsful of water to boiling. Slowly add eight level tablespoonsful of table sugar, stirring the water until the sugar dissolves.

Pour a tablespoonful of the hot solution on ice or packed snow. Describe the appearance of the solution after it has cooled. Compare it with sugar crystals. Discuss the change in terms of order and disorder of the molecules of sugar.

Pour the rest of the solution into two bottles. Put one bottle in a cold place. Put the other in a warm place, such as on a radiator. Examine the contents of the bottles a day later. Describe the products formed.

Veins. Water containing minerals in solution usually fills all the cracks and cavities in rocks.

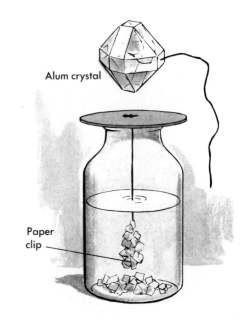

Alum crystal

Paper clip

Quartz veins in sedimentary rock layers

Sometimes conditions change within the rocks; perhaps temperatures fall or pressures drop. The solution may become supersaturated, and atoms of the dissolved chemicals will leave the solution, depositing in cracks and cavities.

Cracks filled with the new minerals are called *veins*. The photograph on page 28 shows veins of quartz in sandstone. Quartz and calcite are two common vein-forming minerals. Two or more minerals may be found in one vein.

Valuable ores such as those of silver, copper, lead, and zinc are usually found in veins. The ores were once dissolved in underground water and were later deposited in concentrated form as veins in rocks.

Crystal-Filled Cavities. Deposition in cavities is commonly so rapid that the resulting crystals are too small to be seen. They are in microcrystalline form. However, sometimes the rate of deposition is very slow, as when a cavity is sealed off. Then visible crystals may form.

Such a cavity is usually lined with a microcrystalline deposit that represents the period of rapid cooling and deposition. Then as the rate of deposition slows down, the atoms arrange themselves in their preferred pattern. Crystals form, growing toward the center. Cavities partly filled with crystals are called *geodes*.

Minerals in Hot Springs. Many minerals such as quartz (silicon dioxide) do not dissolve readily in cold water, but become much more soluble in water at high temperatures. Water that has been heated by passing through hot rocks deep within the earth usually contains dissolved minerals.

Hot Spring Deposits. Water from hot springs and geysers cools quickly when it reaches the surface. The water becomes supersaturated with some of the minerals in solution. These minerals deposit around the openings of the springs.

Calcium carbonate may deposit from hot spring water as *travertine*. Silicon dioxide may deposit as *opal*. A special variety of opal found around geysers in Yellowstone Park is called *geyserite*.

Geode

Old Faithful Geyser in Yellowstone National Park spouts silvery cascades about every 65 minutes. Each display lasts about 4 minutes.

29

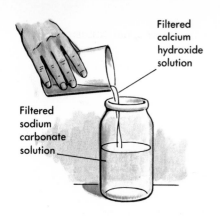

Filtered calcium hydroxide solution

Filtered sodium carbonate solution

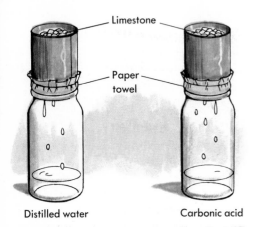

Limestone

Paper towel

Distilled water   Carbonic acid

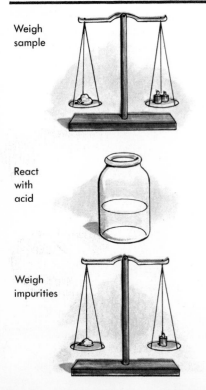

Weigh sample

React with acid

Weigh impurities

Chemical Precipitation. Dissolve a teaspoonful of sodium carbonate (washing soda) in 250 ml of water. Dissolve half a teaspoonful of calcium hydroxide (hydrated lime) in another 250 ml of water. Filter the two solutions, then mix them.

What happens as the chemicals react? Examine the mixture a few hours later. Filter the sediment and test it with dilute hydrochloric acid.

The white sediment is calcium carbonate, which is insoluble. It settled out, or precipitated, as solid particles.

Calcium Carbonate to Calcium Bicarbonate. When carbon dioxide dissolves in water, the solution is known as carbonic acid. Carbonic acid reacts with calcium carbonate to produce calcium bicarbonate:

carbonic   calcium   calcium
acid   carbonate   bicarbonate

$$H_2O + CO_2 + CaCO_3 \rightarrow Ca(HCO_3)_2$$

This reaction is very important in the erosion of limestone because calcium carbonate is not soluble in water. Calcium bicarbonate, however, dissolves readily.

Dissolving Limestone. Punch several holes in the bottoms of two cans. Tie paper towels over the holes. Fill the cans with crushed limestone.

Pour 500 ml of distilled water (rain water) in one can. Bubble carbon dioxide through another 500 ml of distilled water until blue litmus paper turns pink. This shows that the solution is acidic. Pour this slightly acid water through the other can of limestone.

Collect and evaporate the water that trickles through the two cans. Compare the amount of solid material that remains from each. Describe the chemical reaction that took place.

Measuring the Purity of Limestone. Many chemical industries need high purity limestone for their work. Use the following procedure to determine the purity of limestone samples.

Crush a sample of limestone and weigh the powder. Mix the powder with dilute hydrochloric acid, adding more acid as needed until all bubbling stops. Any sediments that remain are impurities. What happens to the calcium carbonate?

Filter out the sediments. Allow them to dry near an open window. Weigh the sediments which were not calcium carbonate. Calculate the

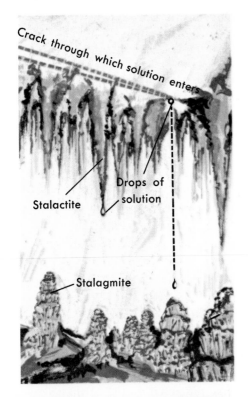

Crack through which solution enters

Stalactite

Drops of solution

Stalagmite

percentage of impurities in the original limestone.

Cave Deposits. Limestone caves often contain beautiful formations. These are produced from mineral solutions trickling into the caves.

Perhaps a drop of saturated calcium bicarbonate solution oozes from a crack in the ceiling of a cave. A few atoms are deposited at this point before the drop can fall. As the process is repeated again and again through the years, an icicle-shaped *stalactite* of calcite grows down from the ceiling.

The drops that fall to the floor also gradually deposit atoms. An upward-growing *stalagmite* is produced on the floor of the cave.

The solution may trickle down a wall or slope. Then calcium carbonate is deposited as *flowstone*.

Explain why stalactites are sharp pointed and why stalagmites are blunt. What conditions might cause a stalagmite beneath a stalactite to join and form a *column*?

Minerals in Solution. Water that seeps through rocks usually contains appreciable quantities of chemicals dissolved from the rocks. One of the most plentiful is calcium bicarbonate, especially in limestone regions. We have seen that calcium carbonate or limestone, is converted by acids to calcium bicarbonate, which is highly soluble.

Water containing dissolved calcium bicarbonate is called *hard water*. Since this chemical can be driven out of solution by heating, the water is also said to have *temporary hardness*.

Two other minerals commonly found in ground water are calcium sulfate (gypsum) and magnesium sulfate (Epsom salt). These chemicals cannot be driven out of solution by heating, and the water is said to have *permanent hardness*.

SUMMARY QUESTIONS

1. What are sediments?
2. How do sediments change into sedimentary rocks?
3. How do stalactites and stalagmites form?
4. What causes bedding in sedimentary rocks?
5. What are the two most common minerals in sedimentary rocks?
6. What is a saturated solution? What happens when it cools?

IGNEOUS ROCKS

Rocks that form when molten rock material cools are classed as igneous rocks. The term "igneous" is not truly suitable for these rocks, because it refers to fire (as in the word "ignite"). Igneous rocks were named when people believed volcanoes were burning mountains.

Igneous rocks may be divided into two groups. The rocks that form at or near the surface, as when lava cools, are called volcanic rocks. Those that form deep within the earth are called plutonic rocks. There are uncertainties about the formation of plutonic rocks, just as there are uncertainties about all processes that take place within the earth.

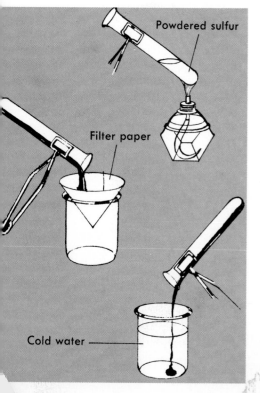

Powdered sulfur

Filter paper

Cold water

Effect of Cooling Rate. The cooling of molten rock materials cannot be studied directly. Conclusions must be drawn from analogies.

Slowly heat a tablespoonful of powdered sulfur until it melts. Pour the liquid sulfur into a paper cone, as shown at the left. Then melt an equal quantity of sulfur and pour it into cold water.

After the sulfur has hardened, break open the masses and examine them with a hand lens. How has the rate of cooling affected the sulfur?

Magma. Geologists call molten rock deep within the earth *magma*. They call molten rock that reaches the surface *lava*. Magma may or may not be the same as lava; probably it is not. Geologists believe that magma contains large amounts of dissolved gases which escape before the magma reaches the surface. Dissolved gases probably affect the chemical and physical properties of magma and make it unlike lava.

Over 40 different kinds of magma have been identified from a study of plutonic rocks. This evidence helps disprove the theory that the center

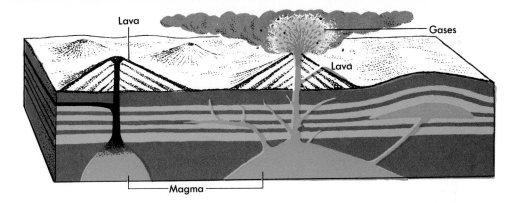

Lava

Gases

Lava

Magma

of the earth is one large body of molten rock from which all igneous rocks are produced.

Movements of Magma. Nearly all that is known about magma has been gained by the study of ancient rocks formed deep within the earth, and now uncovered by erosion. It is believed that magma is pushed upward from great depths, melting its way through bedrock. Sometimes the magma cools to form huge bodies of rock many kilometers across and thousands of meters deep.

Small amounts of magma may squeeze into cracks in bedrock, moving outward or upward. The magma may reach the surface and flow out as lava.

Molten rock which reaches the surface cools quickly and becomes rock within a few hours or days. Magma in cracks may take years to harden. Magma in large bodies far beneath the surface probably requires thousands of years of harden.

The Cooling of Molten Rock. Geologists believe that the crystalline structure of an igneous rock depends partly upon the rate at which it cools. Very rapid cooling freezes the atoms into a state of disorder. The resulting rock is noncrystalline. Slower cooling gives the atoms time to arrange themselves in crystals. The slower the cooling, the larger the crystals. Where would you find igneous rocks with the large crystals?

Volcanic Rocks. Volcanoes produce two types of rock materials. Molten rock which flows out is called *lava*. Pulverized rock produced by explosions is called *volcanic ash*.

Large particles of ash, sometimes called *cinders*, fall near a volcano. These particles may later be cemented to form a rock called *volcanic breccia*. Small particles are often blown by the wind and deposited some distance away. They become a rock called *tuff* if cemented together.

Volcanic Breccia

Volcanic Tuff

33

Lava cools too quickly for crystals to form. Sometimes it hardens into a glassy material called *obsidian*. Usually, however, lava is filled with gas bubbles. Such lava hardens into a porous rock known as *scoria*. Froth on molten lava hardens into *pumice*.

Volcanic rocks also include the rocks formed when lava cools within a volcano or in nearby cracks. Cooling in such locations may be slow enough for microscopic crystals to form. Two groups of this type of igneous rock are recognized. One group made up chiefly of dark-colored minerals is called *basalt*. Another group made up chiefly of light-colored minerals is called *felsite*.

Plutonic Rocks. All rocks classed as plutonic have large crystals. If some of the crystals are much larger than all the others, the rock is called a *porphyry*.

Granite is the most familiar of the plutonic rocks. It is made up chiefly of the light-colored minerals feldspar and quartz. *Diorite* is another light-colored plutonic rock, but it contains little or no quartz.

Gabbro is made up of dark-colored minerals; it contains some feldspar. *Peridotite* is dark-colored but contains no feldspar.

Identifying Minerals in Granite. Use a hand lens to study a piece of granite that has large crystals. Weathered granite should be broken with a hammer to expose fresh surfaces.

Obsidian

Scoria

Pumice

Felsite Porphyry

Basalt

Granite

Diorite

Gabbro

Peridotite

Feldspar breaks smoothly along surfaces at right angles to each other (right angle cleavage). These surfaces give brilliant reflections. The cleavage makes identification certain.

Quartz grains break irregularly and have a glassy look. They reflect less light than feldspar, and so they look both duller and darker.

Mica can be identified by its cleavage into thin sheets. Use the point of a knife to test the cleavage.

Magnetite may be present among the dark grains in granite. Pry out some dark grains and test them with a magnet.

Analysis of Granite. Choose a granite specimen that has a flat or nearly flat surface; a polished piece is excellent. Become acquainted with the mineral in the rock by studying the rougher surfaces.

Lay a square of window screening on the rock, and mark through 100 or more holes with a wax pencil. Using a lens, identify the mineral under each mark, count the number of marks on each kind of mineral, and calculate the percentage composition of the rock. Analyze other specimens of granite in the same way.

Analyzing Other Plutonic Rocks. Feldspar is the most abundant mineral in plutonic rocks. Quartz is abundant only in granite. Dark minerals, such as black mica and magnetite, are abundant in gabbro and peridotite. The proportions of each mineral determine the type of rock.

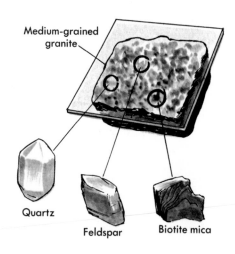

Medium-grained granite

Quartz

Feldspar

Biotite mica

SUMMARY QUESTIONS

1. Some igneous rocks have small crystals and some have large crystals. Why?
2. What is magma?
3. What are two igneous rocks with no or very small crystals?
4. Name two igneous rocks with large crystals.
5. What three minerals are often found in granite?

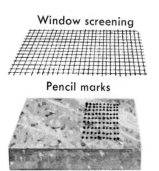

Window screening

Pencil marks

Specimen	Feldspar	Quartz	Mica	Magnetite	Other
Red granite	56%	32%	4%	—	8%
Gray granite	64%	30%	2%	2%	2%

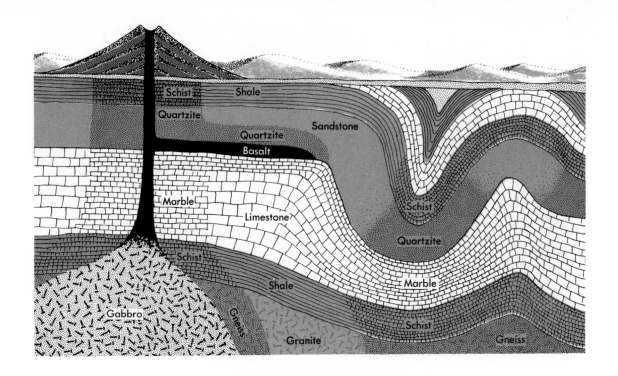

Schist
Shale
Quartzite
Quartzite
Sandstone
Basalt
Marble
Limestone
Schist
Quartzite
Schist
Marble
Shale
Gabbro
Gneiss
Schist
Granite
Gneiss

METAMORPHIC ROCKS

Two classes of rocks have already been described in this chapter—sedimentary and igneous rocks. A third class is made up of rocks that have been formed from rocks of the first two classes.

Heat and pressure cause changes in both igneous and sedimentary rocks. The changes may be physical, chemical, or both. The changes may be slight, or they may be so great that the original rock cannot be recognized. Rocks that have changed from their original form are called metamorphic rocks.

Causes of Changes in Rocks. High temperatures bring about most of the changes that produce metamorphic rocks. High temperatures are often accompanied by great pressures, and sometimes by movements of rock materials.

Magma forcing its way through cracks produces great changes in surrounding rocks. Minerals melt or soften, then react with each other or with chemicals in the magma. Rocks farther from the magma are affected by hot water and gases from the magma. Commonly the gases and liquids deposit new minerals in cracks or pores of rocks through which they pass.

Tremendous sideward pressures cause rocks to twist and fold. Such tremendous pressures and

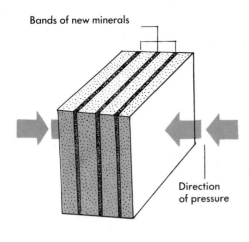

Bands of new minerals

Direction of pressure

Mica foliation

Flaky foliation

movements produce high temperatures. Thus, both chemical and physical changes result.

Types of Changes. Great heat and pressure bring about the following types of changes in rocks:

1. *Densification.* Rocks become more dense, either because the grains are squeezed closer together, or because additional minerals are deposited between the grains.

2. *Loss of Chemicals.* Water of crystallization may be forced from crystals. Oxygen and carbon dioxide may be produced by chemical reactions and escape from the rock.

3. *Recrystallization.* Atoms in softened rock may have additional time in which to rearrange themselves into larger crystals than before.

4. *Formation of New Minerals.* Minerals may break down chemically into elements. These elements may then recombine with each other or with elements from other sources, forming new compounds.

5. *Development of Foliation.* Rocks may develop a tendency to break apart in sheets or flakes because of the arrangement of new minerals with cleavage like that of mica.

Crystal Formation. According to one theory, increased pressure causes atoms to move closer together. The atoms then rearrange themselves into new crystals that take up less space than the original rock minerals. Mica is one of the most common minerals formed in this way.

There is less opposition to crystal growth at right angles to a line of pressure than along the line of pressure. Therefore, new minerals in a metamorphic rock are usually arranged in bands, as shown above. The bands form at right angles to the line of pressure.

Foliation. Metamorphic rocks may contain bands of mica or other minerals which have good cleavage. These rocks tend to split apart through the bands. This form of breaking is called *foliation*.

Foliation may be good or poor depending upon the kinds of minerals and their arrangement. Foliation in fine-grained rocks is usually flat, as in slate. Foliation in coarse-grained rocks is usually flaky. The picture at the left shows flaky foliation that is also wavy and irregular.

Identifying Metamorphic Rocks. Some metamorphic rocks are best identified by their foliation, others by the minerals which they contain. Identification is not always easy, because one type of rock may grade into another, depending upon the amount of change that has taken place.

Limestone becomes *marble* as calcium carbonate recrystallizes into large crystals of calcite. A freshly broken piece of white marble looks like sugar. Dilute hydrochloric acid should be used to test for calcium carbonate.

Sometimes it is difficult to decide whether a rock is a limestone or marble. If the rock is made up of crystals that can be seen without a hand lens, call it marble. If not, call it limestone.

Quartz sandstone becomes *quartzite* after the spaces between the grains have been filled with more quartz. This quartz is usually deposited by hot water flowing through the sandstone.

Quartzite can be identified by its glassy appearance and its hardness. (It scratches glass.) To distinguish between quartzite and sandstone, examine a freshly broken surface with a hand lens. If the rock broke apart through the cement but not through the sand grains, call it sandstone. If the rock broke apart through the grains as well as through the cement, call it quartzite.

Gneiss (pronounced *nice*) is easily recognized by its banding. Foliation is poor, but the rock sometimes breaks along a band.

Changes that produce gneiss are so great that the original rock cannot always be identified. Gneiss sometimes looks as if it had been formed from granite. But probably any rock becomes gneiss when it is exposed to great heat and pressure for a long enough time.

Slate is formed from shale. The amount of change is usually small. Slate looks as fine-grained

White Marble

Verd Antique Marble

Quartzite

Gneiss

as shale. However, slate contains microscopic crystals of mica which lie in flat bands. Therefore, the foliation of slate is flat and usually very good. For this reason, slate has been used for blackboards, flagstones, and shingles.

Slate looks so much like shale that the two are often difficult to tell apart. Slate is usually shinier, harder, and denser than shale. Slate gives off a ringing sound when lightly tapped; shale gives off a duller sound.

The rock called *schist* has usually undergone such great changes that the original rock cannot be identified. Shale, slate, and basalt are known to have changed to schist. Probably any fine-grained rock can become schist.

Slate and schist can be told apart by the size of the crystals. Crystals in slate are microscopic. Crystals in schist can be seen without a hand lens.

Crystals in schist are arranged in bands that are close together. Foliation is good but often irregular or wavy. Large specimens of schist can usually be identified more easily than small ones.

Gneiss and schist grade into each other. Some specimens fit both categories and cannot be classed as one or the other.

UNCERTAINTIES IN SCIENCE. *Science is full of uncertainties. A rock cannot always be identified as shale or slate. Measurements are not exact. Two different theories may explain a condition equally well.*

Many people think that science is always exact, and that it has a positive answer for every question. True scientists, however, recognize that there will always be uncertainties. They know that science can only be made a little more exact and a little more certain.

Red Slate

Gray Slate

Schist with Garnets

Mica Schist

A vein of coal

Coal—An Example of Metamorphism. Coal is one of our most valuble resources. Both hard and soft coal are important sources of energy for industry. Soft coal is also a source of many industrial chemicals.

The coal used today was formed long ago from ancient forests. Trees died, fell, and were buried deep in sediments. Decay could not take place because of insufficient oxygen. The plant materials underwent chemical changes that removed other elements and left nearly pure carbon behind.

Formation of Coal. Changes in deeply buried wood are not easily investigated. However, they seem to be similar to those which take place when wood is heated in a closed test tube. Therefore, the process of coal formation will be studied by analogies.

Weigh two dry, clean test tubes. Set them up as shown at the left. Weigh a sample of wood shavings and heat them in the first test tube over a hot flame.

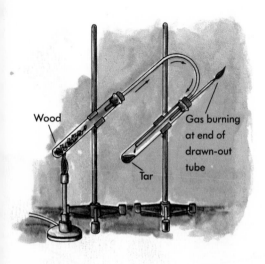

Wood breaks down chemically when it is heated. Some of the products are *volatile* (able to vaporize), and escape as gases. Water and *tar* condense in the cool test tube. The other volatile materials escape from the second tube. Find out if they will burn. The gas which escapes is chiefly *methane*, sometimes called "marsh gas."

When cool, reweigh the test tubes, and calculate the percentage of volatile and nonvolatile substances produced from the wood. Examine the

Peat

Lignite Coal

Bituminous Coal

Anthracite Coal

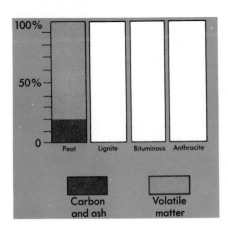

100%

50%

0

Peat Lignite Bituminous Anthracite

Carbon and ash

Volatile matter

nonvolatile substances in the first test tube. What is it called?

Peat. Peat represents the first step in the production of coal. Peat is made up of compressed plant materials. It is usually dark brown in color.

Peat is being formed at the present time in many bogs and ponds, especially in cool climates where decay is slow. Methane is given off during the process. Bubbles of methane can be seen escaping from bogs and marshes. Thus, the name "marsh gas."

Heat a sample of dry peat in a clean test tube. Use the same setup used for the wood shavings. Collect and weigh the products. Calculate the percentages of carbon and volatile material produced. How does peat compare with wood in terms of these products?

Lignite and Bituminous Coal. Pressure of overlying sediments compresses peat into a rock. At the same time the plant remains continue to lose methane and other volatile substances. The product is carbon of increasing purity. The steps are:

peat → lignite → bituminous coal → anthracite

Heat samples of *lignite* and *bituminous coal* in a clean test tube, and calculate percentage compositions, as you did above.

Anthracite. Coal as found in metamorphic rocks may be almost pure carbon. It is called *anthracite*. The heat and pressure that produced the metamorphic rocks also drove most of the volatile materials from the coal.

Determine the percentage composition of a sample of anthracite coal. Compare it to lignite and bituminous coal.

SUMMARY QUESTIONS

1. How do metamorphic rocks form?
2. How does a metamorphic rock differ from a sedimentary rock and an igneous rock?
3. What changes does plant matter undergo as it metamorphoses to hard coal?
4. What metamorphic rocks form from each of the following: shale, limestone, sandstone, soft coal?

Granite

Slate

REVIEW QUESTIONS

1. How was rock No. 1 formed?
2. What processes changed loose sediments into rock No. 2?
3. To what class of rocks does specimen No. 3 belong?
4. What minerals give sedimentary rocks their colors?
5. What is the source of most clay?
6. What is the brownish-red mineral in specimen No. 3 at the left?
7. What is a simple test for calcium carbonate?
8. How does the cooling rate affect rocks formed from magma?
9. How is clay changed into shale?
10. What happens to limestone when it is exposed to high temperatures and pressures?
11. What is methane and how is it produced naturally?
12. What is a precipitate and how may it be formed?
13. What are two rocks formed from the remains of organisms?
14. Why is sandstone commonly composed of quartz sand?

THOUGHT QUESTIONS

1. Why are fossils rarely found in igneous rocks?
2. What was the direction of the pressure that produced the specimen of slate shown at the left?
3. Why is cementation more important in the formation of sandstone than of shale?
4. Under what conditions might graphite, which is nearly pure carbon, be formed?
5. How would you explain limestone beds on a high mountain?
6. What story is told by the rocks in the next to the last picture on this page?
7. What story is told by the rocks in the lowest picture?

3

The Nature of Soils

In many parts of the world, minerals and rock fragments remain on the land above the solid bedrock. With the action of moisture, plants, and animals, these rock fragments slowly turn into useful soil.

The top centimeters of soil provide the people of the world with most of their food supply. In spite of this fact, many people never look at soil carefully.

A quick glance at a handful of soil shows that it is a mixture of many things. Included are minerals, water, air, and biological substances such as roots, small animals, decaying matter, and microorganisms. Because of this variety of substances, biologists, chemists, geologists, and physicists have all added to our understanding of soils. Also, this variety of substances provides many things in the soil to interest the students who study it.

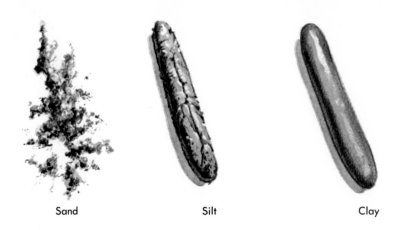

Sand   Silt   Clay

DESCRIBING SOILS

The people who work with soils describe them in different ways. For example, a farmer might say that a soil is too sour to be a good pasture; a construction engineer might describe the same soil as being poorly drained. A chemist would describe the soil as being acidic. Thus, individuals describe a soil in terms of their own training and in terms of the problems the soil presents.

Soil Texture. Break up some dry soil in your hand. Does the soil feel (1) sandy, (2) harsh like clay, or (3) floury? Test soils in other locations.

Break apart all the lumps in small samples of sandy, harsh, and floury soils. Use a hand lens to study the mineral particles. Which soil has the largest particles? Which has the smallest particles?

Soils can be classified into three soil texture groups according to the size of the particles. Particles of *sandy soil* can be seen with the unaided eye. Sandy soils feel gritty.

Clay soil is composed of particles so small that they can be seen only with an electron microscope. Clay soil is harsh when dry and sticky when wet.

Particles of *silt soils* are larger than clay particles but smaller than sand particles. Silt soils feel similar to flour or talcum powder.

Add water to a soil sample until it sticks together. Roll a small ball of this soil between your thumb and forefinger. Compare this roll of soil with those shown above, and classify the soil texture as sand, silt, or clay. Test other soils in this way.

Percentage Composition of Soil. Weigh a liter jar. Add 250 ml of dry soil to the jar and reweigh.

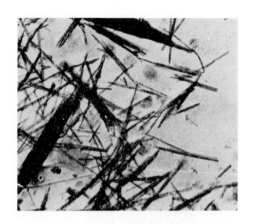

Clay particles magnified 29 000 times with an electron microscope.

Record the weight of the soil in your recordbook.

Add water until the jar is three-quarters full. Stir the soil and water until all the pieces of soil are broken up. Pour everything which does not settle out immediately into a large beaker as shown in picture 1. Do not allow any of the sandy material in the bottom of the jar to escape. Add a little more water to the jar and stir up the soil particles. Again pour the water from the jar into the beaker. Dry and weigh the sand in the jar. Record this weight. Calculate the percentage of sand in the soil.

Allow the soil particles in the beaker to settle for 30 minutes. Pour the water mixture into a wide pan as shown in picture 2. Dry and weigh the silt which settled out in the beaker. Record this weight. Calculate the percentage of silt in the original soil sample.

The minerals in soils are sometimes described as consisting of sand, silt, and clay. *Assume* that your soil consists only of these three materials. How can you calculate the percentage of clay in your soil? Do it.

Classifying Soil. Note point A on the diagram. This point corresponds to a soil which is 40% sand and 20% clay. The other 40% must be silt. Notice from the diagram that a soil of these proportions is called a *loam*.

Use the diagram to classify your soil sample. Test and classify other soil samples.

1

2 Thirty minutes later

Silt

Sand

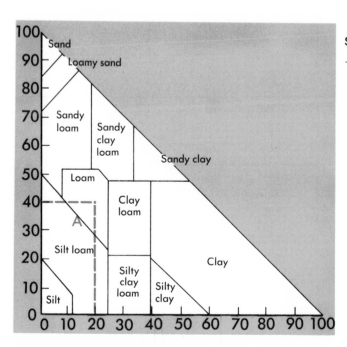

Water in Soil. Weigh an empty can, add a sample of soil, and reweigh the can. Find the weight of the soil sample by subtraction. Record these weights in a table similar to the one shown below.

Heat the can and soil in an oven at a temperature of 107–120°C until all the soil is dry. Calculate the weight of dry soil in the can.

Calculate the weight of water which was in the soil by subtracting the weight of dried soil from the original weight of the soil. Calculate and record the percent of water in the soil.

Organic Matter in Soil. Heat the dried soil from the last activity over a high temperature gas flame to burn out the organic material. Weigh the soil after heating and determine how much weight was lost. Record these weights in a table. Calculate the percent of organic matter in the original soil.

Repeat the experiment using different soils, such as woodland and garden soil. Compare the percent of organic material in different soils.

Decayed organic matter forms a useful substance in soil which is called humus. The dark color of topsoil is usually due to humus. Humus and other organic matter are changed by heat into gases, such as carbon dioxide (CO_2) and water vapor (H_2O).

WATER IN SOIL

Soil Type	Mr. Harris's Garden	
A. Weight of soil and can	385 g	
B. Weight of can	88 g	
C. Weight of soil (A–B)	297 g	
D. Weight of dry soil and can	325 g	
E. Weight of dry soil (D–B)	237 g	
F. Weight of water (C–E)	60 g	
G. % Water $\left(\dfrac{F}{C} \times 100\right)$	20%	

Cutting ridge off top of can

Cutting a soil sample

① ②

Removing sample
of soil

③

Soil Air. Use tin shears to remove the ridge from around the top of a metal can. Place the can on the ground, sharp edge downward. Turn the can, forcing it downward into the ground. Lift out the can filled with soil.

Fill a second can of the same size with water. Slowly pour the water into the soil in the can until the soil is unable to hold any more water. Calculate the volume of water added to the soil. Calculate the percent of air in the original soil by assuming that the water fills up the air spaces in the soil.

Compare the amount of soil air in soils of different textures.

Assumptions in Measuring Soil Air. In the last activity it was assumed that when water was added to the soil, the water would only fill the air spaces between the soil particles. Soil scientists have discovered that this assumption is correct for sand and silt, *but not for clay.* Water reacts with clay particles to produce new substances which swell and increase the volume of the material. Thus, the calculated volumes of soil air are accurate for sand and silt, but not for clay soils.

ASSUMPTIONS. Conclusions in science experiments are often based on assumptions. For example, the determination of the amount of air in soil is based on the assumption that the soil particles don't change size when water is added to the soil. Sometimes the assumptions are later found to be incorrect and the conclusions have to be changed. Much of your own thinking is based on assumptions. Look for assumptions in your thinking and try to find out if any of these assumptions are incorrect. Incorrect assumptions may lead to wrong conclusions.

SUMMARY QUESTIONS

1. Describe two different ways of determining the composition of a soil sample.
2. How can you determine the volume of air in a sample of sand?
3. What is humus?
4. What is an assumption? Give an example.

PHYSICAL CHANGES IN SOILS

Soil is constantly changing. Soil responds to many forces which tend to pack it or loosen it, to temperature changes, and to changes in the available water. The way a soil is changed by some or all of these factors determines the usefulness of the soil.

To the trained observer, the changes which a soil undergoes in a few seasons are quite noticeable. They are similar to the changes in living things, as they develop and increase in vigor, or grow old and begin to decay.

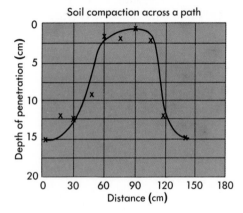

Measuring Soil Compaction. People often take shortcuts across a lawn or field. The grass soon dies where these people walk. Study how tightly the soil is packed along a line across such a path.

Mark equal spaces on a pencil as shown above. Push the pencil point into the soil, about 60 cm from the path, pressing with the palm of your hand. Record the depth to which the pencil is pushed when the pressure becomes uncomfortable.

Move 15 cm closer to the path, and again test the soil this same way. Repeat this testing every 15 cm until you are 60 cm off the path on the other side. Plot your observations on a graph. Compare your graph with the one shown here.

Soil compaction is the term used to describe the amount that a soil is packed. How did walking on the soil affect soil compaction? How did walking on the soil affect plant growth?

Use a calibrated pencil to study the compaction of soils where different species of plants grow. What species of plants grow best in soils which are only slightly compacted? Which plants can survive in highly compacted soil?

Depth and Soil Temperature. Use a thin metal rod and a hammer to make holes in the ground 15, 30, and 45 cm deep. Measure the temperature at the top of the soil and at these depths. Plot the temperature and depth data on a graph.

48

Repeat the measurements at different times of day and under different weather conditions (such as sunny and cloudy days). Plot this information on the same graph. Does soil conduct heat well? Explain your answer.

Animal Life and Soil Temperature. The picture at the left shows the underground corridors and chambers of an ant colony. What changes in temperature would you expect to find at different depths in an ant colony on a warm day?

People who study ants have noticed that young adult ants move ant larvae to greater depths on warm days. What protection would this give the larvae?

The 17-year locust (cicada) spends most of its life in the soil. When the eggs hatch, the young cicada burrows into the ground. For over 15 years, it migrates upward in the soil in the summer and downward in the winter.

Soil Temperature in Sun and Shade. Compare the soil temperature in the sun and shade near the school. Compare the soil and plants in the lawn which is shaded much of the day with the lawn which is never shaded. How might soil temperature be a factor in explaining the observed differences?

Sloping Land and Soil Temperature. Measure and compare the temperatures of soil on different sides of a hill. Do north-facing or south-facing slopes have the higher soil temperature on sunny days? Why?

Study the picture below. Which stream bank faces south and which faces north? On which slope will the soil be hotter and drier next summer?

Water Penetration in Soil. Cut both ends from a metal can, and push the can into the soil as shown above. Pour a known volume of water into the can in the soil. How long does water take to penetrate into soil?

Test soils of other textures in the same way. Through which soils does water penetrate quickly? Through which soils does water penetrate slowly? Why is this type of data important to septic tank builders?

Penetration Rate of Water. Tie a cloth over one end of each of four transparent cylinders or glass columns. Fill each tube three-quarters full with one of these dry soil materials: sand, silt, clay, and humus. (Use dry peat moss if humus is not available.) Gently pack the soil as you fill the columns.

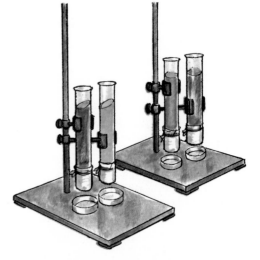

Add water to one column. Measure the distance the water penetrates every 15 seconds. Plot the distance and time on a graph. Add water and plot the data for other types of soil. Label each line on the graph with the name of the soil texture. Compare the penetration rate of water in the different types of soil.

The Cause of Erosion. Water does not penetrate quickly into clay soil. Thus, when rain falls on bare clay hillsides, the water flows along the surface instead of penetrating into the soil. As the rainwater picks up speed, the water erodes away the clay, producing gullies such as the one shown here.

Soil conservationists suggest planting the slopes of such hills to reduce erosion. How could plants reduce the speed and amount of water running off a hill after a rain?

50

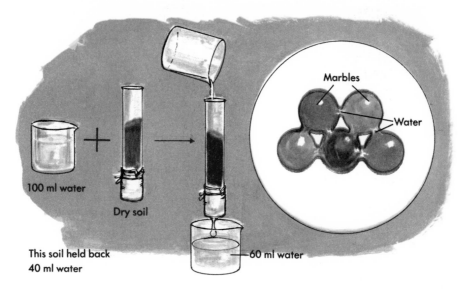

100 ml water

Dry soil

This soil held back
40 ml water

60 ml water

Marbles

Water

Water Retention. Fill a transparent cylinder with dry soil. Add known volumes of water until some comes out the bottom of the column. Collect and measure the water which seeps out. Calculate the volume of water still held by the soil particles.

Measure the volume of soil in the column. Divide the volume of water remaining in the soil by the soil volume. This ratio is a measure of the *water retention*, or water holding power of the soil.

Measure the water retention in different types of soils. In which soils is water retention high?

Adhesion of Water to Glass. Fill a small jar with glass marbles. Cover the marbles with water. Hold back the marbles while pouring the water out of the jar. Notice the water which remains. Where is it located?

The attraction of two unlike substances, such as water and glass, is called *adhesion*. The adhesion of water to soil particles is an important factor in the water retention of soils.

Water Retention in Different Soils. Note the different conditions of the two lawns shown at the right. Both soils receive the same amount of rainfall and lawn sprinkling. However, the differences in soils cause one lawn to be green and the other to be brown. Explain the differences in terms of water penetration and water retention.

How is water retention knowledge useful to (1) home builders, (2) septic tank installers, (3) farm pond builders, and (4) irrigation farmers?

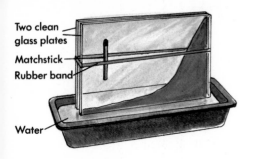

Two clean glass plates
Matchstick
Rubber band

Water

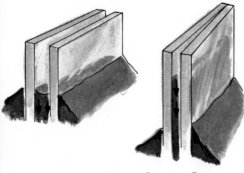

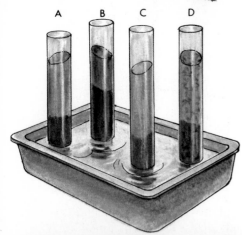

A B C D

Capillary Action. Set up the apparatus shown here. Observe the rise of water between the glass plates. How does the height of the water change as the distance between the plates increases?

Water is attracted to glass by the force of adhesion. The forces are shown at the left by colored arrows. Adhesive forces lift the water until the weight of water (black arrows) pulling down balances the adhesive force pulling up. Why is the water level higher where the plates are closer together? This effect is called *capillary action*.

Capillary Action in Soil. Place four transparent cylinders with different types of dry soil in a pan of water. Observe what happens to the level of water-moistened soil. Compare the levels of water moistening in the cylinders after several hours. In which soil does the water rise highest? Compare the size of spaces between soil particles and the height the water rises by capillary action.

Soil Water Classification. Scientists recognize three kinds of soil water: (1) drainage water, (2) capillary water, and (3) combined water.

Drainage water is present in soils just after a rain. Drainage water moves downward under the influence of gravity.

Capillary water adheres to soil particles and is used by plants. Capillary water gives soil its moist appearance when soil is spaded on a dry day.

Combined water is chemically united with soil minerals. Plants are not able to use water in this condition.

The clay soil below has lost much of its combined water. As the water is lost, a new mineral is formed which takes up less space than the hydrated mineral. Thus the soil shrinks and cracks. During the next rain, the clay minerals will react with the rainwater and swell.

SUMMARY QUESTIONS

1. How does soil compaction affect plants?
2. What happens to soil temperature with increasing depth? Is this true all year?
3. What effects does slope have on soil?
4. Describe the behavior of water in soil?

CHEMISTRY OF THE SOIL

Soil minerals do not dissolve readily. Otherwise, the soil would have dissolved long ago. However, only the small fraction of each soil mineral that does dissolve in soil water can be absorbed and used by plants. The other chemically useful part of the soil is composed of bits of soil only slightly larger than those which dissolve in water. These are called colloidal particles. Thus, the two most important parts of the soil, dissolved minerals and colloidal particles, are both too small to be seen with the unaided eye.

ESSENTIAL PLANT ELEMENTS

1. Carbon
2. Hydrogen
3. Oxygen
4. Phosphorus
5. Potassium
6. Nitrogen
7. Sulfur
8. Calcium
9. Iron
10. Magnesium
11. Boron
12. Manganese
13. Copper
14. Zinc
15. Molybdenum
16. Chlorine

Dissolved Chemicals. Use clean jars to collect samples of rainwater, well water, stream water, and rainwater which has passed through various types of soil. Place these open jars where the water will evaporate. Observe the jars after all the water has evaporated. Compare the amount of material that has dissolved in rainwater with the amount dissolved in the other water samples.

Phosphorus starved   Calcium starved

Potassium starved   Nitrogen starved

Elements in Plants. The minerals which a plant needs must first be dissolved in soil water before plant roots can absorb these chemicals. Each chemical element in the list at the left has been found necessary for plant growth. More than 90 different elements have been found in plants, but many of these do not seem to be essential.

Element Deficiencies in Plants. Plants must absorb enough of the essential elements to grow normally. Plants which lack one essential element are shown on this page. How does a lack of nitrogen affect the growth of corn? How does a lack of phosphorus affect the growth of corn?

There are two major causes for deficient soils. Some soils develop from rocks which do not contain the necessary elements. Other soils become deficient in one or more elements when the same crop is grown year after year. Eventually the plants use up all of one or more essential elements.

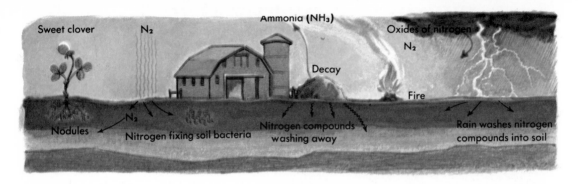

Sweet clover   N₂   Ammonia (NH₃)   Oxides of nitrogen   N₂

Decay

Fire

Nodules   N₂   Nitrogen fixing soil bacteria   Nitrogen compounds washing away   Rain washes nitrogen compounds into soil

Nitrogen in the Soil. Plants use nitrogen in the form of nitrate ions (NO_3^-), nitrite ions (NO_2^-), and ammonium ions (NH_4^+). Unfortunately, soil compounds containing these ions dissolve readily in water and wash out of the soil. Thus, nitrogen is often the one element that a soil lacks in order to produce maximum plant growth. For this reason, the study of how nitrogen compounds are replaced in topsoil is an important part of soil chemistry.

Nitrogen Cycle. The processes by which nitrogen enters and leaves the soil make up the nitrogen cycle. Many of the steps of the nitrogen cycle are shown in the picture above.

Nitrogen from the air is shown entering the soil at the left of the picture. This elementary nitrogen is plentiful in soil air, but most plants are not able to use it. A few soil bacteria can change this nitrogen into compounds useful to plants. Many of these nitrogen-fixing bacteria live in bumps or nodules in the roots of plants such as clover, alfalfa, and soybeans. As these bacteria die, the nitrogen compounds in their cells are used by other plants.

Pull up a clover or alfalfa plant. Compare the roots with those in the photograph.

Oxides of nitrogen are produced when nitrogen and oxygen combine during lightning storms. Other nitrogen compounds enter the atmosphere from fires and from decay processes such as in manure piles. These nitrogen compounds dissolve in rainwater and are washed into the soil.

Soil Testing Kits. Chemical tests for elements such as nitrogen, phosphorus, and potassium require special chemicals not readily obtained. For this reason, soil testing kits are sold which contain these special chemicals along with directions for using them to test soils.

Visit a farm store or garden shop that sells soil

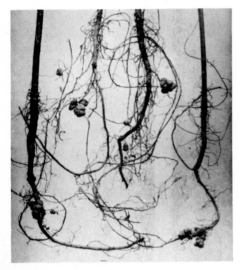

Nodules in legume roots

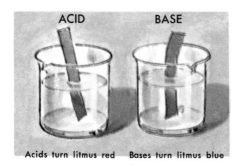

ACID　　　　BASE

Acids turn litmus red　　Bases turn litmus blue

SOIL pH FOR BEST
PLANT GROWTH

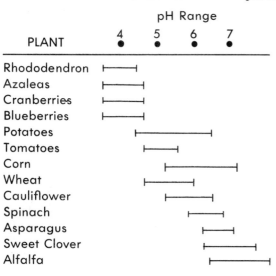

PLANT	pH Range
	4　5　6　7
Rhododendron	⊢—⊣
Azaleas	⊢—⊣
Cranberries	⊢—⊣
Blueberries	⊢—⊣
Potatoes	⊢———⊣
Tomatoes	⊢—⊣
Corn	⊢———⊣
Wheat	⊢——⊣
Cauliflower	⊢——⊣
Spinach	⊢—⊣
Asparagus	⊢—⊣
Sweet Clover	⊢——⊣
Alfalfa	⊢——⊣

testing kits. How many of the essential plant elements can these kits detect? Bring a soil testing kit to class and use it to test soils.

Testing for Acidic and Basic Soils. Acidic, neutral, and basic solutions can be identified by using litmus paper. Blue litmus paper turns red when dipped in acidic solutions. Red litmus paper turns blue when dipped in basic solutions. If neither red nor blue litmus turns color when moistened, the solution is neutral.

Use litmus paper to find out if a soil is acidic, basic, or neutral. If the soil is dry, add several teaspoonsful of distilled water. Then moisten the litmus with the wet soil.

Test soils in different locations. In what locations do you find soils which are acidic? Are these soils usually wet or dry?

Soil pH. The acidic, neutral, or basic property of soils can be described more accurately in terms of pH. Neutral soils have a pH of 7. Acidic soils have pH values less than 7. The more strongly acid the soil, the lower the pH value.

Basic soils have pH values greater than 7. The more basic the soil, the higher the pH value.

An Example of the Effect of pH. The hydrangea plants shown below are both of the same species. The conditions under which they grew were identical except that iron sulfate was added to the soil of the plant on the right.

Add some iron sulfate ($FeSO_4$) to moist, neutral soil. Test the soil with litmus to find out how the pH of soil changes when iron sulfate is added.

Plant Growth and Soil pH. Grazing animals prefer to eat plants growing in well-drained soils where clover grows. Such soils usually have pH values of 7 or greater, and are called sweet.

Studies show that plants have different pH preferences for best growth. Thus, the types of plants which grow well in acidic soils usually do not grow well in basic soils. Some examples of the soil preferences of plants are given on page 55.

Liming Acidic Soils. Use litmus to test water which has lime in it. Is lime acidic or basic?

Dig up a cube of acid soil 10 cm on each edge, and place it in a pail. Moisten the soil. Add a teaspoonful of lime to the soil, and test the soil with litmus. Repeat this process until the soil is neutralized. How much lime is needed to neutralize this amount of soil? How much lime would be needed to neutralize a square meter of soil to a depth of 10 cm?

The photographs show the effect of pH on the growth of alfalfa. Lime was added to an acid soil, resulting in a better crop. Note the pH preference of alfalfa in the table on the previous page.

Colloidal Soil Particles. Add several teaspoonsful of clay soil to a jar of water and stir it well. Observe the jar after several hours.

The small bits of soil which keep water cloudy, even after many hours, are called *colloidal particles*. Colloidal particles are composed of crystalline clay particles and noncrystalline humus particles. The colloidal clay minerals in the photograph below have been magnified about 30 000 times. Colloidal particles are especially important in soils because useful chemicals are attached to the colloidal particles.

Colloidal particles are believed to have negative charges on their surfaces. Positively charged ions, such as calcium (Ca^{++}), magnesium (Mg^{++}), potassium (K^+), ammonium (NH_4^+), and hydrogen (H^+), are attracted to the surfaces of the colloidal particles. Notice how the positive ions are attached to the colloidal particles shown on the next page.

Rainfall and Soil Differences. The map on page 57 shows the locations of two main types of soils called pedocal and pedalfer soils. *Pedocal* (ped–soil, cal–calcium) soils contain relatively large amounts of calcium ions. The important calcium ions are missing from *pedalfer* (ped-soil, al–aluminum, fer–iron) soils.

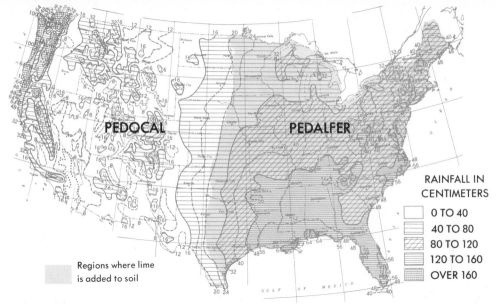

RAINFALL IN CENTIMETERS

	0 TO 40
	40 TO 80
	80 TO 120
	120 TO 160
	OVER 160

Regions where lime is added to soil

Compare the rainfall in regions where pedocal soils developed with regions where pedalfer soils developed. Make a hypothesis to explain how rainfall might have affected the development of these soils.

Colloidal particles are believed to play an important part in the development of soils. The role of these particles is pictured in the diagram.

Colloidal soil particles attract positively charged ions, such as calcium ions (Ca^{++}). In regions of high rainfall, some of the soil water breaks up into hydrogen ions (H^+) and hydroxyl ions (OH^-). These positively charged hydrogen ions (H^+) replace many of the other ions on the colloidal surfaces. The positive ions, such as calcium (Ca^{++}), which were removed from the surfaces of the colloidal material, wash out of the soil.

The hydrogen ions in the soil particles react with litmus paper in the same way that hydrogen ions in acids do (acids contain hydrogen ions). Thus, where rainfall is heavy, soils tend to be acidic and low in calcium.

Note regions on the map where lime is often added to soils. Why is lime added to pedalfer soils?

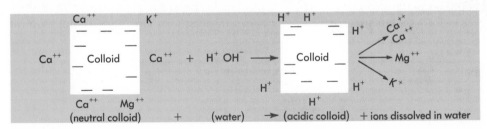

57

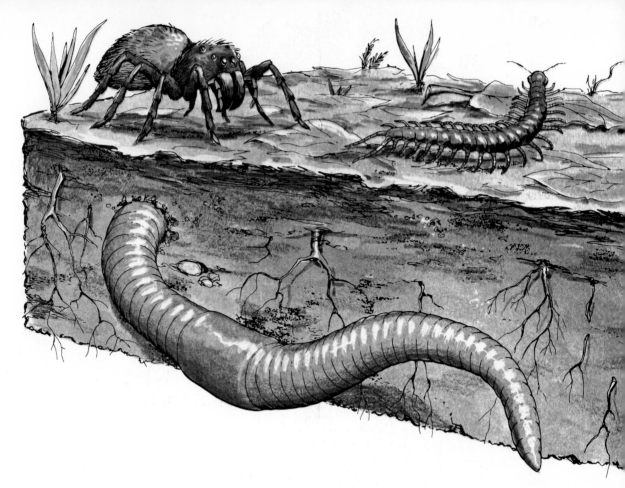

SOIL LIFE

Each spadeful of soil contains millions of plants and animals, most of which are microscopic. These organisms have many different sources of food. Some digest dead plant and animal material found on top of the soil. Others exist on roots and other plant parts in the soil. Some soil animals are predators, that is, they search for, attack, and eat other soil organisms.

Dead, green plant material on top of the soil is probably a major food source of soil life. This material passes through or becomes a part of many different soil plants and animals. Usually nongreen plants called fungi are the first to digest and use the material in green plants. Fungi, in turn, serve as food for many different types of bacteria and small soil animals. Eventually, much of the original plant material ends up as humus; but the exact process is not understood. This once-living material improves soil in many ways.

Animals in Soil. Carefully lift out one cube of woodland soil about 10 cm on each edge.

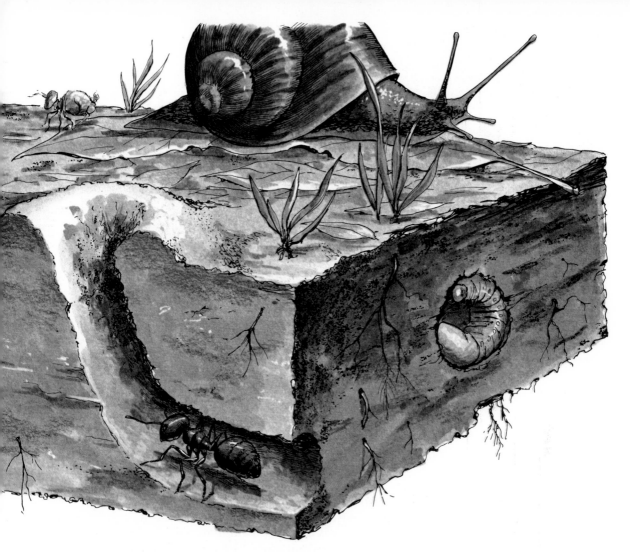

Spread the soil on a newspaper. Remove any small animals that you find in the soil and place them in jars. Use the picture on these pages to identify each animal. Keep a record of the type and number of organisms in this cubic decimeter (1 liter) of soil.

The number of small animals in the soil depends on many factors, such as food supply, moisture, and temperature. A study of soil in a New York forest revealed over 600 insects, mites, and other small animals in a soil sample about this size.

Animals such as worms, termites, springtails, millipedes, and snails feed on plant material, thus aiding in the process of plant decay. Other animals, such as spiders, ants, and centipedes, are predators. They feed on the plant-eating animals in the soil.

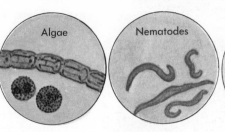
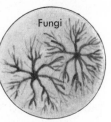

Rotifer Protozoa Algae Nematodes Fungi

RECIPE: Add 3 grams of agar and 1 boullion cube to 100 ml of water. Boil for 20 minutes. Add mixture to sterile Petri dishes. Allow to harden before using.

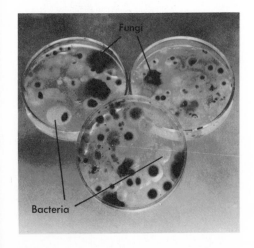

Fungi

Bacteria

Dobruson/Boston

60

Microscopic Soil Life. Boil some grass and leaves in water for 20 minutes. After the mixture cools, add two teaspoonsful of soil. Cover the jar and put it in a warm place.

After several days, examine some of the drops of the culture through the low-power lenses of a microscope. Use higher magnification to aid in drawing some examples of the different forms of life. Identify algae, fungi, protozoa, rotifers, nematodes, and other organisms from the pictures above and in other reference sources.

Biologists have calculated that 99.999996% of all the organisms which live in the soil are microscopic. Microscopic fungi compose about half of the living organisms in woodland soil. Bacteria and protozoa make up most of the other half. The bacteria and fungi are responsible for most of the decay which occurs in the soil. The protozoa feed on the bacteria and fungi.

Culturing Soil Microorganisms. Prepare several sterile dishes containing agar. (See the recipe.) Sprinkle a pinch of soil from different places in each dish. Keep one dish unopened as a control.

After several days, compare the condition of the control dish with the dishes to which soil was added. Use the photograph to aid in identifying bacteria and fungi in the culture.

Making Humus from Compost. Many gardeners collect lawn clippings, leaves, and other plant remains, and make *compost* piles of the material. The compost is digested by soil organisms and broken down and changed into colloidal particles of humus.

Use a sharp-bladed shovel to cut vertically through a compost pile. Observe the particles at the top and bottom of the pile. Which particles have been decaying longest? What changes in particle sizes have occurred? What changes in color have occurred? Make a list of the organisms visible in the

A-horizon / topsoil

B-horizon

Subsoil

C-horizon

Fractured rock

Bedrock

compost. Gardeners return the humus to the soil as a way of improving their crops. In spite of the importance of humus, its exact method of formation and its chemical composition are still not understood.

Soil Profiles. One useful way to summarize your understanding of soils is to examine a soil profile. Some examples of different soil profiles are shown below in the photographs. Humus regions are usually dark in color. This dark upper region of the soil is called *topsoil,* or the A horizon.

Below the topsoil is a region called *subsoil,* or the B horizon. Clay and minerals in the topsoil may wash down into this region. In some parts of the country, capillary action brings dissolved minerals up into the subsoil forming a white mineral layer in the profile.

The C horizon is the region just above the bedrock. This region is usually composed of broken pieces of bedrock material.

Visit an excavation near your school and examine a soil profile. Try to identify the A, B, and C horizons. Suggest what processes are operating on this soil to change it.

Collect samples from each horizon of the same profile. Set up an experiment to compare the growth of plants in the soil from these three horizons.

Northeastern United States forest soil

Northern Great Plains soil

Southern Great Plains soil

Central Mississippi Valley soil

REVIEW QUESTIONS

1. What four different types of matter make up soil?
2. Which soil has the largest particles: sand, silt, or clay? Which has the smallest?
3. Where is there an example of highly compacted soil near your school?
4. What effect does moisture have on the temperature of soil during a warm day?
5. Which side of a hill would have the highest soil temperature? The lowest soil temperature?
6. Through which soils does water penetrate quickly? Slowly?
7. Why is the penetration rate an important factor in the erosion of hillsides?
8. Why is the retention of water by soils important?
9. Why is capillary action important in soils?
10. What is the difference between drainage water, capillary water, and combined water?
11. What are some of the essential plant elements?
12. Of what importance are colloids in soils?
13. What is the difference between pedocal and pedalfer soils?
14. Of what importance to soils are each of the following: fungi, insects, bacteria, protozoa?
15. What is a soil profile? What are the major regions of a soil profile?

THOUGHT QUESTIONS

1. Why is distilled water used when testing soils for different elements?
2. Many soils of the eastern United States rest on bedrock of limestone, yet lime must be added to these soils. Explain.
3. Shaded lawns and sunny lawns have different types of weeds. What other factors besides temperature might be responsible for this difference?
4. Why would a hillside with plants on it have less water runoff than the same hill without plants?
5. Why would a farmer cut a soil profile before buying farm land?

Investigating on Your Own

The study of rocks and minerals provides students with many opportunities for doing science projects. Projects which make use of the rocks in your own region are especially interesting. Some of the projects suggested in this chapter require students to plan, carry out, and think about their results in the same way that research scientists do. In this way, you can find out whether a career in some field of science has interest for you.

Many of the activities suggested in this section require more time than one class period, or special materials not found in many schools. For these reasons, you should identify one or more activities from these pages to carry out on your own, outside of class. Ask your teacher to help you plan your work in class so that you will be able to get the materials you need to carry out your research. Plan to share your findings with others during later class periods.

CONTROLLED EXPERIMENTS

Science had its beginnings when people first observed the world around them and tried to figure out what they saw. However, modern science is often said to have begun with the discovery of the method of conducting controlled experiments. The next page gives suggestions for controlled experiments you can do on your own.

How to Do a Controlled Experiment. A geologist would be doing a controlled experiment when finding out how the rate of cooling affected the size of mineral crystals forming from molten granite. The geologist might melt granite a number of times, cooling it quickly and then slowly. Each time, the experimenter would probably measure the sizes of many of the crystals which formed.

An experimenter usually changes one factor or *variable* at a time. The variable that the experimenter changed in the above experiment is the rate of cooling. This variable is called the *independent* (or manipulated) *variable*. What is the independent variable in each problem below?

Usually, the experimenter observes how the independent variable acts on one other variable. The other variable in the experiment above is the size of the crystals which formed. Because the size of the crystals is thought to *depend* on the cooling rate, this second variable is called the *dependent* (or responding) *variable*. What is the dependent variable in each problem below?

Imagine an experimenter who carried out the experiment above and observed different sized crystals. How could you be sure the differences were caused by differences in cooling rate, and not some other variable? Such variables might be the

Problem	Independent Variable	Dependent Variable
Does the size of a piece of gypsum affect how much water you can drive off?	Size of pieces of gypsum	
Does sand near the water's edge contain more quartz than sand farther up on a beach?		Percent of quartz grains
Does CO_2 in water make limestone dissolve faster?		

type of container used, the pressure, the types of minerals present, and so on.

When an experimenter plans an experiment so as to control all other variables, the experimenter is doing a controlled experiment. What variables should be controlled in the problems at the bottom of page 64?

IDEAS FOR INVESTIGATION

1. Set up an experiment to compare the growth of plants in humus soil with plants grown in soil composed of crushed rocks, such as shale.

2. Set up an experiment to determine what effect sulfur has on the pH of soil.

3. Set up an experiment to compare how much water remains in clay, sandy, silty, and humus soils when a bean plant starts to wilt in each type of soil.

4. Twist round metal cans into the soil to obtain soil samples. Set up the apparatus shown here to test the soil air for carbon dioxide. Compare the amount of CO_2 in different soils.

5. Find out if sterilizing soil affects the growth of plants. Sterilize one sample of soil in an oven at 150°C for an hour.

6. Study the stones used for monuments in a cemetery. Prepare a circle graph showing the percent of each type of stone used.

7. Study the monument stones in a cemetery. Has one type of stone been the most popular since the cemetery began? If not, show what stones were most popular in which times. Propose hypotheses to explain your findings.

8. Study the stones in a cemetery. Which types are most resistant to weathering?

9. Test expanded vermiculite as a heat insulator. Fill two bottles with the same amount of hot water. Put thermometers through one-holed stoppers into the bottles. Bury one bottle in vermiculite; leave the other in open air. Keep a record of the temperature changes, and plot your observations on a graph.

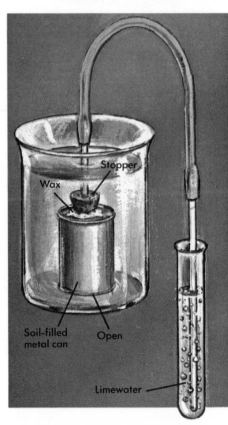

Stopper

Wax

Soil-filled metal can

Open

Limewater

MINERAL AND CRYSTAL STUDY

Few areas of science provide as many hobbies for nonscientists as do minerals and crystals. Hundreds of clubs exist throughout North America where "rock hounds" regularly meet. At these meetings speakers sometimes talk to the group. At other meetings, members swap samples of different minerals or fossils, exchange information about places to collect, or organize field trips to mineral collecting sites. Most such clubs would welcome junior high students.

Faceted stones

Mineral crystals are classified into six groups according to their shapes. Imaginary lines called *axes* (singular-axis) help to classify each crystal. Crystals belonging to the *cubic* group have three axes at right angles to each other. Each axis is the same length.

Crystals with different shapes can belong to the same general group. Do the following:

Carve a cube of raw potato or balsa wood. Cut a little from each corner as shown at *B*. Halite sometimes exhibits this type of crystal shape.

Continue cutting the same amount from each corner. The final shape will be like *C* (magnetite). Compare the number of faces and corners in the new shape to the cube you started with. What happens to the shape, its faces and corners, if you now cut off each corner?

Start with another cube. This time carve off the edges as at *D*. Continue until the new faces meet. The shape is that found in garnet crystals.

Cubic

Three equal axes at right angles to each other.

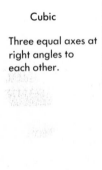

A. Halite

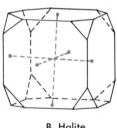

B. Halite

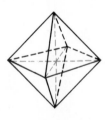

C. Magnetite

D. Garnet

Tetragonal

Three axes at right angles but one is longer than the other two.

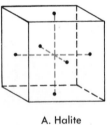

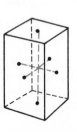

Zircon

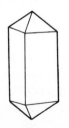

Cassiterite

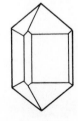

Vesuvianite

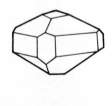

66

Orthorhombic

Three unequal axes at right angles to each other.

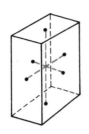

Barite

Pyroxene

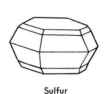

Sulfur

Monoclinic

Three unequal axes, two at right angles and the third at some other angle with the two.

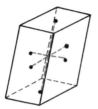

Gypsum

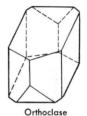

Orthoclase

Triclinic

Three unequal axes, none at right angles with the others.

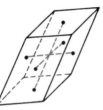

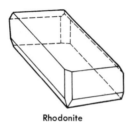

Rhodonite

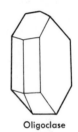

Oligoclase

Hexagonal

Three equal axes at 120° angles and a fourth longer than the others and at right angles to them.

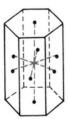

Beryl

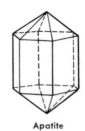

Apatite

Quartz

Calcite

Cubic   Tetragonal

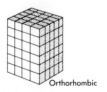

Orthorhombic

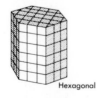

Hexagonal

Monoclinic

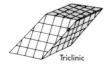

Triclinic

LEARNING ACTIVITIES WITH CRYSTALS AND MINERALS

1. Carve figures representing members of the other five crystal shapes. Use the same procedure of cutting off edges or corners until a different shape is formed.

2. Prepare a set of paper crystal models from lined paper. Such models clearly show the lengths and angles of the crystal axes. In laying out the patterns, be sure to provide flaps for pasting the sides together.

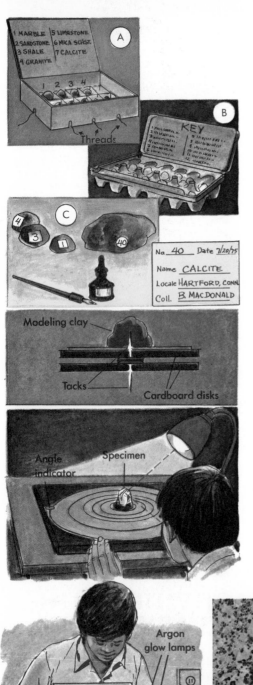

3. Make a collection of crystals, both natural and artificial. Classify the crystals according to the crystal system to which each belongs.

4. Arrange a trip to a museum where minerals are displayed. Find out which ones can be found locally and how they are used.

5. Make a mineral (or rock) collection. Look for mineral specimens at ledges, quarries, mine dumps, and lake shores. Keep specimens in boxes or cartons, each by itself. Number each specimen using indelible ink on a spot of quick drying enamel. Keep records which give the name, place and date collected, and name of the collector.

6. Measure the angle between cleavage faces or crystal faces with the device shown at the left. Paste a sheet of polar coordinate paper onto a stiff piece of cardboard. Insert a thumbtack through the center of the disc. Mount a mineral specimen in a piece of clay on top of the thumbtack. Ask someone to read the angle through which you turn the disc as you sight a reflection from one face, then the next face, of a mineral specimen.

7. Test minerals with ultraviolet light. Mount one or two argon glow lamps in a box. Compare the way the minerals look in white light and in light from these lamps. Some minerals glow (fluoresce) in ultraviolet light. (**CAUTION:** Do not look directly at an argon glow lamp; ultraviolet radiation from such lamps harms the retinas of the eyes.)

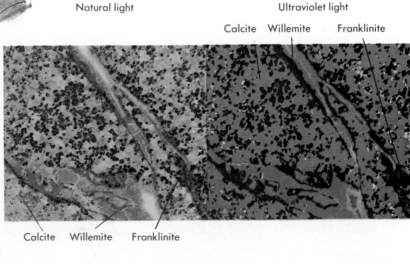

Natural light   Ultraviolet light

Calcite   Willemite   Franklinite

Calcite   Willemite   Franklinite

OTHER INVESTIGATIONS AND PROJECTS

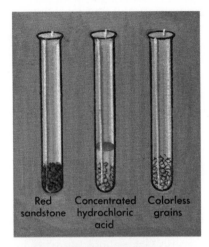

Red sandstone   Concentrated hydrochloric acid   Colorless grains

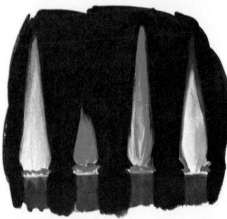

Sodium   Strontium   Copper   Potassium

1. Visit an exposure of bedrock. Identify the types of rocks. Develop a brief geological history of the area based on such clues as the kinds of rocks, bedding, folding, and foliations.

2. Determine the amount of magnetite in a rock by crushing the rock into a fine powder and collecting the magnetite with a strong magnet.

3. Determine whether the red color of a rock is due to red iron oxide (hematite). Crush the rock, put a pinch of the powder into a test tube and cover it with concentrated hydrochloric acid. (**CAUTION:** Concentrated hydrochloric acid causes severe burns.) If the red coloring is hematite, it will react with the acid to produce a brownish-yellow liquid. The powder remaining will be colorless after it is rinsed with water.

4. Determine the percent of carbon and carbon compounds in a sample of black shale. Crush the shale, treat the powder with dilute hydrochloric acid to remove any calcium carbonate which may be present, dry the powder, and reweigh it. Then heat the powder red hot to oxidize the carbon and carbon compounds. Weigh the powder again, and calculate the percentage of weight lost by oxidation.

5. Use flame tests to detect the presence of certain elements in minerals. Get a length of platinum or nichrome wire sealed in the end of a glass rod. Clean the wire by dipping it in strong hydrochloric acid and then holding it in a gas flame until no change in color of the flame can be seen. Touch the wire to a bit of powdered mineral which has been moistened with the acid and then hold it in the flame. Colors produced by four elements are shown at the right.

6. Calculate the approximate weight of a large rock, using its volume, and specific gravity. Decide whether the shape of the rock is closest to that of a sphere, cylinder, cone, pyramid, or rectangular block. Select from a mathematics book the formula for determining the volume, and make the necessary measurements. Chip off a sample of the rock and determine its specific gravity. (See page 8.)

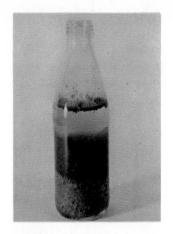

7. Add equal amounts of different clay soil samples to each of several bottles. Fill the bottles with water. Shake the bottles well and let them stand for a week or two. Observe the bottles daily. Keep a record of the rate of settling of any apparent layers in the mixture. Each layer represents a large fraction of particles of the same size, all settling at the same rate. Compare the slowly settling layers in each bottle. What does this show about the clay samples?

8. Visit the county agricultural agent. Ask the agent what soil problems are of importance locally. Report your findings to the class.

9. Invite a county soil agent or interested farmer to speak to your class about local soils.

10. Prepare a photographic scrapbook showing examples of good and poor soil conservation practices in your area.

11. Make a study of the different types of small animals found in a decaying log. Report your findings.

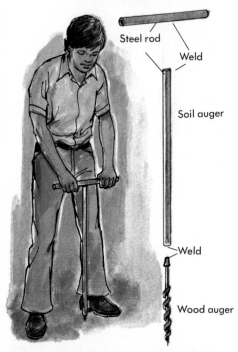

Steel rod

Weld

Soil auger

Weld

Wood auger

12. Prepare a display of products which are made from soil minerals; for example, glass and china.

13. Prepare a report about loco weed and other plant conditions caused by an excess of one element in the soil.

14. Make a soil auger by welding two pieces of steel rod to an old wood auger, as shown here. Use the soil auger to study the texture and appearance of soil at different depths. Study soil profiles using this device. Prepare a map showing the area studied, indicating the regions which have similar soil profiles.

15. Do an experiment to find out what is the best amount of fertilizer (or lime) to add to a soil. Add different amounts of fertilizer to each of several pots containing the same kind of soil. Grow the same species of plant in each pot. Make graphs showing the growth of the plants for each amount of fertilizer added.

16. Use Hydrion paper to test the pH of worm castings. (See the picture at the left.) Compare the pH of the castings with the pH of the nearby soil.

Castings

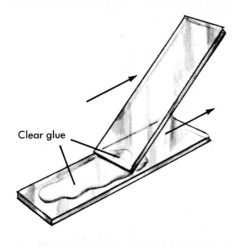

Clear glue

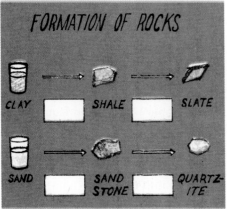

FORMATION OF ROCKS

CLAY   SHALE   SLATE

SAND   SAND STONE   QUARTZITE

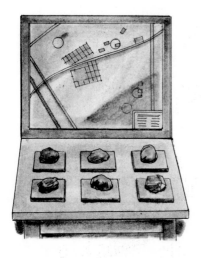

17. Study particles of soil with a microscope. Spread a thin layer of clear glue on a microscope slide as shown. Sprinkle sandy soil particles on the slide. Prepare slides of particles of other soil types. Examine the slides with a microscope. List the minerals you identify in each soil type.

18. Make an exhibit of rocks having economic importance. List ways in which the rocks are used.

19. Report on the rock and mineral resources of your local region or your state. Obtain information from chamber of commerce bulletins, state publications, journals, and encyclopedias.

20. Make an exhibit showing how rocks are formed and transformed. For example, display clay, shale, and slate with arrows showing the stages in the formation of slate. Use labels which describe the changes.

21. Make a map of the region around your school. Mark on the map the location of all exposures of rock. Collect samples and identify the rocks. Write the names of the rocks on the map. Prepare an exhibit of specimens to accompany the map, and mount it on the bulletin board.

22. Prepare a report on the raw materials and processes used in the production of portland cement. Collect pictures to illustrate your report.

23. Make glass as follows: Mix equal volumes of clean quartz sand, sodium carbonate, and lead dioxide in a metal can. Set the can in a hot fire (outdoors or in a furnace) for one hour. Then pour the contents of the can on several thicknesses of paper on the ground.

24. Make a list of the different kinds of building stones used for public and private buildings in your community.

25. Visit a local industry that uses rocks as raw materials in a manufacturing process. Prepare an exhibit of specimens and pictures which describe the process.

26. Prepare a brief history of glassmaking through the ages.

27. Investigate the process of brickmaking and the types of bricks produced.

71

Charge

Air heater

Air

Air

Slag

Iron

Slag

Iron

1. A flint-and-steel set contains: 1. Punk, the material to catch the spark. 2. Flint or quartz. 3. File or jackknife. 4. Tinder, such as cedar bark.

2. Hold flint and punk between fingers. Hold wick 2mm from edge of flint. With back of blade, strike glancing blows, vertically against the edge.

3. Aim blows of steel so sparks will fly into punk. When spark is caught and ember starts glowing, place punk in tinder. Blow with soft blows.

28.   Make a model of an oil well or a coal mine.

29.   Weigh limonite before and after heating to determine the percent of water of crystallization driven off by heating.

30.   Produce copper from cuprite (an ore of copper) by use of a reducing flame and a blowpipe, as described for galena on page 18.

31.   Make a diagram of a blast furnace like that at the left. Find out the purpose of each part, and the chemical changes that take place within the blast furnace during the reduction of iron ore.

32.   Prepare a report on the way in which mercury is obtained from its ore.

33.   Use a soil test kit to test some garden soil. Prepare a report about any mineral deficiencies you discover.

34.   Prepare a report on the methods by which an important metal such as copper is extracted from its ore.

35.   One method by which early men put sharp edges on their flint tools is shown at the left. Find a piece of flint and a deer's antler or other pointed instrument and try to produce a good cutting tool.

36.   Visit an amateur gem cutter and watch gems being cut. Prepare a report on gem cutting.

37.   Prepare an exhibit of important minerals and their products.

38.   Find out from a scout manual how pioneers used flint and steel to start fires. Practice the method, and demonstrate it to your class.

① ② ③

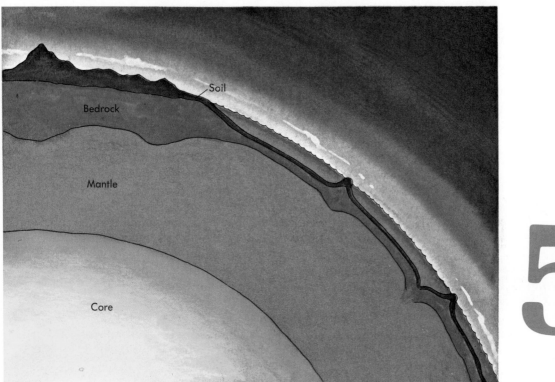

Soil

Bedrock

Mantle

Core

What's Below the Bedrock?

Many scientists have remarked that we know more about some distant star or the sun than we know about the inside of our own earth. Perhaps there is a reason. For example, the sun continually sends information about itself in the form of sunlight. By analyzing the light, we can learn much about its source, the sun.

The earth's interior sends no such regular information about itself. An occasional volcano sends up cinders or molten rock material, and earthquakes shake the ground below us. Caves and mines permit us to look just below the surface, and holes drilled in rock allow us to lower thermometers to learn a bit more. In spite of these limited bits of information, earth scientists are piecing together a very interesting picture of the inside of the earth.

THE TEMPERATURE INSIDE
THE EARTH

The temperature inside the earth would be useful to know. If the temperature is very high, the material inside the earth is probably molten. If the temperature is low, the material is probably solid.

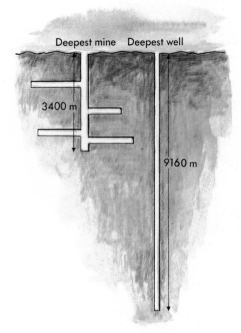

Deepest mine   Deepest well

3400 m

9160 m

DEEPEST WELL

Year	Meters
1859	22
(Drake's first oil well)	
1895	366
1909	1730
1924	2230
1935	3890
1948	5440
1955	6550
1972	9160

The Temperature Below the Earth's Surface. Can you recall when you were much smaller? Perhaps you once dug into the earth to see what was down there. When you got tired, you may have noticed how cool the dirt in the hole was. From this observation, you may have concluded that the deeper you go, the cooler the earth's temperature is.

Another way to learn about the temperature inside the earth is to go down into a cave. Cave visitors often comment on how cool it is down there.

Both of these observations are biased. That is, they do not give an accurate idea of temperatures below the earth's surface. The reason for the bias is that such observations are usually made in the summertime. People who dig holes in the ground or who visit caves in the winter usually observe that the temperature inside the earth is higher than it is at the surface.

Geologists have observed that the temperature of the earth about 15 meters below the surface remains the same all year long. The heat of summer and the cold of winter do not affect the temperature at this depth. The temperature remains the same as the average temperature at the surface throughout the year. Why are caves in New Mexico warmer than caves in Kentucky?

Temperature in Deeper Holes. Mines and wells both provide data about temperatures inside the earth. The deepest mines in the world are in South Africa where diamonds and gold are mined. One such mine is over 3400 meters deep. At this depth, the temperature is 51°C.

The deepest holes into the earth have been made by oil well drillers. In 1972, an oil well was drilled in Oklahoma to a depth of 9160 meters. Geologists estimate that the temperature increases about 25°C for every kilometer (1000 meters) of depth below the earth's surface. Estimate the temperature in the bottom of the well in Oklahoma.

The table on page 74 shows how improvements in equipment and know-how have permitted the drilling of deeper and deeper wells. Make a graph of these data. Mining engineers in Russia have announced a program to drill a well 15 000 meters deep. Use the graph to estimate the year this might happen. What factors might change to upset your prediction?

Geysers. The temperature below the earth's surface increases 25°C for every 1000 meters of depth *on an average*. There are, however, many regions where the temperature does not increase at an average rate. Regions with active volcanoes, hot springs, and geysers are not average. Here, the temperature increases at a much faster rate.

North America's most famous geyser is shown at the left. Old Faithful is located in Yellowstone National Park. This geyser got its name because it faithfully spouts up hot water and steam about every 65 minutes.

The diagram at the left shows a type of geyser action found in kitchens. The coffee percolator is heated at the bottom. Bubbles of water vapor form at the bottom and are directed upward through the small tube in the center. The rising gas bubble pushes water up ahead of it. The "boiling" water splashes against the top of the percolator and drains back downward, passing over the ground coffee.

Does the action of a percolator prove how geysers work? Explain. Use a percolator's action to propose how the geyser below might work.

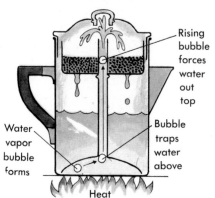

Rising bubble forces water out top

Water vapor bubble forms

Bubble traps water above

Heat

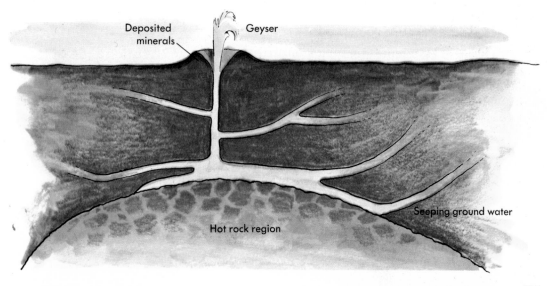

Deposited minerals

Geyser

Hot rock region

Seeping ground water

THE REGIONS OF THE EARTH'S INTERIOR

The distance from the earth's surface to its center is about 6380 kilometers. The deepest we have ever been inside the earth is about 3 km. The deepest we have drilled and brought out pieces of rock is 9 km. Thus, our ideas about the earth's interior below 9 km must be made without ever having been there or having samples to look at.

Although no one has ever seen the earth's interior, many ideas about its composition, state, and temperature have been proposed. Most of these ideas come from records made by energy waves which spread outward from earthquakes. These waves travel to distant parts of the earth's surface. There is evidence that some of these waves travel through the earth's interior. By studying the records of these waves, scientists have proposed hypotheses about the material through which the waves pass.

Seismograph. The seismograph pictured here shows a beam of light reflecting from a mirror onto a photographic film on a slowly rotating drum. When the earth shakes in the direction shown by the double arrow, the seismograph and recorder drum move. But the pendulum weight has so much mass that it moves very little. As a result, the light beam traces a wavy line on the photographic film as the recorder's drum slowly turns.

Seismograph Data. Three earthquake records, called *seismograms*, are pictured on page 77. They were made by seismographs at three different stations. Seismograms usually show three different types of earthquake waves. The first group of waves to reach station A are called *P* waves. The *P* waves

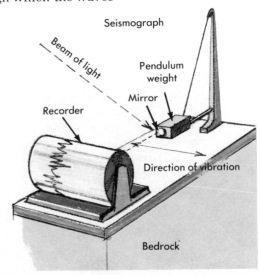

Seismograph

Beam of light

Pendulum weight

Mirror

Recorder

Direction of vibration

Bedrock

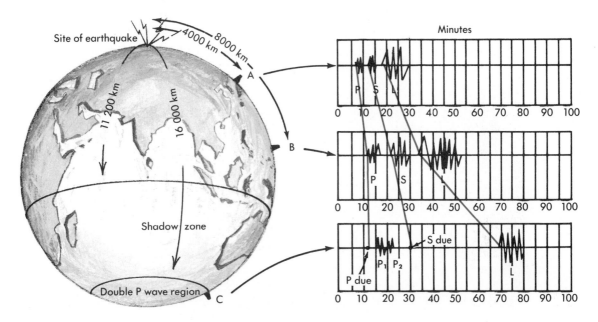

Minutes

Site of earthquake

4000 km

8000 km

A

11 200 km

16 000 km

Shadow zone

Double P wave region

C

B

P   S   L

P   S   L

P due

P₁ P₂

S due

L

took 8 minutes to travel to station A. How long did the P waves take to reach stations B and C?

The second group of waves reached station A 13 minutes after the quake occurred. These waves are called S waves. How long did the S waves take to travel to station B? To station C?

The third set of waves on a seismogram are called L waves. How long did the L waves take to reach each station?

Station B above is twice as far from the earthquake as is station A. The L waves take twice as long to reach the seismograph at B as to reach the one at A. This is to be expected if the L waves travel at a constant rate along the surface. However, the S and P waves take less than twice as long to arrive at B as at A. What does this suggest?

Seismographs 10 500 km from an earthquake may receive strong S waves, but stations at distances greater than 11 000 km never receive S waves. Is there an S wave on seismogram C above?

The P waves which are received in the region 11 000–16 000 km from an earthquake, arrive later than expected. For these reasons, the region 11 000–16 000 km from an earthquake is called the *shadow zone* .

In the region 16 000–20 000 km from the earthquake, the P wave appears to arrive late and at two different times. These waves are labeled P_1 and P_2 in seismogram C.

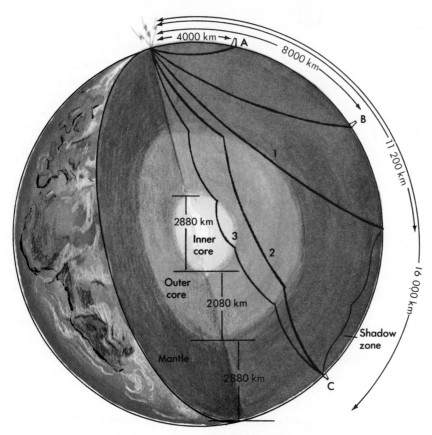

Interpreting Seismograms. Seismograph B on the preceding page is twice as far from the earthquake region as is A. P and S waves take slightly less than twice as long to reach B as A. Scientists assume that P and S waves travel *through* the earth instead of along the surface.

The shadow zone region can be explained if there is a sudden change in the earth's composition at a depth of 2900 km. Scientists believe so strongly in this change in composition that they call the region below a depth of 2900 km the core. The next outer region of the earth's interior is called the mantle. P waves would bend as they pass from mantle to core, and bend again as they pass from the core into the mantle again. This is shown by the waves labeled 2 above.

S waves must be unable to pass through the core because they never arrive at seismographs further than 11 300 km from an earthquake. One way to explain this is to suggest that the core is molten, and that S waves cannot pass through molten material.

The two *P* waves which arrive at the earth's surface opposite an earthquake can be explained if the inner part of the core changes again. Perhaps the inner core is solid. Under these conditions, two different waves, such as waves 2 and 3, traveling through matter in two different states, would travel at slightly different speeds. Therefore, these waves would arrive at seismograph C at slightly different times.

Discontinuities. There are other regions in the earth's interior at which earthquake waves are believed to bend or to reflect. These regions are called discontinuities. Each discontinuity is believed to represent a change in the composition of the earth's interior. One important discontinuity occurs only a few kilometers below the surface. It is called the Mohorovicic discontinuity, or the Moho. The material above the Moho is called the earth's crust. The region below is called the mantle.

The diagram below shows the Moho under a continental region and under the ocean. Note that the Moho is much closer to the earth's surface under the oceans than under the continents. This discontinuity exists at a greater depth under mountains than under flat lands.

The New Geological Revolution. Most geologists have studied a broad range of topics in the earth sciences. Geologists, however, tend to be interested in and gather data about some specialized area of geology. For example, a geology professor may teach different courses, but carry out research in the area of mountain formation. A mining geologist may be paid to know where minerals are located inside the earth, but also be studying ways of drilling down beyond the Mohorovicic discontinuity. In the past, these researchers usually explained their observations using a model of the earth in which the continents were considered as land masses fixed at definite places on the earth's surface.

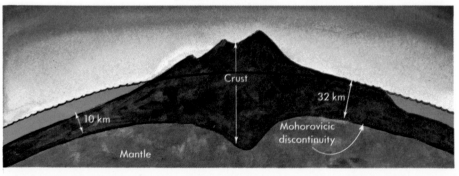

Crust

32 km

10 km

Mohorovicic discontinuity

Mantle

In the late 1960's, geologists studying the floor of the oceans made observations about the rocks there which could only be explained if Europe and North America were moving apart. Fossil geologists noticed that the locations of fossils in Europe and North America, and those in Africa and South America, could be best explained by using this same idea. This idea now has the name of *continental drift*. All geologists, including those studying the earth's interior, must now explain their observations in terms of an earth's surface with drifting continents. A basic change in the way scientists think about their area of study is called a *scientific revolution*.

REVIEW QUESTIONS

1. What evidence supports the idea that the earth gets hotter as you go deeper?
2. What is a geyser? How does it work?
3. What is a seismograph? How does it work?
4. What are the different kinds of earthquake waves? What do they tell us about the earth's interior?
5. What is a discontinuity?
6. Describe the variations in the thickness of the earth's crust.
7. What evidence suggests that the earth has a core of solid iron surrounded by molten iron?
8. What evidence supports the idea that the molten part of the earth's core is in motion?

THOUGHT QUESTIONS

1. What are some ideas which caused scientific revolutions in the past 300 years?
2. Why do we need to know about the earth's interior?

6

Using the Earth's Resources Wisely

Our lives and happiness are closely bound to the earth. Our houses, cars, and other things in our daily life all come from rocks and minerals in the earth, or from plants growing on the earth. In addition, nearly all of the energy for heating our houses and moving our cars comes from coal and oil in the earth.

In past centuries, the world population has been much smaller. In addition, each person used smaller amounts of the minerals from the earth. In those days few people believed that the minerals would ever be used up. Today, however, the population of the world is increasing rapidly. Also, each person (particularly in the United States, Canada, and western Europe) requires increasing amounts of energy (gas, oil, electricity) as well as increasing amounts of metals for cars, trailers, plumbing, radios, and televisions. Most people who think about it today believe that the earth cannot continue to supply us with minerals as it has in the past. In this chapter, some ideas are presented to assist you in thinking about this problem which is very important to you as well as to your future children.

ENERGY NEEDS AND ENERGY SUPPLIES

The top two photographs show two people using energy. The person in the top photograph is rapidly using food energy supplied by the body. The one in the next photograph is using skills and equipment (technology) to do most of the work. Why does the person who consumes more human energy probably take home less pay? Why does the technological worker, who pollutes the air and spoils the land to secure energy, take home more pay?

For many years, people have been saying that countries which use lots of energy are technologically advanced and rich; those which don't are technologically backward and poor. As a result, billions of dollars have been spent supplying less advanced countries with electric power plants, trucks, and highways so that these less fortunate countries can enjoy greater prosperity. At the same time, our needs for energy increase to supply our electric toothbrushes, air conditioners, golf carts, lawnmowers, and exercise machines.

Recently, people have begun to realize that energy consumption leads to increased air and water pollution. They realize that the world's sources of energy may not be able to satisfy our rapidly increasing demands. Certainly there is not enough fuel to supply the world's population with the amount of energy per person we use in our country. Keep these ideas in mind as you study our energy needs and energy supplies.

Human Energy. People use energy whenever they move, or their arms or legs move something else. The upper photograph shows someone tossing alfalfa. How do you know energy is being used?

The human body can do work for many hours because of stored food energy within itself. Foods containing sugars, starches, fats, or oils can be used by the human body to resupply it with energy. When a person runs out of energy, the supply must be renewed by taking in food.

The Source of Human Energy. The photograph at the left shows one example of sources of human energy. Where did the energy come from which is stored in this food?

Some types of energy can be restored quickly by natural processes. Such energy sources are called *renewable*. Is the energy in plants a renewable or a nonrenewable energy resource? Explain your answer.

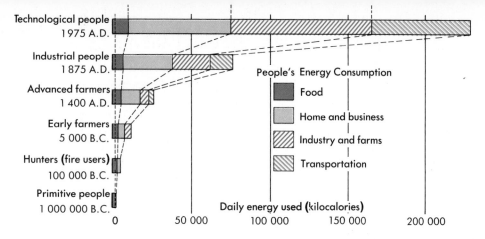

People's Energy Consumption

- ■ Food
- ■ Home and business
- ▨ Industry and farms
- ▨ Transportation

Daily energy used (kilocalories)

A typical grain field is able to convert about 1% of the sun's energy striking it into food energy. In 35 years the world's population is expected to double, because there are more new births each year than deaths. Propose two possible solutions for growing enough food for the world's population 35 years from now.

Who Uses the World's Energy? Energy is measured in units called *kilocalories* (sometimes called food calories). A kilocalorie *(kcal)* is the amount of heat energy needed to raise the temperature of one liter of water one Celsius degree. A typical American eats about 2500–3500 kcal per day, depending on the amount of activity.

A million years ago, smaller primitive humans probably used about 2000 kcal per day of energy from the world's supply. All this was renewable energy in foods. Later generations discovered the uses of fire, domesticated animals, water and wind power, steam, electricity, and the internal combustion engine. Today, as the diagram here shows, people in highly technological countries use an average of 230 000 kcal of energy per day.

Study the chart above. Use the ideas in the previous paragraph to explain the various energy uses of people at each stage of development. What are some ways by which Americans use the world's energy? Where today can you find examples of people consuming energy at each rate?

A small part of the increase in kcal consumed between primitive tribes and modern societies is caused by a better diet of 2500–3500 kcal actually eaten. The diagram at the left shows where the rest of the energy goes in supplying technological nations with food.

10 000 Kilocalories to Feed Technological Man

Usable plant energy = 2200 kc

Processing wastes

Feed for Pigs and Cows

Usable meat energy = 900 kc

3100 kc

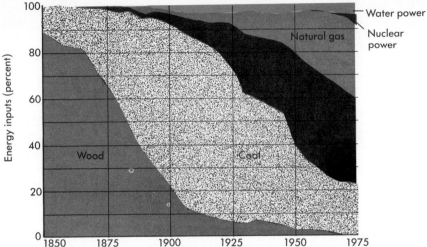

Changing Energy Sources. The diagram above shows the changes in nonfood energy sources in the United States in the past 125 years. What was the major source of energy in 1850? Why would it be impossible to supply 90% of today's energy needs with wood? Study the types of fuels used in 1975 and suggest how these fuels were used by industrial people as shown in the chart on the previous page.

Which two energy sources in this diagram are renewable? How much of our energy is produced from these sources? Make a list of the 1975 energy sources in order of decreasing use.

Graphs help us to predict what will happen in the future. Predict which source of energy will play a more important role by the year 2000.

Coal as an Energy Source. Many of your grandparents can remember when coal was the only common way of heating homes and producing power. The soot-blackened older buildings in cities remind us of the once widespread use of coal. Today coal is the one energy resource which is not in short supply. Today we know of coal in the ground in the United States which will last for 400 years based on the present rate of use. Unfortunately, much of this coal produces black smoke and air pollutants such as sulfur dioxide. As a result, it cannot be used as a fuel in large cities where air pollution is a problem.

Many scientists believe that if enough money were spent on research, techniques could be developed to release the energy in coal without polluting the air. Because of these problems, the use of coal is limited to large factories and power generating plants having large smokestacks and

expensive equipment to reduce the pollution from this fuel.

Petroleum and Natural Gas Formation.
Methane and other kinds of *natural gas* are produced during the chemical breakdown of any plant or animal material. *Petroleum*, on the other hand, seems to have been formed only from marine sediments, not from land organisms. There are many uncertainties about the formation of petroleum.

A lot of the natural gas and petroleum formed in the world has escaped to the surface and been lost. However, some has been trapped in porous rocks between layers of nonporous rocks. Wells drilled into the porous rock make the natural gas and petroleum available. The diagram on this page shows one arrangement of rock layers that allows large quantities of gas and petroleum to collect in small areas.

Petroleum must be refined before it can be used. During refining, it is separated into useful energy products such as gasoline, kerosene, diesel oil, heating oil, motor oil, grease, paraffin wax, and road tar.

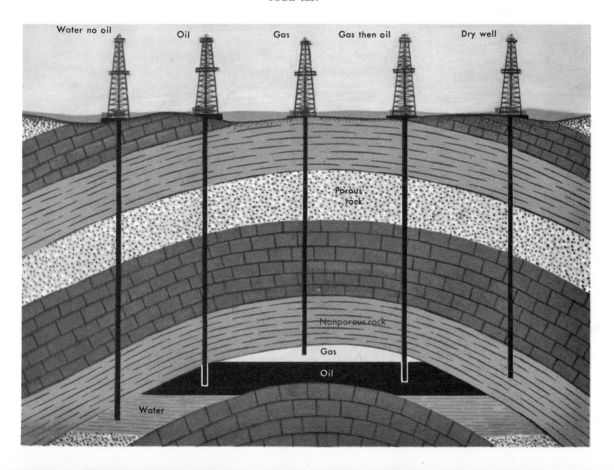

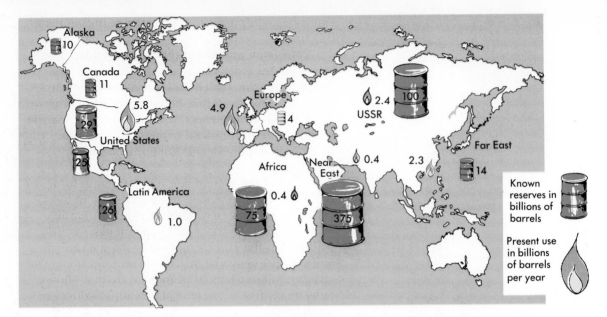

Known reserves in billions of barrels

Present use in billions of barrels per year

Users and Suppliers of Oil and Gas. The map above shows which regions use and which supply oil. All the figures on the map are in billions of barrels of oil. One barrel contains 160 liters. What regions of the world use large amounts of oil? What regions are major suppliers?

If Europe did not import oil, how long would its own oil reserves last? If North America did not import oil, how long would its oil reserves last? How long could the Far East continue without importing oil?

Such predictions can only be accurate if (1) the use of oil does not change, and (2) the known reserves do not change. What might happen in future years to change either or both factors? Discuss in class how changes in either of these factors will affect your answers above.

What would happen to Europe or North America if the Far East and Africa decided to increase the price of oil to ten times its present price? Discuss the possibilities if the Far East and Africa decided not to export oil at all to Europe and North America. Suppose some Far East nation got in an argument with a South American nation and the United States was asked to help get these countries to agree on a peaceful solution. Which country might our government favor in its decision? Why? Suggest a reason why each government would like a 100-year supply of oil reserves.

Generally, oil and gas are found in the same

regions. Thus the same countries have excesses or shortages of both gas and oil. Unfortunately, the worldwide picture is worse for gas than for oil. Experts predict that the known stocks of oil in the world may last for 30–35 years, but the supply of gas will last only about 17 years.

Hydroelectric Power. Many early towns in North America were located near a waterfall. Some of the water could easily be made to turn a water wheel. The turning wheel was used to grind grain, cut lumber, or run machines. Today, tall dams are used to provide energy. Giant turbines and electric generators produce electricity which is sent hundreds of kilometers to homes and industries.

In North America about 8% of the available rivers are dammed, and used for power. What are some arguments for constructing more dams? What are some arguments against such construction? What is the source of the energy which causes water molecules to evaporate before they condense as rain and form the streams which provide hydroelectric power? Is this a renewable energy source?

Nuclear Fission. During World War II, a group of scientists, including Albert Einstein, showed that immense amounts of energy could be released by splitting the element Uranium-235. When the nuclei of Uranium-235 atoms split apart (*nuclear fission*), new elements form. Great quantities of heat and light energy are released in the process. As a result of these studies, the atomic bomb was developed.

Today, a similar process of releasing energy is carried out under carefully controlled conditions so that the energy is released more slowly. Such furnaces for releasing nuclear energy are called *nuclear reactors*. Nuclear reactors are an important new source of power. Nuclear power stations usually use the heat energy produced to make steam; the steam then turns turbines which drive electrical generators.

One important pollutant from this process is heat. Vast quantities of water are needed to carry the heat away. This water is heated a few degrees, but not enough to be useful as an energy source. Suggest why nuclear reactors are built near large sources of water. How might these reactors affect the life in these waters? For what other reasons might people not want nuclear power stations near their homes?

Water cooling towers at a nuclear power station.

A nuclear reactor.

Controlled Fusion. Hydrogen bombs release vast amounts of energy when hydrogen is turned into helium. The process of fusing the nuclei of atoms together is called *nuclear fusion*.

Today, scientists with very expensive apparatus are trying to fuse hydrogen nuclei into helium nuclei under controlled conditions. So far, they have not been successful, but the research goes on. Only 1% of the world's supply of the special type of hydrogen needed for this reaction would produce 500 000 times as much energy as all the people in the earth's history have used to date.

Geothermal Energy. Recently, geologists have accepted the idea that the continents slowly move about on the earth's surface. The energy for this process comes from the heat within the earth. So, why not harness this heat?

The photograph at the left shows one example of a power generating plant located in a region of hot springs in California. Other such plants are found in Iceland, Italy, and New Zealand, which also have hot springs. Experiments are now being conducted to investigate the possibility of drilling very deep wells into the earth's surface. The heat in the rocks at these depths might be used to power electric generating stations.

Tidal Power. The sketch at the left shows one idea for harnessing the power of the twice-daily tides. Water would be permitted to enter bays as high tides approach. After high tide, dams across the mouths of bays would be closed. Then the water behind the dam would be used to turn water turbines which would be connected to electrical generators. What are some problems which would have to be solved before this method of obtaining energy could be used?

New Technologies and Discoveries. Geologists know that many layers of porous shale in the western United States and Canada contain vast amounts of stored energy in the form of oil. However, it is very difficult and costly to extract the oil from shale. As long as foreign oil was available at a low price, getting oil from shale was too expensive to be practical. But with the rapid rise in the price of oil, researchers are investigating the possibility of getting oil from this abundant source. What are some of the problems that must be solved in extracting oil from shale?

The size of a country's fuel reserves depends

Harnessing
tidal power

on how much money is spent looking for future supplies. There are no doubt other sources of oil not yet discovered. However, most geologists think that future discoveries will be of smaller and smaller oil fields in a time of greater and greater demands for oil. How might these discoveries affect the energy shortages of the world?

Setting Up an Energy Budget. The energy of the world can be divided into renewable and nonrenewable sources. Renewable forms of energy are renewed by the sun's energy which reaches the earth. Sunlight, food, wind, waterpower, wood, and muscle power are examples of renewable sources of energy. How does the sun renew each kind?

Most of the energy reserves now being used in the world are in the forms of natural gas, oil, and coal. These are called *fossil fuels*. How did the sun produce these nonrenewable energy resources? Uranium-235 is another nonrenewable resource.

Some people compare nonrenewable and renewable energy resources to the resources in dollars of a family budget. Suppose a family has $2000 in savings (nonrenewable) and $200 a week income from working (renewable each week). How long can they spend $2200 a week before they run out of money? $1200 a week? $200 a week?

As we have seen, scientists are trying to find ways to release slowly the energy given off when a type of hydrogen changes into helium in the process of nuclear fusion. If these efforts are successful, the results will be the same as if the family above suddenly received a gift of $50,000 to add to their savings. How long could they now spend $2200 a week? Are this family's immediate problems of spending $2200 a week solved by the $50,000 gift? Are their problems solved forever? How would success in fusion research affect our energy crisis?

Scientists using a compressor in a fusion research project.

SUMMARY QUESTIONS

1. Which source of energy is in shortest supply?
2. Which source of energy is most abundant?
3. Which nations of the world use the most energy?
4. Which nations of the world have the greatest reserves of oil? The fewest reserves?
5. What are some energy sources which did not begin with sunlight?

METALLIC AND NONMETALLIC RESOURCES

Many of the things in our daily life come from the earth's crust. The steel in cars, the copper in wires, and the gypsum in plasterboard all started as minerals in the earth. When metallic minerals occur in amounts large enough to be mined, they are called ores. Iron, copper, and aluminum ores make up some of our most important resources. Is iron ore a renewable or nonrenewable resource? Explain your answer.

Technological nations demand greater quantities of metallic and nonmetallic minerals each year. This demand causes at least two problems: (1) how to supply the increasing demand, and (2) how to get rid of the materials when they are no longer useful. This second problem is one of solid waste disposal.

What raw materials are used in making steel?

Iron. The United States has been a major source of iron ore for many years. Originally, small iron ore deposits were mined in the northeastern United States. More recently, larger deposits in Alabama and northern Minnesota have been mined. Today, huge deposits in eastern Canada are important suppliers.

The process of changing minerals to metals, such as iron ores to iron and steel, is called smelting. Iron smelting requires other minerals in addition to such iron minerals as hematite, magnetite, and limonite. The most important other mineral is coal. In order to produce one metric ton of steel, 2 to 3 metric tons of iron ore, 1 metric ton of coal, and ⅓ metric ton of limestone are consumed. In addition, 4 metric tons of air, and 40 kiloliters of water are used, mostly for cooling.

Aluminum. The ore of aluminum is called bauxite. Bauxite contains aluminum oxide. Large deposits of bauxite are found in Alabama, Arkansas, Georgia, and Tennessee.

Aluminum has many uses because of its valuable properties of strength, light weight, resistance to corrosion, and good electrical conduction.

Aluminum is a lot more expensive than iron per kilogram. One important reason is the high cost of separating the aluminum from the oxygen in aluminum oxide. This requires large amounts of electrical power. For this reason, most aluminum producing plants are located near sources of cheap electrical power, such as large hydroelectric stations.

Bauxite

Copper. Although copper is sometimes found as the pure metal, most copper is smelted from one of several common minerals. These include chalcopyrite ($CuFeS_2$), chalcocite (CuS_2), and cuprite (Cu_2O). Copper's properties of excellent electrical conduction, heat conduction, and ability to be shaped make it a very useful and valuable metal. About half the copper mined in the U.S. comes from mines in Arizona like the one shown above. Other copper ores are mined in Utah, New Mexico, Montana, Nevada and Michigan.

Lead and Zinc. Minerals containing these two metals are often found in the same region. Frequently the minerals are found in mountains or once mountainous regions where the rocks are highly metamorphosed. Such deposits are found in New York, New Jersey, Idaho and Utah. Other important deposits are found in the sedimentary rocks of Missouri, Kansas, Illinois and Wisconsin.

Lead is used in the production of electrical storage batteries, solder, paint, and in chemically resistant pipe. Zinc is used as a plating on iron (galvanized iron), to make alloys with other metals (brass, bronze, solder), and in pigments. The United States is the leading producer and consumer of these two metals.

Chalcopyrite

Chalcocite

OTHER METAL RESOURCES

Name	Uses
Tin	Plating iron in cans; in solder
Nickel	Alloying (mixing with other metals)
Gold	Jewelry
Silver	Jewelry, photography
Uranium	Nuclear reactors, medicine

Cuprite

91

The Case for Nonmetallic Minerals. Most of us have seen at least a few movies describing the excitement and glamour of mining metallic minerals such as gold and silver. Other films show the drama and wealth awaiting those who discover mineral fuels such as oil or uranium. For some reason, book writers and film-makers rarely consider the search for nonmetallic resources as worthy of attention. In spite of this lack, nonmetals play an important part in our lives. Study the picture above. How many different nonmetals can you find? Use the table on the next page to learn more about these nonmetals in your daily life.

The Economics of Nonmetals. Nonmetals include such different substances as diamonds and gravel. Some nonmetals, such as diamonds, are rare and have very special properties found in no other substance. As a result, diamonds are shipped all over the world and are very expensive. Another nonmetal, such as gravel, may vary greatly in its

properties, is heavy, and is required in huge amounts. As a result, gravel is generally trucked less than 30 kilometers. Larger travel distances increase the shipping price so much that a new, closer supply is usually used. Fortunately, gravel (or its substitute, crushed stone) is available widely. Although diamonds may seem more glamorous, many more dollars are spent and many more people work in the gravel industry.

SOME NONMETALLIC RESOURCES

NAMES	USES

Abrasives:
  garnet, corundum,
  emery
	cutting aid

Asbestos	cement pipe, sheets, insulation, brake linings
Borax	glass, fiberglass, solar batteries, nuclear reactors
Clay: kaolin, fire clay	bricks, tiles, pipes, absorbent, drilling mud
Crushed stone	mixed with sand, cement, and water to make concrete; road base, fill, more expensive substitute for gravel
Diamond	gem stone, hardest known abrasive
Dimension stone	granite—monuments, buildings, curbing slate—roofing, chalkboards, flagstone, pool tables marble and limestone—buildings, monuments sandstone, quartz, quartzite—buildings, high purity sand
Fluorspar: fluorite	making steel, aluminum, toothpaste; in freon
Graphite	heat-resistant containers, lubricant, pencils
Gravel	same uses as crushed stone
Gypsum	plaster, wallboard, sculpture, fertilizer, cement
Limestone: calcite	lime, cement
Mica: muscovite	heat and electronic applications
Phosphate	fertilizer ingredient
Potash	fertilizer ingredient
Pumice	lightweight concrete, abrasive/polish
Salt: halite	raw material in chemical industry, seasoning
Sand (construction)	concrete, fill, road base, bricks, pipe, plaster
Sand (special) quartz	glass, blasting, molding, furnaces, abrasives
Sulfur	sulfuric acid, steel, oil, rubber, chemicals
Talc: soapstone	lubricant, ceramics

SOLID WASTES

Increasing use of metals, nonmetals, and energy minerals has resulted in increasing amounts of waste. Experts in our government estimate that 230 billion kilograms of waste are produced in the United States each year. This includes waste from homes, factories, businesses, and institutions. Most waste is in a solid form. Estimate the average waste produced each day. Estimate the average waste produced each day by each person.

COMPOSITION OF TRASH

Type	Percent	Percent Increase Per Year	Heat Energy kcal/kg
Glass	10	3–4	36
Metal	10	3–4	410
Paper	51	4–5	4400
Plastic	1	10	6400
Leather, Rubber	2	3–4	5600
Textiles	3	4	4400
Wood	3	1	4800
Food waste	18	1½	4700
Miscellaneous	2	?	1900

Solid wastes are still collected and dumped at sea.

About ten years ago, each American created only about half as much solid waste per day as now. Many experts predict that today's waste production will double in ten more years. List some changes in living styles in the last ten years which have added to the waste disposal problem. What are some changes in the last 50 years which have increased waste production?

Composition of Solid Waste. Some groups of people have kept their trash for several days to find out what is in it. The table at the left shows one group's results.

Recall some of the recent trash from your own home. What are some examples of things of each material you have recently thrown away?

The third column on the chart shows how the composition of trash is changing. Which trash material is increasing at the fastest rate? Which one is increasing at the slowest rate?

The fourth column shows the amount of heat energy in each material. Which material will give off the most energy per kilogram if the trash is burned? What are some possible uses for this energy?

Disposing of Solid Wastes. Today, 78% of the United States population has a regular way of getting rid of trash. These people include nearly everyone in urban and suburban areas. 12% of the American people have no trash disposal service. Generally, these people live in rural areas and must dispose of their own trash, frequently by burning or open dumping on their land.

The most common type of trash disposal today is to collect trash and dump it in some other place. There are 14 000 sites in the United States where trash is dumped. In order to reduce disease and pollution, regulations have been established for these sites. These regulations require that all trash be covered each night with dirt. Government inspectors have estimated that only 6% of the 14 000 disposal sites meet these standards. Some 75% of the sites make no effort to cover any trash. Open burning is often observed which contributes further to our air pollution problems.

Many regions with large populations have built *incinerators* to burn trash. In this process, much of the matter in trash is converted to carbon dioxide and water which escapes into the atmosphere. As a result, the total volume of trash is reduced to an ash with 5% of its original volume. This ash is then spread on the ground at a dump. The ash is odorless, does not blow around, and does not serve as a breeding site for rats.

There are over 400 incinerators in use today in the United States. Unfortunately, half of these are not completely burning the material in the trash. As a result, much unburned waste is dumped out of the incinerator, and air polluting substances escape from the chimneys. The photograph at the left shows a modern incinerator which pollutes the air and ground with its waste. When expensive government efforts to reduce pollution fail, the designer often blames the builder, the builder blames the men who operate it, and these men blame the designer. Suggest some reasons why it is difficult to prove who is responsible in such cases.

Recycling. *Recycling* means using a resource over again. Recycling has been going on for many years. Early pottery-makers reused the waste clay left over, much as doughnut makers reuse the holes to make more doughnuts. Steel makers have always trimmed sheet steel and recycled the waste back into their steel furnaces. Similarly, tin can makers sell their scrap back to steel mills. In both cases, the companies find that the cost of recycling the metal is less than the cost of disposing of the waste in any other way.

The problem of recycling tin cans from home trash is much more complicated. First, the tin can has to be separated from the other trash, then the waste food has to be separated from the can, and the label has to be separated from the clean can. Usually the cost of these tasks is much greater than what the can is worth to a recycling mill. As a result, cans accumulate in the disposal area. How might this change as the cost of leaving the can in a disposal area increases because we are running out of disposal sites?

A trash incinerator.

SPEED LIMIT

1000 kg — 100 km/h

1500 kg — 90 km/h

2000 kg — 75 km/h

The Changing Picture in Conservation. What might happen to our fuel and mineral reserves if:

1. The government assisted oil companies by giving them bigger tax refunds for the money they spent looking for oil?

2. The government substantially increased the tax on gasoline and gave the money to cities to build quick, clean rapid transit systems?

3. A low speed limit was set (and enforced) for all cars which don't average good gas mileage?

4. All tolls on highways and bridges were eliminated for people going to work in cars with three or more people in them?

5. New car buyers were taxed according to the ability of their new car to last 300 000 km?

6. The Environmental Protection Agency forced small towns to cooperate in their trash disposal, and to build large efficient sorting devices for separating trash?

7. All federal agencies could buy materials only from companies which recycled materials?

8. All businesses were taxed 120% of the cost of the government clean-up of *all* their waste materials?

QUESTIONS FOR DISCUSSION

1. Propose an outline of a letter to your representative in congress suggesting changes in our laws to reduce the energy shortage.

2. Propose an outline of a letter to your representative in congress suggesting areas of research the government should consider funding to help solve the energy shortage.

3. Propose a solution to the solid waste problem where you live. What are some arguments against your proposal?

THOUGHT QUESTIONS

1. In 1925, 4% of our energy came from hydroelectric power plants. Although many new hydroelectric power plants have been built since then, today only 4% of our energy is produced this way. Explain why.

2. Study the figures at the left. What are some reasons for the large differences in the percentages of each material being recycled?

Material	% Recycled
Paper	20
Aluminum	25
Lead	65
Steel	50
Copper	50
Zinc	65

THE EARTH'S SURFACE

Topographic Maps

When did you last use a map? What did you use it for? What kind of a map was it? If this is your first year in this school building, you might have used a map of the building to find your way around. On an automobile trip, a road map is a necessity, if the places you go are new to you.

For geologists interested in landforms on the surface of the
earth, there are maps which show the highlands and the lowlands.
They show the steepness of the hillsides, the shapes of the valleys,
and many other characteristics of the landscape.

You will learn about making, reading, and interpreting
topographic maps in this chapter.

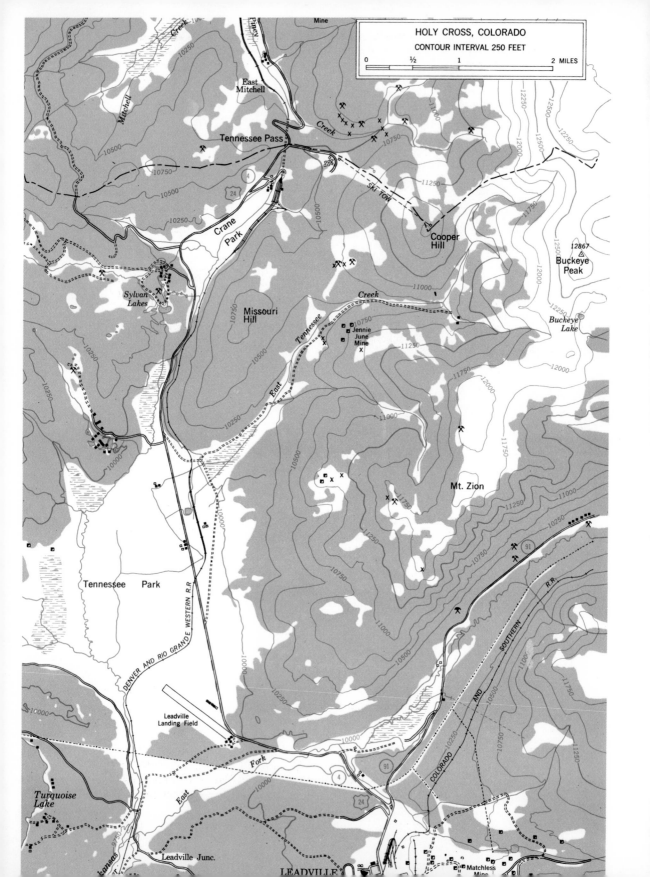

Hard surface, heavy duty road, two or three lanes............................. ▬▬▬▬
Hard surface, medium duty road, two or three lanes......................... ▬▬ ▭▬ ▬▬
Improved light duty road... ════════
Unimproved dirt road—Trail.. ┼┼┼┼┼┼
Railroad: single track—multiple track................................. ╪═╪═╪┼═╪
Tunnel: road—railroad .... →═→ ═══←
Power transmission line... ·─── ·── ·──
Telephone line, pipeline, etc.. ▬ ▬ ▬ ▬ ▬
School—Church— ....■ ▪   Barn...........□   Dwelling...........▪
Open pit, mine, or quarry—Prospect..................................... ⚒ ──── X
Shaft (Mine)—Tunnel entrance.. ◩ ──── Y

Perennial streams ....⌇〜〜   Intermittent streams ....〜⌐〜〜

Marsh (swamp) ....⟦🌿⟧   Sand area ....⟦░⟧

Depression contours ....⟦◎⟧

MAKING AND READING MAPS

Trips to volcanoes, glaciers, canyons, beaches, and high mountains make the study of landforms more interesting. However, such trips are not always practical. Instead, these landforms can be studied with the use of topographic maps, such as the one on the opposite page. Topographic maps are designed to show accurately the shapes of hills and valleys of a particular region.

Topographic Map Symbols. Many different symbols are used on topographic maps. A few important symbols are shown above. Study them and then answer the following questions to learn about the region shown on the map.

1. How do you think Leadville got its name? What evidence can you find for your hypothesis?

2. What is the distance from Leadville to Tennessee Pass as measured along a straight line? What is the distance between these two places measured along U. S. Highway 24? Why was the road not built along a straight line?

3. What is the general direction of the Denver and Rio Grande Western Railroad on this map? Describe the route of the railroad between Tennessee Pass and East Mitchell.

4. Topographic maps use contour lines to show the slopes of landforms. Each contour line is composed of points having the same elevation in feet above sea level. Every fifth contour is usually a darker line. Where is the highest elevation?

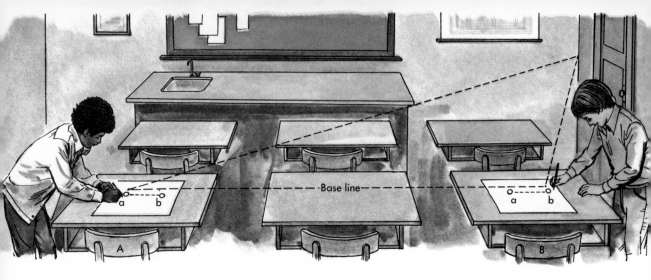

Base line

5. Forested areas are printed in green on colored topographic maps. Locate two large areas on the map which are not covered with forests. Why do you think trees are not growing in these areas? What evidence can you find for these hypotheses?

6. Locate five swamps, ten small lakes, a race track, a landing field, a power line, a ski tow.

Mapping the Classroom. The two students in the picture above are making a map of their classroom. When they finish, their map will represent a scale model of the classroom. They are using a method called *base line mapping*.

The students set up a base line in the classroom. They label the ends of the base line *A* and *B*. Next, they draw a line on their paper for the base line of the map. The ends of the scale base line are labeled *a* and *b*. The room base line is 3 meters long, and the map base line is 12 centimeters long. Find the ratio:

$$\frac{\text{base line on map}}{\text{base line in the room}}$$

This ratio is called the *scale* of the map. The scale of this map is ¹/₂₅, or 1 cm equals .25 meter.

The paper is then placed so that point *a* on the map is directly over point *A* on the base line, and the scale base line points directly at point *B* in the room. With the map in this position, the directions to both corners of the room are sighted accurately and marked on the map.

After drawing and labeling these sighting lines, the students move to position *B*. They place point *b*

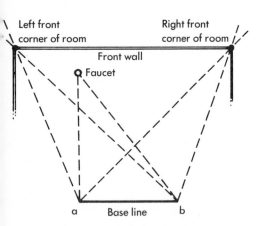

Left front corner of room

Right front corner of room

Front wall

Faucet

a   Base line   *b*

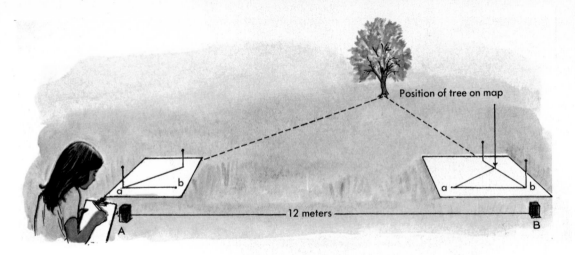

Position of tree on map

a   b

12 meters

A   B

on the map directly above point *B* of the room, and orient the map's base line so that point *a* on the map points toward point *A* of the room. Again, the directions to both corners of the room are accurately sighted and marked on the map.

The map is completed by extending the sighting lines until pairs of lines pointing to the same corner intersect. (Note: Tape other sheets onto the map, as needed.) The point of intersection represents the corner of the room. Lines drawn on the map between the corners of the room represent the walls of the room.

Make a map of your classroom as described here. When your map is completed, compare the scale length of a classroom wall with the length of the wall as measured with a meterstick. Compare your map with maps made by other students. Discuss possible sources of errors. From your discussion, decide whether base lines should be long or short to improve accuracy.

Outdoor Mapping. Select a region near your school to map. If possible, choose an area which is not perfectly level. Before going outside to map the area, tape a large sheet of paper onto a board. Take a chair with you to support your surveying board.

Decide where to set up your base line in the field. Pound two stakes into the ground to represent the ends of the base line. Use a steel tape to measure the length of the base line.

Choose a scale so that the map will fit onto the paper. Draw the base line on the paper to the correct scale, and complete the map following the procedure used in mapping the classroom. Draw the

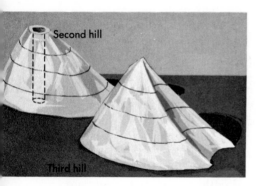

outer boundary of your map, and locate hilltops, buildings, large trees, and other objects of interest.

Compare your completed map with the maps of your classmates. Discuss reasons for errors, and suggest methods which would improve the maps.

Contour Lines. Cut a potato in the shape of a cone 6 cm in diameter and 6 cm high. Use a cork borer or large nail to cut a hole from the top of the cone to the base, as shown above. Draw a small triangle on a sheet of paper, and place the potato on the paper so that the hole in the potato lines up with the triangle on the paper. Trace the outline of the base of the cone onto the paper. Label this line, *0 cm*, to represent the elevation of the base of the cone.

Place a pencil on a wooden block so that the point of the pencil is exactly 2 cm above the desk top, as shown above. Use the pencil to draw a line around the potato at an elevation of 2 cm above the base. Slice off the bottom 2 cm of the potato, and place the remaining cone-shaped part on the paper. Center the potato over the triangle again. Trace the outline onto the paper. Label this line, *2 cm*.

Continue the process of slicing off 2-cm sections of the base and tracing the outline on the paper until the remaining cone-shaped section is less than 2 cm tall.

Carve a second hill from a potato. Make the slope on one side much steeper than on the other. Construct another topographic map as before, and compare the two maps. How do contour lines of steep slopes differ from those of gentle slopes?

Study the topographic map on page 100. Locate (1) a cone-shaped hill, (2) a steep slope, (3) a flat area, and (4) hills with valleys on the sides.

104

Contour Intervals. The difference in elevation ween two contour lines next to each other is led the *contour interval*. The contour interval of maps in the last activity is 2 cm. If one contour e represents an elevation of 4 cm, the next uphill ntour line represents an elevation of 6 cm. What the contour interval of the map on page 100?

Making a Topographic Map. Make the sighting el shown below, using 2 pieces of wood and a ter-filled bottle. The distance from the top of the hting level to its base is 50 centimeters.

Set the sighting level at the lowest place on the ld that you mapped earlier. Look along the hting device as a friend holds it level. Locate a int that is 50 cm higher than where you are nding. Aim the sighting level in other directions, d locate a number of points on the slope at the ne elevation. Drive stakes at these points. Tie a l cloth around each stake.

Move the sighting level to one of the l-flagged stakes. Again locate a number of points the field which are 50 cm higher. Drive in stakes these higher points. Label each of these stakes th a white cloth. Repeat the sighting procedure ing different colored cloths for each elevation til the highest point in the field is reached.

Use the base line mapping method to locate the sition of each red-flagged stake on the map nich you made earlier. Draw a line connecting ese points to represent the 50-cm contour line. peat and draw other contour lines on your map.

To test the accuracy of your map, locate the gion where the contour lines are closest together. mpare with the steepest part of the field.

Air bubble
50 cm contour
Lowest point of elevation is 0 meters

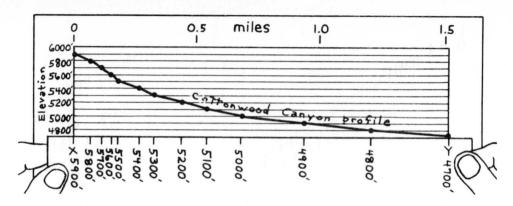

MAPS OF STREAMS

Streams shape many features of the earth's surface. The kinds of features they form depend upon how fast they flow. Two factors which affect stream flow are slope and amount of water. How can you tell from reading a map whether a stream is flowing quickly or slowly?

Stream Slopes. The map on page 107 shows a region in Colorado. Locate some flat areas. Locate some steep slopes. What caused them?

Note the two contour lines at points X and Y that cross the stream bed in Cottonwood Canyon. Use the elevations of these two lines to determine the direction this stream flows. Do contour lines point upstream or downstream?

Stream Profiles. Two streams and stream valleys are shown on page 107. How does the slope of the stream bed in Cottonwood Canyon compare with the slope of the Souris River?

Cut a strip of paper slightly longer than the space between X and Y on the map. Mark a point with an X along the edge at one end of the paper strip. Place this point over X on the map, as shown here. Bend the strip to match the bends in the stream bed. Mark the strip at places where contour lines cross the river. Label these marks with the elevations of the contour lines.

Continue this process until you reach point Y on the map. Then compare your paper strip with the one shown above. Use the data on the strip to make a graph showing the slope of the stream bed. This type of graph is called a *stream profile*.

Count the contour lines which cross the Souris River. Describe a stream profile of this river.

Cross Profiles. Make a *cross profile* of Cottonwood Canyon and the Souris River Valley

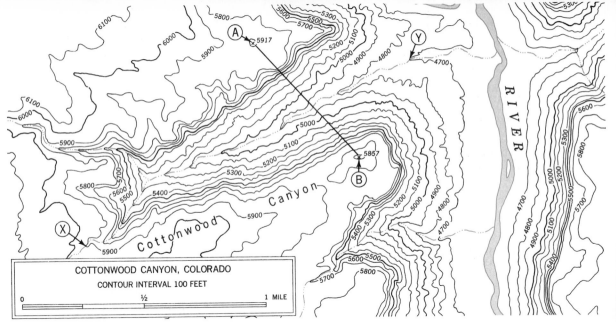

COTTONWOOD CANYON, COLORADO
CONTOUR INTERVAL 100 FEET

0   ½   1 MILE

from points *A* to *B*. Compare the cross profiles of
these two stream valleys. Did the V-shaped valley
develop where stream velocity was high or low?

Note the flat region along the Souris River. This
feature is called a *flood plain*. Flood plains are
composed of silt and clay particles. Suggest how
floods might aid in developing these plains.

Note point *C* on the lower map. How does the
width of the valley change at this point? Note the
locations marked *D*. These are *depression contours*.
Describe the landforms they represent.

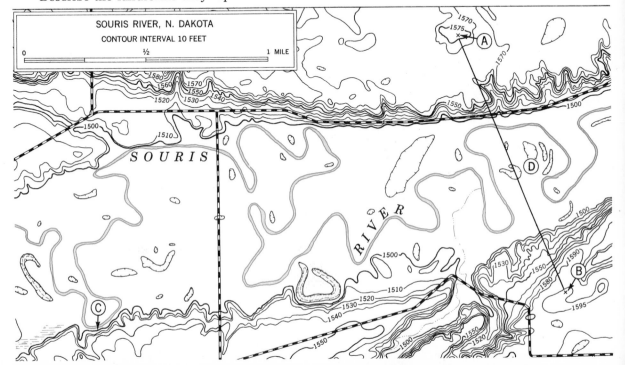

SOURIS RIVER, N. DAKOTA
CONTOUR INTERVAL 10 FEET

0   ½   1 MILE

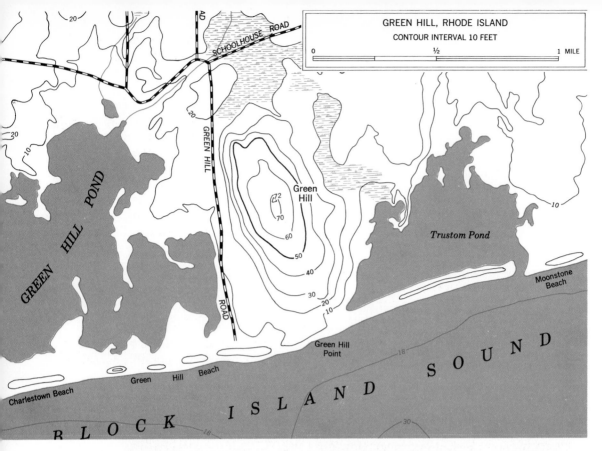

INTERPRETING TOPOGRAPHIC MAPS

The general shape of an ocean shoreline is usually determined by whether the level of the sea has been rising or falling in the past thousands of years. Shorelines are often straight when the level of the sea has been falling (or where **the land** has been rising). These are called shorelines of emergence. Where **the land** has been slowly dropping, the resulting shoreline of submergence is **often** very irregular. By studying the topographic maps of an area, it is **possible** to interpret the geological changes that have taken place.

Interpreting Contour Lines. The Atlantic Coast region of Rhode Island is shown on this page. What is the contour interval of this map? What is the elevation of the small oval contour line near the top of Green Hill? Describe the shape of Green Hill.

Locate a 20-foot contour line along Green Hill Beach. What is the smallest possible height of this hill? How do you know the hill is not more than 29 feet high?

Place a ruler along the Block Island Sound shoreline. What point bulges slightly outward from the otherwise straight shoreline? Suggest a possible reason why the shoreline bulges here. Suggest a cause for the bulge in the 18-foot depth contour line offshore from Green Hill Point.

Trace the shoreline if the land submerged 20 feet. Trace the shoreline if the land emerged 18 feet. Which shoreline is straighter? Does the straight shoreline of this region suggest that the land is submerging or emerging?

Submerged Shoreline. The bottom of Mobile Bay on the map below is believed to be part of a river valley which is now below sea level. Dog River is now a salt water bay. Describe the possible appearance of Dog River before this region was submerged.

Locate Rattlesnake Bayou on the map. Suggest how this region differed when the river in Mobile Bay was above sea level.

Speculate about some possible causes for shorelines of submergence. Examine a map of the eastern United States and locate possible shorelines of submergence and emergence.

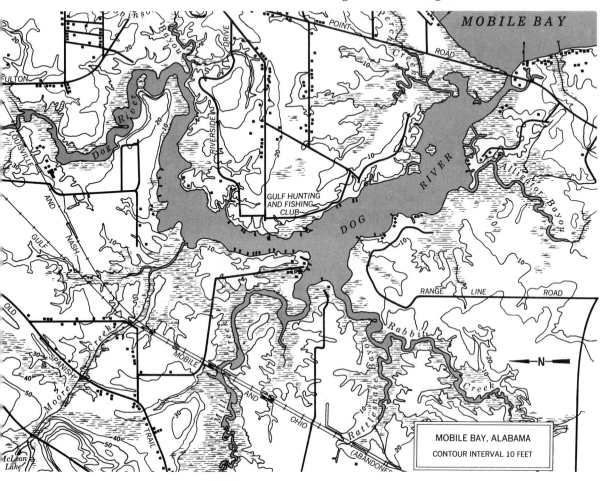

MOBILE BAY, ALABAMA
CONTOUR INTERVAL 10 FEET

Topographic Map of Your Area. The United States Geological Survey has surveyed most of the United States. Study a USGS topographic map which shows your school and community. By reading the map, determine the following:

1. Where are the highest and lowest points?

2. Identify any rivers or streams, and trace their paths. Choose the largest and determine its profile.

3. What is the straight-line distance from your home to school? The shortest walking distance?

4. Locate an area which you think should be preserved for recreational purposes.

REVIEW QUESTIONS

1. How do features *A* and *E* differ in the map below?
2. Propose how the flat land at *C* and the steep slope next to it were formed.
3. Which way does Peterson Creek flow?
4. Part of Peterson Creek is safe to canoe, another part is not. Locate each section.

THOUGHT QUESTIONS

1. Why are 1-foot contour lines used in topographic maps of flat areas and 250-foot contour lines used in maps of mountains?
2. How is a shoreline of emergence different from a shoreline of submergence?

PETERSON CREEK, N.Y.

CONTOUR INTERVAL 10 FEET

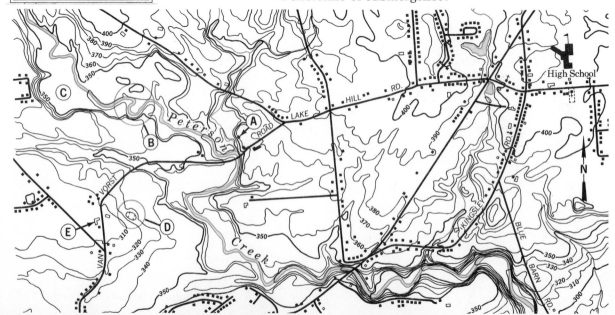

Erosion and Its Effects

    The rocks of the earth are under constant attack wherever they lie exposed. Chemicals in air and water weaken rocks, and frost splits them into fragments. Wind, ice, and moving water grind the fragments against each other and against unbroken rock. Gradually, the high portions crumble and are carried away.

    These processes of erosion are generally very slow. Few people notice any differences in the size of a hill or the shape of a valley during their entire lifetimes. But the person who watches closely sees many tiny changes that gradually remodel the landscape.

Weathered
granite

WEATHERING

A rock begins to change as soon as it is exposed to the weather. The changes may be chemical, as when magnetite unites with oxygen. The changes may also be physical, as when a rock is split apart by frost action.

Rocks differ greatly in the rate at which they are affected by weathering. Some contain chemically active materials. Some are weak. Others are strong and chemically inactive. All rocks are affected by the number of cracks and pores into which water and air can seep.

Weathering rock

Feldspar is a mineral found in granite. It can be identified by its flat surface, and its edges which form right angles when they break.

Chemical Weathering. Most igneous rocks are strong and tough when they are first formed. However, some of the minerals in igneous rocks are easily affected by chemicals in air and rain. Chemical reactions weaken a rock in two ways: (1) Mineral grains may increase in size by uniting with oxygen or water. This increase in size wedges other grains apart and loosens them. (2) The products formed by the reactions may be soft or soluble. Removal of the products by wind or rain leaves the remaining grains with less support.

Changes in Feldspar. Feldspar reacts slowly with water and carbon dioxide. The products of the reaction include clay, silica, and one or more carbonates. One form of feldspar (orthoclase) reacts with water and carbon dioxide as follows:

feldspar + water + carbon dioxide →
                    clay + silica + potassium carbonate

Potassium carbonate is soluble and is soon removed by water. Clay crumbles and is washed or blown away.

Feldspar reacts slightly with carbon dioxide and water in the laboratory. Grind a piece of

feldspar into a fine powder. Put a pinch of the powder into each of two test tubes. Fill one test tube with distilled water (rainwater) and the other with carbonated water (water solution of carbon dioxide). After 10 minutes test each with litmus paper. A slow change of red litmus to blue indicates that potassium carbonate, or some other basic chemical, has been produced.

Changes in Iron Oxides. *Magnetite* is a black oxide of iron; it is common in many igneous rocks. When magnetite unites with oxygen from the air it becomes the red oxide of iron called *hematite*. Hematite in turn unites with water and becomes the yellow-brown oxide of iron called *limonite*.

$$magnetite + oxygen + water \rightarrow limonite$$

Combining with oxygen and water increases the size of magnetite grains so that neighboring grains are wedged apart and loosened. The end product, limonite, is soft and easily washed away by rain. This leaves spaces into which more oxygen and water can enter.

Weathered granite showing hydration and oxidation of its minerals

Examine weathered granite and other igneous rocks for brown stains of limonite. Break open the rocks and look for particles of unweathered magnetite. Use a magnet to identify the magnetite.

Weathering of Basalt. Basalt, like that shown in the picture above, contains several iron compounds. These compounds may unite with chemicals in the atmosphere, producing limonite as one of the end products. In moist climates, basalt is soon stained brown with limonite.

113

Exfoliated granite

Joint cracks and bedding

Exfoliation. The photograph at the left shows an effect of weathering called *exfoliation*. To find out how exfoliation occurs, a research geologist carefully put a polished block of granite under a lamp. The lamp was timed to go on for five minutes and off for ten minutes. While the light was off a fan cooled the rock. The temperature of the surface of the block ranged over 60C° with this treatment. The geologist repeated the work 89 400 times, and there was no apparent change in the block. Then she began spraying a fine mist of water on the rock while the light was off. In just ten days of this treatment, the granite block lost its polish, the crystals of feldspar became clouded, and exfoliation cracks appeared. Geologists think that the water combines chemically with the crystals in the outer portion of the rock. The crystals grow larger with the added water and cause the outer layers to peel off somewhat like the layers of an onion.

Frost Action. Freezing and thawing are very important in the weathering process. During warm periods, water seeps into cracks and pores. During cold periods, the water freezes and expands. Forces great enough to split rocks are exerted by the expanding water.

The middle photograph shows blocks of granite pushed apart by freezing water. How does frost action help to speed up other weathering processes?

Joints. Most rocks are divided by cracks which geologists call *joints*. These cracks are usually too small to be seen when a rock is first exposed. But weathering eventually widens them so that they appear similar to those shown in the middle photograph.

Joints provide pathways into which air and water can seep. They also weaken a rock so that it breaks apart more easily.

Bedding. Sedimentary rocks usually show layers which geologists call *bedding*. The bedding shown in the bottom photograph is nearly level; the joints run up and down.

Bedding can be especially important in the weathering process when some of the layers are much weaker than others. The weak layers crumble, leaving the remainder of the rock without support. Large pieces then break off because of their own weight.

Erosion of a Cliff. The cliff shown above is continually exposed to the weather. Discuss the erosion of this cliff.

Explain why the rock at the top of the cliff is divided up into blocks. What causes these blocks to separate? What part does chemical weathering seem to be playing in the destruction of the cliff?

Describe the processes which might move a large block to the edge of the cliff. Is there any evidence that blocks have already fallen? What may happen to a block when it lands at the base of the cliff?

Talus

Talus. Pieces of rock that have been loosened from a cliff drop down and pile up around the bottom of the cliff. This pile of loose materials is called *talus*.

Study the photographs of talus on this page. Are the pieces of rock smoothly rounded or are they sharp edged? Explain why this is so.

Are the pieces nearly the same size or are they different in size? What effect does the type of joints have on the shape and size of the pieces of rock in the talus? What effect does the bedding have?

Visit a cliff and look for talus. Study the pieces of loose rock and explain their shape. Find pieces of rock that are about to fall from the cliff, and describe what may happen to them. If you cannot visit a cliff, make a model that shows how talus is formed.

Bottom of a talus slope

Angle
of
rest

Angle of Rest. Talus piles up at the bottom of a cliff until a fairly steep slope is formed. Thereafter, additional materials do not remain on the slope, but roll or slide to the bottom.

The angle of steepest slope upon which loose material can remain is called the *angle of rest*. This angle depends upon the type of material being piled up. Some materials can remain on steeper slopes longer than others.

Measuring the Angle of Rest. Pour dry sand on a flat surface. Watch the way the sand piles up. Note when the angle of rest is reached. What happens as additional sand is poured on the pile? What happens if part of the pile is cut away?

Fasten two sticks together with a screw or bolt. Use this device to measure the angle of rest of the sandpile.

Comparing Angles of Rest. Measure the angle of rest for a number of different materials, such as dry garden soil, dry forest soil, road dust, round pebbles, and crushed stone. Make a chart which shows the angle of rest for each of these materials. Test the effect of mixing materials of different sizes.

Water and the Angle of Rest. Crush dry clay or garden soil into a fine powder. Pour the powder onto a flat surface, and measure its angle of rest.

Mix the powder with one–eighth of its volume of water. Pour out the mixture, and measure its angle of rest. Repeat the measurements using increasing amounts of water each time.

Dry sand

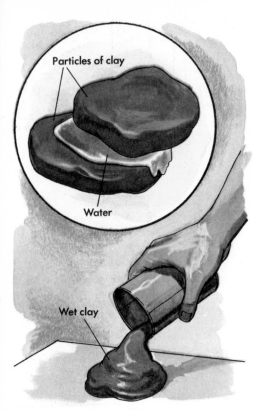

Particles of clay

Water

Wet clay

Make a graph of the results of this experiment. Plot the percent of water against the angle of rest. Explain what your graph shows.

Test the effect of water on the angle of rest of other materials, such as sand and gravel. Compare the results with those obtained from clay.

Lubrication of Clay. The finest particles of clay are too small to be seen with an ordinary microscope. More powerful microscopes show that clay particles are thin flakes, much like the artificial snow sold at Christmas time.

The friction between dry particles of clay is very high. A bank of dry clay can be cut so that its walls are straight up and down.

Water lubricates particles of clay so that they slip past each other easily. The same effect can be noted when dry and wet microscope slides are rubbed together. Explain why water affects the angle of rest for clay.

Mudflows and Slumping. Visit a road cut or other excavation in clay soil after a heavy rain. Look for places where the soil has slumped. Look for *mudflows*—places where the soil flows almost like water. Explain the slumping and any mudflows you observe.

Landslides. Loose materials are able to remain on a slope as long as the frictional force is greater than the downward pull. Any decrease in the frictional force, or any increase in the weight of the materials, may result in a landslide.

Most landslides take place in the spring or after heavy rains. Explain how rain can affect the frictional force in a pile of loose materials. Explain how rain can affect the weight of the pile.

The photograph at the left shows a landslide. The enormous amount of material that rushed down the mountain slid some 100 meters up the other side, then settled back like a liquid being sloshed in a large basin. The landslide formed a dam about 70 meters high across the valley, creating a lake almost eight kilometers long. Spring floods in the following year raised the water over the lip of the dam. As the water flowed over the dam, it caused such down-cutting that the level of the lake was lowered 15 meters in five hours. The resulting flood below the dam caused much damage. Several people lost their lives. Such slides are more common than most people realize. What part do small landslides play in the erosion process?

The photograph below shows a huge landslide that flowed down a mountainside and dammed a small river. What part do large landslides play in the erosion of a region?

Creep. Loose materials often move slowly down a slope instead of rushing downward suddenly as in a landslide. Geologists call this slow movement *creep*. It operates steadily in most regions, and is probably more important than landslides in the total picture of erosion.

Examples of creep can be seen on many steep hillsides. Telephone poles, fence posts, stone walls, and gravestones are often tilted as the soil beneath them moves. They may need resetting every few years to keep them from tipping over.

Trees may also be affected. Trees tend to grow in the opposite direction to the pull of gravity. As creep occurs, the tree is tilted. With continued growth the tree trunk bends as it responds to gravity. Study the diagrams which show a tree growing as soil creep is occurring. How does the tree in the photograph show that the soil is slowly moving down the slope? Very young trees are apt to turn upward as rapidly as they are tilted, as shown in the diagrams at the left. The resulting curve may remain in the trunk during the lifetime of the trees.

Make a study of trees, posts, walls, and gravestones on steep hillsides. Use a carpenter's level or a plumb bob to discover how much various objects are affected by creep.

SUMMARY QUESTIONS

1. In what ways may chemical action weaken a rock?
2. What happens to feldspar when it is exposed to water and carbon dioxide?
3. Give four examples of nonchemical weathering.
4. Describe the erosion of a cliff, using the words talus, frost action, and angle of rest.
5. How does water affect the angle of rest of clay?
6. What is soil creep?

EROSION BY STREAMS

Swift streams are among the most important agents of erosion. Their action is chiefly mechanical, although there are some chemical effects as well.

Swift water may strike against jointed or bedded rock with enough force to loosen blocks along the cracks and lift out the blocks. The loose pieces become agents of erosion in their turn, grinding against each other and the stream bed as they are swept downstream.

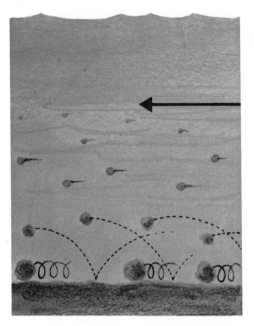

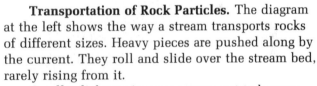

Transportation of Rock Particles. The diagram at the left shows the way a stream transports rocks of different sizes. Heavy pieces are pushed along by the current. They roll and slide over the stream bed, rarely rising from it.

Smaller lighter pieces are more apt to bounce along, being carried some distance by the current before they strike the stream bed and bounce again. The swifter the current, the larger the pieces the stream can carry.

Very small pieces may not strike the bottom at all if the current is swift. These pieces cause the muddy color in many streams.

Abrasion. The wearing away of rocks as they rub against each other is called *abrasion*. Abrasion takes place in streams as rock particles are swept along by the current. Abrasion also takes place when wind blows sand against rocks.

The diagram at the left shows how abrasion takes place in streams. Rocks wear away small particles from each other and the bedrock as they roll and slide along. Bouncing rocks may chip off pieces each time they strike each other. What will happen to the size and shape of the rocks after much abrasion?

121

Producing Abrasion. Break up a soft rock with a hammer. (**CAUTION:** Wear goggles to protect your eyes.) Choose 25 pieces of rock having about the same size and shape. Set aside three pieces for later comparisons.

Put the remaining pieces in a jar half full of water. Shake the jar 100 times. Remove three pieces and label them. Shake the jar another 100 times. Remove and label three more pieces. Continue until all the pieces have been removed.

Compare the shapes of the specimens removed from the bottle at different times. How has abrasion changed them? Make a record of the changes by tracing around each specimen. Which kinds of rocks abrade more quickly? Observe the water in which you have been shaking the rocks. What has happened to it? Test other rocks in the same way. Use the idea of resistance to abrasion to explain the differences you see.

	None	100 shakes	200 shakes	300 shakes	400 shakes	500 shakes
Limestone						
Shale						
Granite						
Quartz						

Gorge through sandstone

A pothole in a gorge

Gorges and Canyons. Abrasion gradually wears down the bed of a swift stream. Each rock tumbled along by the current removes a bit of the bedrock. Within a few thousand years, a deep canyon may be formed.

Visit a gorge or canyon. Look for signs of erosion (such as round rocks, smooth walls, and places where the banks have been undercut). Make diagrams to show what the region may have looked like at different times in the past. Make a diagram of what the region may look like a few thousand years from now.

Potholes. The beds of many streams contain round pits called *potholes*. Study the diagram here, and explain how a pothole is formed.

Why are potholes found only in the beds of swift streams? Why are the walls of a pothole smooth? Why are the stones in a pothole usually round and smooth? What may happen to the stones in a pothole after a few years?

Hose siphon

A

20 cm
40 cm
60 cm
80 cm
100 cm

Piece of wood

Roof gutter

B

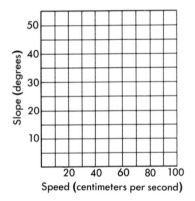

Slope (degrees)

50
40
30
20
10

20   40   60   80   100

Speed (centimeters per second)

Slope and Rate of Flow. Set up the apparatus shown above, working outdoors if possible. Keep the water level nearly constant in pail A so that the siphon delivers a steady flow of water.

Measure the speed of the water in the trough by dropping a bit of wood in the water and timing its movement from one point to another. (A stopwatch is useful for this.) Make several measurements, and calculate the average speed.

Measure the angle of the slope with a protractor and a carpenter's level.

Change the slope of the trough and measure both the slope and the speed of the water. Try other slopes. Make a graph of speed against slope. What does your graph show?

Volume of Water and Rate of Flow. Experiment to find the effect on the rate of flow of changing the volume of water. What variables are changed in this experiment? What variables should be kept constant?

Increase the volume of water by using two or more siphons. Measure the speed as you did before. Measure the water collected in pail B to determine the amount of water flowing per second. Graph your results.

Waterfalls. When water flows over some types of land, waterfalls will form. Try the following activity to see how some types of waterfalls can form. Put sand in the upper end of the trough so that the sand fills half the trough. Begin a gentle

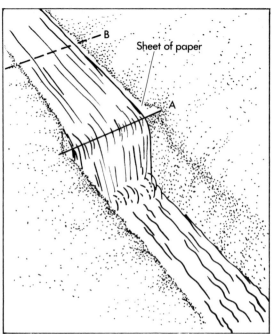

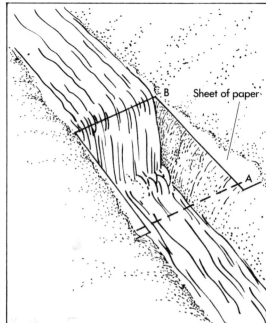

flow of water over the sand and observe what happens. If a waterfall does not form, what might you do to cause one to form?

Set the trough up as you did earlier with the sand at the upper end. This time bury a piece of paper so it is flat, and start the water flowing again over the paper. What happens as the water flows over the edge of the paper? What kind of formation would you find in nature that might cause the same effect as the piece of paper?

What action causes a cave to form behind the waterfall at the edge of the paper? As the cave gets bigger, cut or tear a slot into the paper and continue running the water over it. Each time the edge of the paper is about to collapse, cut out another slot. Notice the gorge which forms downstream from the waterfall.

Compare the aerial drawing of the waterfall with your own waterfall. Notice that the waterfall in the drawing has a gorge downstream. Explain how this gorge might have formed.

Some kinds of rock are more resistant to abrasion than others. Whenever a resistant rock lies over a weak rock in a stream bed, waterfalls are apt to form. The more resistant rock will often form a ledge. As the water flows over the ledge, it cuts into the less resistant rock layer below.

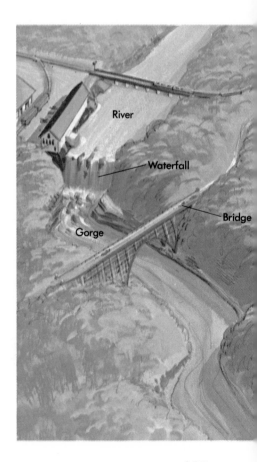

Limestone

Shale

Limestone

Sandstone

Shale

Shale

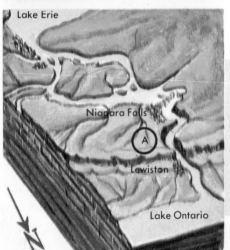

Lake Erie

Niagara Falls

A

Lewiston

Lake Ontario

N

The diagram at the left shows the formation of Niagara Falls. The limestone in the top layer is much more resistant to erosion than are the weak shales underneath. What different processes of erosion take place below the falls?

Every few years a block of limestone breaks from the edge of the falls. Why does this happen? What happens to the pieces of rock that break off?

A diagram of the area around Niagara Falls is also shown. Geologists believe that the falls were at the point marked A about 10 000 years ago. If this is true, why are the falls in their present position today? Where may the falls be 10 000 years from now?

Make a study of a waterfall near your school. Explain the formation of the falls, and describe the probable history of the falls.

SMALL-SCALE EXPERIMENTS. Conditions in small-scale experiments are commonly much different from conditions in full-scale experiments. For example, the force with which water clings to objects greatly affects a trickle of water, but it has little effect on the flow of a river. Conclusions drawn from small-scale models and experiments may not be completely correct for actual situations.

Age of River Valleys. The diagrams here show a river in three different stages. The young stream loses elevation rapidly and flows fast. Such a stream erodes its bed and cuts a canyon. Its high speed usually produces a straight valley.

As the young stream erodes downward to nearly the elevation of its outlet, its rate of flow decreases. As a result, this mature, slower moving stream cannot erode its stream bed. Instead, materials deposit on the stream bed, turning the water aside, and causing the stream to twist. The walls of the canyon erode back, and the valley widens. During floods, the riverbanks overflow, and sediments deposit on either side of the river.

By the time a stream reaches old age, it has flooded many times. Sediments have been deposited on both sides of the river forming banks. The land slopes down away from the river, and swamps form.

Stream Slopes and Valleys. The map below shows a plateau region in Colorado. Using your skill at reading topographic maps, how would you classify the valleys on the map? Note the two contour lines that cross the stream bed at points *X* and *Y* in Johnson Canyon. Use the elevation of these lines to find the direction this stream flows.

Note the scale of the map. What is the distance between points *X* and *Y*? What is the difference in elevation between these two points? What is the drop in elevation per mile between these points?

The change in elevation per mile is called the *slope* of a stream bed. Determine the slope of the stream bed in Johnson Canyon between 5750 feet and 6000 feet, and between 6250 feet and 6500 feet.

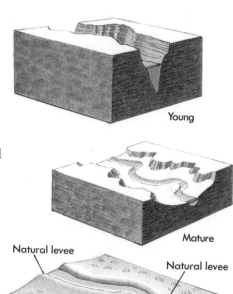

Young

Mature

Natural levee

Natural levee

Swamp

Old age

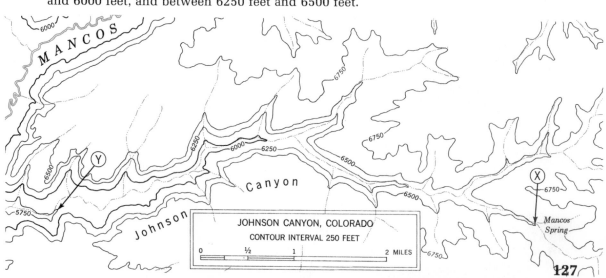

MANCOS

Canyon

Johnson

JOHNSON CANYON, COLORADO
CONTOUR INTERVAL 250 FEET

0   ½   1   2 MILES

Mancos
Spring

Mountain Streams and Valley Streams. Most mountain streams lose altitude rapidly, dropping over falls and tumbling down steep slopes. The swift water has much energy and can do a great deal of work, carrying rocks and eroding the stream bed.

On the other hand, streams in broad valleys lose altitude slowly. The water flows slowly, has little energy, and can do little eroding.

Examine the pictures on this page and discuss the streams shown. Compare the speed of the water, the energy of the water, the rate of erosion, and the size of the particles that can be carried.

Visit several streams. Compare them with each other and the streams shown in the pictures. Look for signs of erosion, and discuss the possible history of the streams.

Seasonal Changes in Streams. On this page are shown two pictures of the same stream taken at different times of the year. When do you suppose each picture was taken? Give reasons for your statements.

Compare the speed and energy of the water in the pictures. At what season can the stream do the most work? When is the stream able to erode its bed most rapidly?

What evidence is there that the stream in the pictures sometimes carries very large pieces of rock?

Make a study of a nearby stream throughout the year. Measure the speed of the water at different seasons. Look for seasonal changes in the rate of erosion.

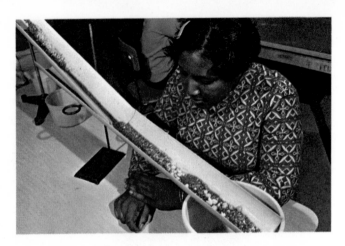

Effect of Decreased Rate of Flow. Set a short trough in a longer one, as shown above. What happens to the rate of flow of the water going from one trough to the other?

Put pebbles of various sizes in the first trough. Watch them as they are washed downward. Explain why some particles stop moving sooner than others, and why some do not stop moving at all. Where might you see the same effect in natural streams?

Sediments. Rock particles that are dropped from water (or from wind or ice) are called *sediments*. Sediments may range in size from microscopic particles to boulders weighing several metric tons. Gravel and boulders are carried only by swift streams, but mud can be carried by the slowest of streams.

Put a handful of soil in a jar of water and shake the jar several times. Then watch the mixture. Note the size of the particles that settle out immediately. Look at the jar a few hours later. Have more particles settled out? If so, what are their sizes? Has the water become perfectly clear?

Suspensions. Microscopic sediments may not settle from water for months or even years. These tiny particles are said to be in suspension. A mixture of water and tiny particles is called a *suspension*.

Scientists explain a suspension in terms of the belief that water molecules are in constant motion. Very tiny particles of clay or other sediments are bumped first one way and then another by the water molecules. The particles are unable to fall directly to the bottom and, if small enough, may never reach it.

Streams on Gentle Slopes. Swift mountain streams cut their beds downward without making many curves. However, streams on gentle slopes are more likely to erode the sides of their banks and thus produce many curves.

Make a study of a stream at a curve. Stretch a cord from one bank to the other. Measure the depth of the water, and the height of the bank at 25-cm intervals. Make a cross-section diagram of the stream to scale.

Measure the speed of the stream at different places across it. Drop in bits of wood and find out how far each goes in one second. Write the speeds on the cross-section diagram.

Which bank is being eroded? How do you know? Why is it being eroded? Why is the other bank being built up?

Drive a series of small stakes into the river bed at a distance of one meter from the banks of the stream. Return after heavy storms and floods, and note if erosion has taken place at the curve. Where has the most erosion taken place? Draw a sketch of how the erosion near the stakes would appear from above.

Valley Widening. Study the diagrams of the changes in a valley shown below. Discuss each stage, and describe how the changes were brought about.

Meanders. A stream flowing along a nearly level valley develops wide sweeping curves called *meanders.* Sometimes these curves continue to develop until nearly complete circles are produced. It is possible to walk a few meters between two points on a stream that are separated by kilometers of travel by boat.

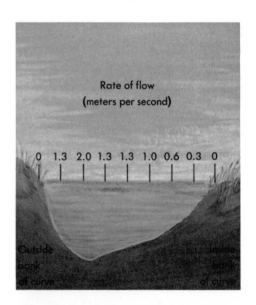

Rate of flow
(meters per second)

0 1.3 2.0 1.3 1.3 1.0 0.6 0.3 0

Outside bank of curve   Inside bank of curve

A   B   C

131

A meandering
stream

Unless something happens to stop it, a meander
continues to develop until the circle is complete
and the narrow neck is broken through. The stream
now takes the shorter path. The long meander is left
without a current.

Oxbow Lakes. After a meander has been cut
off, sediments from the main stream fill in the ends.
A curved lake remains, shaped like the bow of a
yoke once used with oxen. Therefore, these lakes
are commonly called *oxbow lakes*.

Study the diagram below. Tell what is
happening or has happened at each of the lettered
places.

Meandering Streams. Henrys Fork, a curving or
meandering stream, is shown on the next page.
What is the contour interval of this map? What do
the dotted contours show? Describe the slope of
Henrys Fork. Describe the probable rate of flow of
this stream.

Note the features labeled *A* on the inside banks of several meanders. Use the chart of symbols on page 101 to identify these features. Do such features result from erosion or deposition? Compare the regions of deposition along Henrys Fork with those in the photograph above.

Which side of a meandering stream has the greater rate of flow, the outside or the inside of the curves? How does the difference in rate of flow affect the shape of meanders?

Note the neck of land labeled *B* on the Henrys Fork map. Describe the change in the stream route if this neck of land erodes. How did the features labeled *C* form? What will happen after many years? (Hint: These features occur in flood plains.)

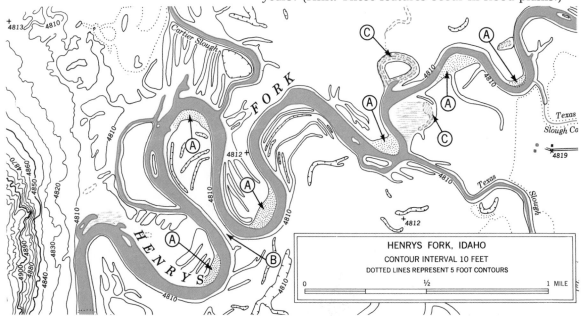

HENRYS FORK, IDAHO
CONTOUR INTERVAL 10 FEET
DOTTED LINES REPRESENT 5 FOOT CONTOURS

0   ½   1 MILE

Natural Levees. The map on this page shows a section of the Mississippi River. What is the width of the river in this region? Identify the features labeled A and B.

How far is point A from point C? What is the elevation at each point? Calculate the slope in feet per mile from C to A.

Note the dark straight lines about 3 miles long near A. These lines represent drainage ditches. In what direction does water flow in these ditches? Describe the shape of a profile drawn from A to B.

The plain on either side of the Mississippi River is made of sediments from flood waters. More sediments deposit on the flood plain near the river every time the river overflows and loses velocity. As a result, a raised flood plain develops along the river's edge. This feature is called a *natural levee*.

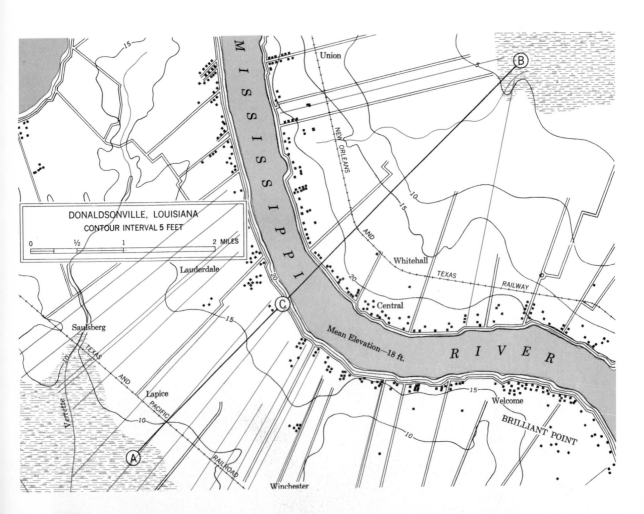

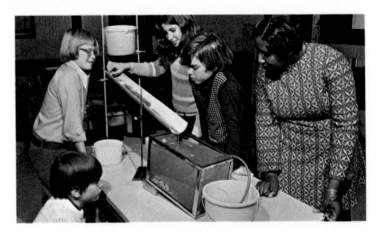

Deposits in Quiet Water. Fill a trough nearly full with a mixture of sand and gravel. Prop up the trough so that one end rests on an aquarium of water as shown above. Fill the aquarium with water, and start a gentle stream flowing down the trough. Let the water flow over the edge of the aquarium if the experiment is done outdoors. Otherwise, use a siphon to remove water from the aquarium as rapidly as it flows in.

Watch what happens to the sediments carried into the aquarium. Where are the large particles deposited? Where are the small particles deposited?

Keep the stream flowing until the deposit of sediments reaches the top of the aquarium. Add more sand and gravel to the trough as needed. Study the way the deposit builds out into the water.

Beds of Sediments. Change the rate of flow of water in the trough. Note that there are changes in the sizes of sediments deposited in one place. A layer of coarse particles may be dropped on a layer of fine particles. Such layers are called *beds*.

Divide an aquarium with a sheet of metal or a pane of glass to produce a narrow space along one side, as shown in the photograph below. Let sediments from the trough fall in this space. Change the speed of the stream from fast to slow. Note how the beds build up. Are all the sediment beds level?

Colorado River entering
the Gulf of California

Deltas. The velocity of a stream slows almost to a stop as soon as the stream enters a pond, lake, or ocean. Sediments carried by the water drop around the mouth of the stream and form a *delta*.

Heavy particles settle out immediately. Smaller particles are carried farther out into the body of water. Fine particles in suspension are carried much farther before they are deposited.

Study the diagram of a delta. Note the bedding. Why are some beds slanting, others nearly level?

Study the topographic map of the region around Taughannock Falls State Park in central New York. What is the land called which juts out into Cayuga Lake at the mouth of Taughannock Creek? What material makes up this feature?

Note the houses along the shoreline in the southeast part of the map. Why are these houses built next to stream outlets?

Construct a profile along line *AB*. Note that the depth contours are given in feet below the level of the lake (382 ft). Thus, the 100-ft-depth contour represents the lake bed at an elevation of 282 feet above sea level. Add a dotted line to the profile to show the level of Cayuga Lake.

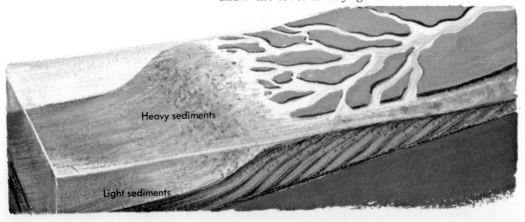

Heavy sediments

Light sediments

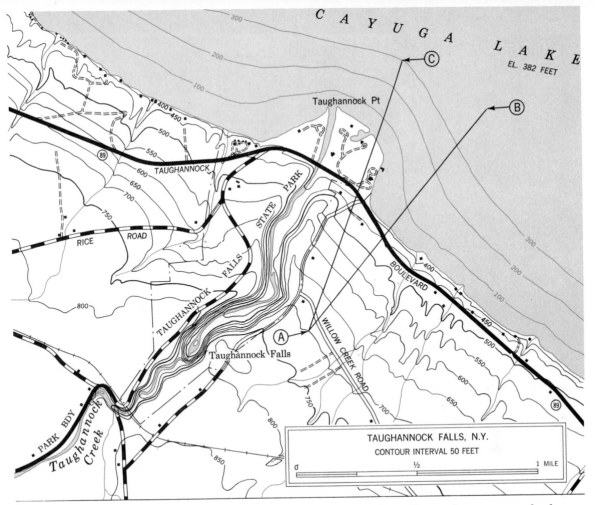

Draw a second profile on the same graph along line *AC*. Shade in the part of the graph which represents the delta of Taughannock Creek.

Estimate the height of the gorge just below Taughannock Falls. Estimate the height of the gorge above the falls. Use these two figures to estimate the height of Taughannock Falls.

Alluvial Fans. If a stream enters a broad valley from a gully on a mountainside, it slows down suddenly. The water can no longer carry the load of rock materials which it picked up on the steep slope.

The rock materials are dropped around the mouth of the gully. Gradually the deposit builds up in the form of a cone. The deposit is called an *alluvial fan.*

Alluvial fans are common along the base of a hill or mountain. Look for them wherever a steep gully ends in a level valley or plain.

137

An alluvial fan from a cliff

Alluvial fan

A small alluvial fan in a road cut

Small alluvial fans are also common at the base of a pile of dirt or road cut after a heavy rain. Look for such deposits near your school. Produce an alluvial fan by sending a gentle stream of water down the side of a pile of loose soil.

SUMMARY QUESTIONS

1. How do streams cut down through the stream bed?
2. Describe how potholes are formed.
3. How does the speed of flowing water affect the amount and kind of sediment carried by a stream?
4. How do young streams differ from mature ones?
5. How are deltas formed?
6. In what forms may streams carry materials?
7. How are beds of sediments deposited?

WIND EROSION

Wind is an agent of erosion in regions where there are fine sediments that can be blown about, as on deserts and beaches. Wind also causes erosion along coast lines by dashing waves against the shores. But wind is much less effective than water in eroding the surface of the land.

Transportation by Wind. Moving air transports sediments much as moving water does. This includes rolling larger particles, bouncing smaller particles, and carrying the finest particles suspended above the surface. Moving air, however, transports only comparatively small particles. This is because: (1) moving air has much less energy than moving water, and (2) air has little buoyant force compared to water. Water may support as much as a third of the weight of some of the common rock particles.

Consequently, only dust and very fine sand are carried by even the strongest winds. Coarse sand is rolled along and sometimes bounced by extremely strong gusts. Most of this transportation takes place close to the ground.

Plants effectively check wind erosion by reducing wind velocity near the ground. Thus, wind has little effect in regions that are covered with trees or grass. Wind erosion is most effective in deserts or along beaches where plants are few or absent.

139

Wind Abrasion. Wind alone has no abrasive power, but the sand it carries may have. Sand usually contains quartz, a very hard mineral. This sand is often carried at high speeds. An automobile windshield exposed to a sandstorm on a desert is quickly scratched.

The effect of sandblasting on rocks depends greatly upon the nature of the rocks. A hard rock of uniform structure is worn smooth. All its corners and bumps are more exposed to the action of the sand and are eroded more quickly. The resulting surface becomes very smooth, and takes on a waxy luster.

A rock that is not uniform in structure is eroded in the soft regions more than in the hard regions. Sandstones, for example, usually have soft spots; these wear away, leaving pits. A granite made up of large crystals is also pitted. The crystals of feldspar wear away faster than the crystals of quartz.

Wind Sculpturing. Many arid regions contain sculptured rocks like that shown above. At one time these were believed to be a product of abrasion by wind-driven sand. However, such rocks are rarely highly polished as they would be if sculptured by sandblasting.

It is now believed that these features are chiefly the results of chemical weathering that attacks some portions more than others. Wind plays a part by blowing away the loosened particles. Probably a limited amount of sandblasting removes some of the softer products of weathering, and smooths off corners.

Wind-Cut Pebbles. Pebbles that have been exposed to sandblasting often develop peculiar shapes like the one shown at the left. The exposed end of a partly buried pebble is cut away, forming a face that slopes upward away from the wind. If the pebble is dislodged, a new face develops. The number of such faces varies. Often, there are three, and the pebble is shaped somewhat like a Brazil nut.

Sand Dunes. Sand being carried by the wind is dropped wherever the wind is slowed. Often a bush or a stone acts as a windbreak behind which the sand is deposited. Such a deposit is called a *dune*.

Dunes have many shapes, depending upon the wind direction, the land surface, and the type of sand. A common form of dune is a long ridge at

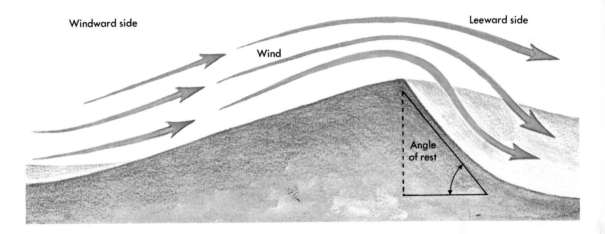

Windward side

Leeward side

Wind

Angle
of rest

right angles to the direction of the wind. Note the
dune on page 140. From which direction does
the wind seem to blow most of the time?

When a dune is small, the wind flows evenly
across it, blowing sand up one side and down the
other. A small dune has about the same slope on
both sides.

After a dune becomes a meter or two high, the
wind produces a low pressure area on the rear side,
as shown above. Sand blows up the front side and
is dropped after it reaches the top. The dune
develops a long, gentle slope on the front side and a
steep slope on the rear side. The rear slope has the
same angle as the angle of rest of dry sand.

Migrating Dunes. Some dunes form because of
fixed obstructions such as rock ridges. Others form
independently in the open. These are apt to move
across open spaces without changing their basic
shape.

A sand dune moves when wind picks up sand
from the windward side and drops it on the
leeward side, as shown in the diagrams. These
dunes may move several meters a year. If large
enough, they may bury buildings and trees. They
stop moving when the speed of the wind is
reduced.

Dune movement can be checked by providing
an obstruction that reduces wind velocity. For
example, grass can be planted on dunes in areas of
ample rainfall. The grass slows down the wind
close to the ground. Artificial obstructions must be
built where there is not enough rainfall to support
vegetation.

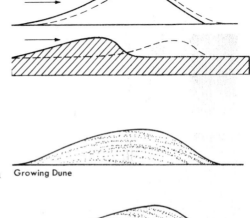

Growing Dune

Migrating Dune

Bedding in Dunes. A dune that is growing has two sets of beds. There is a gently sloping bed on the windward side, and a steeply inclined bed on the lee side, as shown in the diagram on page 141.

A migrating dune has only one set of beds, a steeply inclined set on the lee side. These beds are likely to be destroyed as the dune advances.

Any change in the direction of the wind changes the form of a dune. Earlier beds are partly cut away, and may be covered with new beds that have different slopes and directions.

This type of bedding which has slopes at different angles is called *cross-bedding*. Cross-bedding also takes place in water. However, the beds in a dune are rarely horizontal as they are in water-laid deposits.

Sand dunes have sometimes been transformed into sandstone. How might the origin of such sandstone be determined from the nature of the bedding, as shown on page 141?

REVIEW QUESTIONS

1. Describe some of the processes by which rocks weather.
2. How has a stream been able to cut the deep gorge shown above?
3. What types of erosion are probably taking place on the sides of the cliffs that line the gorge?
4. Why is there no talus at the base of the cliffs?
5. What conditions produce a waterfall?
6. What are two variables that determine the velocity of a stream?
7. Why are geologists interested in the speed of streams?
8. How does a meander form?
9. Why are the sediments in a delta in layers?

THOUGHT QUESTIONS

1. Which of the stones at the left was collected from the base of a cliff; which from a beach; which from a pothole; and which from a windswept desert?
2. Why does chemical weathering take place more rapidly in warm regions than in cold regions?

3

Record of the Rocks

Our present knowledge of the history of the earth is based chiefly upon the study of the rocks that are found at the earth's surface. Unfortunately, many of these rocks are so twisted and crumpled that their histories are not clear. Other rocks have eroded away and their histories are lost forever. In addition, there were probably long periods when no rocks were formed. Therefore, present ideas about the earth's history include many speculations about the meanings of the relatively few facts that have been discovered. Geologists agree on the general pattern of historical events, but they often disagree about the details.

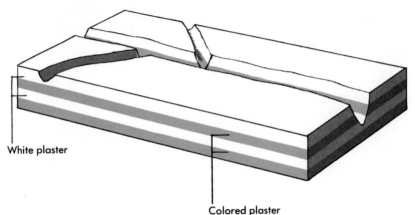

White plaster

Colored plaster

INFORMATION FROM ROCK LAYERS

Sedimentary rocks provide most of our present knowledge about the geologic past. The types of sediments making up the rocks, together with the way they were deposited, tell us something about conditions at the time of deposition. If there are fossils, they provide clues to the types of life and the climate during the period when the sediments were laid down.

Tracing Sedimentary Beds. Mix dry plaster of Paris with dry poster paint or colored chalk dust. Spread a thin layer of the mixture in a mold. Add a second layer of white plaster, then another layer of colored plaster. Continue until there are four or five layers of alternating colors in the mold. Dampen the plaster thoroughly. When set, remove the mold.

Imagine that this block of plaster represents sediments that were deposited in a delta beneath thousands of meters of other sediments. Because of the great weight, the sediments were squeezed together and cemented. Later, the delta was raised above sea level and the upper layers eroded away.

Pretend that streams are flowing across the block. Use a knife to cut the channels that such streams might dig into the rock layers.

Trace the beds along the stream channels. If these were real rocks, how might you know that you were tracing the same bed at all times? Why might some beds be more difficult to trace than others?

Geologists often claim that certain beds are buried beneath the surface even though the rocks cannot be seen. What is the basis for such claims? What assumptions have the geologists made? Why might their assumptions be wrong?

144

Comparing Ages of Rocks. Which beds in this cliff were probably deposited first? Which bed is oldest?

Fossils such as trilobites, like those at *A*, are found in the upper beds; trilobites like those at *B* are found in beds below the line. Which type of trilobite seems to have lived earlier in history?

Suppose that a layer of rocks several kilometers away contains trilobites like those at *A*. What inferences might you make about the age of this layer?

Geologists believe that dinosaurs lived long before elephants, and that the first birds appeared about the time that dinosaurs appeared. What evidence might provide the basis for these beliefs?

Rock Series. Study the photograph of the Grand Canyon on page 143. Which layers are probably youngest? Which are oldest? How do you know that conditions changed during deposition?

How can layers in the canyon walls be used to compare the ages of fossils found some distance apart? How can fossils be used to compare the ages of rocks found some distance apart?

The lowest rocks in the Grand Canyon contain no fossils. The next higher group of beds contain fossils of ancient sea animals. What inferences can you make from this?

The highest beds shown here contain fossils of sea animals that are different from those found below. What inferences can you make?

About 300 kilometers from the Grand Canyon are higher layers of rock. These contain fossils of large reptiles that probably lived on land. What inferences can you make?

Determining the Relative Age of Rock Layers. Geologists believe that the rock layers deposited first are covered by the layers deposited later. This basic idea is called the *principle of superposition*.

Another fundamental idea is that fossils differ from rock layer to rock layer. The comparison of fossils among rock layers is called *fossil correlation*.

Geologists have noted that the processes which change the surface of the earth today leave the same features that can be seen in very old rock formations. This observation leads to a third fundamental idea, that the earth's surface has been changed in the past in the same way it is being changed today. This idea is often referred to as the *principle of uniformity of process*.

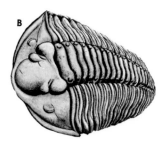

Inferences — conclusions based on observations

145

Relative Time and Measured Time. By using the principle of superposition and fossil correlation, geologists can tell whether one rock layer is older than another. To find how much older, we have to know how much time has passed. We need a measurement of the time that passed between the formation of the old and of the more recent layers.

Natural Timekeepers. There are several natural timekeepers, two of which are illustrated here. How can you tell the number of years a tree has been growing? How are tree rings formed? What do tree rings tell about the changing climate in which the tree was growing?

Thin sedimentary layers in the photograph occur when streams deposit their load in still water. The color and texture of the deposits vary with the seasons. Thicker layers of light coarse sediments are deposited in the spring when more water flows in the streams. What else do these layers show?

Dating by Deposition. Three meters of sediments were deposited around monuments on the Nile Delta in 3000 years. If the delta is at least 1000 meters thick, estimate its age. What conditions might affect your estimate?

Geologists estimate that it takes 13 000 years to deposit 1 meter of sandstone or shale, and 30 000 years for 1 meter of limestone. How can these figures indicate the length of geologic eras?

Radioactive Dating. Scrape a little paint from a clock dial marked with paint that glows in the dark. Examine the paint in darkness with a microscope.

The paint contains a radioactive substance. Atoms of this substance break up, causing fluorescent materials in the paint to glow. Each flicker seen is caused by the breakup of an atom.

About half of an amount of radium breaks up in 1600 years, changing to a form of lead. Half of the remaining radium becomes lead in another 1600 years. Radium has a *half-life* of 1600 years.

Suppose that a scientist someday finds a buried clock. How old is the clock if its paint contains one part of lead for each part of radium?

Geologists use other elements with longer half-lives for dating rocks. One type of uranium has a half-life of about 4.5 billion years, changing to a form of lead. One type of radioactive potassium with a half-life of about 1.3 billion years changes to argon. The oldest rock thus far dated has an age of about 3.75 billion years.

146

Fossil footprints in shale

ANCIENT SHORELINES AND DELTAS

Note the tracks and the impressions of raindrops on this slab of sandy shale. Where might these sediments have been deposited? What inferences can you make about nearby streams and other conditions? Discuss the possible source of the sediments.

What happened to the water level after the sediments were deposited? What might have caused the change in water level?

What inferences can you make from the tracks? Which animal seems to have walked by first? What inferences can you make about the climate at the time the sediments were deposited?

Mud Cracks. Mud exposed to air loses water, shrinks, and develops cracks. Sedimentary rocks sometimes show similar cracks. What conditions can be inferred from rocks containing such cracks?

Make a mixture of mud and water about 3 cm deep in a pan. Set it aside to dry. Note the time needed for cracks to develop. Discuss whether mud cracks in a sedimentary rock might form in a period between high and low tides. If not, what time interval is needed? What conditions might have caused the formation of mud cracks?

Cracks in mud

Fossil mud cracks

Coral reef

Fossil Coral Reefs. In what parts of the world are coral reefs being formed today? What is the water temperature in these places?

What inference can you make about the depth of the water over the reef shown above? Do you think that rivers are depositing sediments near this reef? Explain your answer. What types of sediments would you expect to find around the base of this reef?

The picture below shows an ancient coral reef in limestone. Notice that the reef has been buried deeply by lime mud which turned to rock, and was later exposed by erosion. What inferences can you make about the climatic conditions during the time the reef was being formed?

Dry Land Deposits. Many sedimentary rocks are dark in color because they contain carbon from decaying organisms. It is believed that such rocks were deposited in deep water, or in other situations where oxygen could not reach and oxidize the carbon.

Sediments deposited above the water level might have lost their dark color because oxygen was able to unite with any carbon in them. They might have become red at the same time iron compounds oxidized to form hematite.

Not all red rocks are dry land deposits. Some dark rocks become red on the outside, where they are exposed to oxygen. Other red rocks may have been formed from red sediments. However, most red rocks were once open-air deposits.

How can fossils aid a geologist in deciding whether a red rock was deposited beneath the sea or above it?

Salt Deposits. The Great Salt Lake in Utah is one place where table salt (halite) is being deposited. Where are some other similar places? What other conditions are necessary for the minerals in water to be deposited?

149

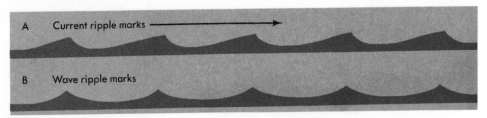

A Current ripple marks ⟶

B Wave ripple marks

Dissolve as much table salt as possible in a liter of water. Stir in a cupful of mud. Pour the mixture into a tray, and let it stand while the water evaporates. Describe the deposits that form in the tray. What conditions can be inferred from deposits of halite found between layers of shale?

Ripple Marks. Ripple marks are produced in sand or mud as currents of water move the sediments. The direction of movement can often be inferred from the shape of the ripple marks.

Waves in shallow water also produce ripple marks, but of a different shape. Study the diagram above. Describe the difference in the shape of the current and wave ripple marks. Explain how the direction of a current can be inferred from a set of ripple marks.

The photographs show ripple marks that have been preserved in sedimentary rocks. Which were formed by waves? Which were formed by a current? What inferences can you make about conditions at the time the rocks were deposited?

Steady winds also produce ripples in sand. Describe the shape of these ripple marks. How might you tell whether ripple marks in sandstone were formed by wind or by water?

Locating Ancient Deltas. Materials that erode from mountains are carried elsewhere and deposited. Some may be dropped near the foothills as alluvial fans, some deposited at the mouths of rivers as deltas. See the top of page 151.

What is the size of the particles usually dropped nearest the shoreline? What size is dropped next? What size is carried farthest from land? What types of rock may these three types of sediments become?

The photographs on page 151 show two kinds of sedimentary rocks. Where might the rocks have been deposited? Suppose that the pictures were taken a few kilometers apart. Which rocks might have been closer to an ancient shoreline? Which might have been in the center of an ancient delta?

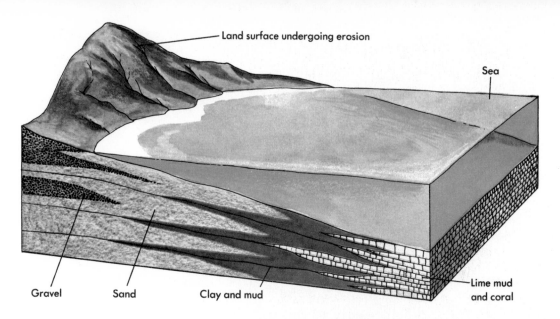

Land surface undergoing erosion

Sea

Gravel   Sand   Clay and mud

Lime mud and coral

Conglomerate cliffs

Sandstone cliffs

SUMMARY QUESTIONS

1. How do geologists trace sedimentary rock layers from one place to another?
2. What is the principle of superposition?
3. How do fossils help to identify a rock layer?
4. How does relative time differ from measured time?
5. Give two examples of natural timekeepers and explain how they keep time?
6. Beside keeping time, what other information can be deduced from natural timekeepers?
7. What clues indicate the existence of an ancient shoreline?
8. Describe the kinds of rock formed from an ancient delta.

151

ROCK MOVEMENTS

The notion of a solid earth has no place in geology. Marine fossils on high mountains provide evidence that great sections of the continents have moved upward. Folded and twisted rocks like those in the photograph are evidence of powerful forces acting upon rock masses, changing their shape and position.

Several theories have been proposed to explain the causes for rock movements, but direct proof of these theories has not been obtained. Much more must be learned about the interior of the earth before complete explanations are possible.

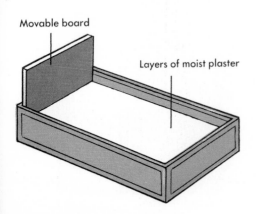

Movable board

Layers of moist plaster

Producing Folds. Cut a small board to fit loosely in one end of a cigar box. Spread layers of white and colored plaster of Paris in the box as described earlier in this chapter. Moisten the plaster but do not make it fluid. Force the board along the mold until the top of the plaster bulges about two centimeters. Let the plaster harden and remove it from the mold. This is a *fold*.

Synclines and Anticlines. Some folds produced in rocks arch upward; these are called *anticlines*. Some folds curve downward; these are called *synclines*.

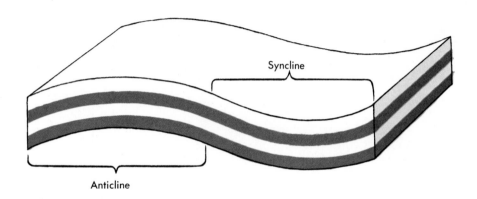

Syncline

Anticline

Examine the folds produced in your plaster model. Find an anticline and label it. Find a syncline and label it.

Study the folded rocks shown on the opposite page. Which type of fold is shown?

Metamorphism. Make another plaster model, squeezing the layers until they fold into tight curves. Sometimes you can produce both anticlines and synclines. Do any of the folds tip over? Do any of the layers break?

Great forces are needed to bend rocks into folds like those shown by the model. Geologists believe that such great forces heat the rocks, often softening them slightly. Some of the minerals may change chemically, often uniting with nearby minerals to produce entirely new compounds, usually in crystal form.

Thus, greatly folded rocks are usually metamorphic rocks. Shale may be changed to slate or even to schist. Into what does limestone change? What are some other types of metamorphic rocks?

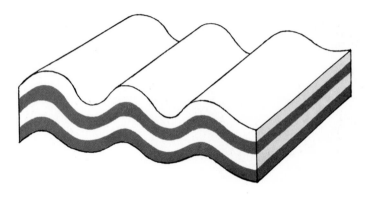

Mountain Ranges. The great mountain ranges of the world, such as the Rocky Mountains, show evidence of folding. We rarely see the tops of the original folds; erosion has usually removed the upper layers. Nevertheless, there seems little doubt that great sideward forces play an important part in the formation of all great ranges.

Study the photograph above. Trace the layers on thin paper placed over the picture. Then draw dotted lines to show the possible position of the complete beds before erosion removed the upper portions. What type of fold is shown?

The great mountain ranges of the Andes, Rockies, Alps, and Caucasus are all composed of thick layers of sedimentary rock. Within these layers are found fossils of animals that are known to have lived in shallow ocean water. Mountains thousands of meters high have shallow water fossils even at their base. Explain how shallow water fossils can be found thousands of meters below other shallow water fossils.

Geosynclines. Some ancient delta deposits are thousands of meters thick, but fossils in them show that even the bottom layers were never very far below sea level. Geologists have suggested that the great weight of a delta may push portions of a continent downward as fast as sediments are deposited.

The diagram shows how a syncline might form beneath a delta as sediments are added. This type of very large syncline is called a *geosyncline*.

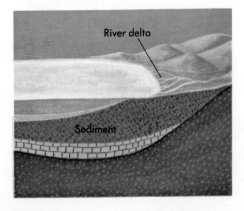

River delta

Sediment

It is believed that geosynclines were formed under many of the seas on the North American continent during geologic history. Where such geosynclines were formed, we often find mountain ranges today.

Unconformities. Make a model of folded beds as described earlier. Before the plaster is completely hard, remove it from the mold and cut off the tops of the folds with a knife. The model now represents folded beds with the upper portions removed by erosion.

Put the model back into the mold and add more layers of plaster. Moisten the plaster, let it set, and then remove it from the mold. Compare the model with the photograph below.

Geologists explain the photograph much as you might explain the model. How did the beds probably appear when first deposited? What happened to them before erosion took place? What did erosion do to the beds? What must have happened to the region before the top beds were formed?

A break between two sets of beds is called an unconformity. An unconformity shows that there was a period when no sediments were being deposited. Explain how an unconformity represents a break in the geological record.

Dating Earth Movements. Fossils in the lower beds of the photograph below are older. Fossils in the upper reddish-brown beds are younger. What inferences can you make about earth movements of the region shown here? Explain your answer.

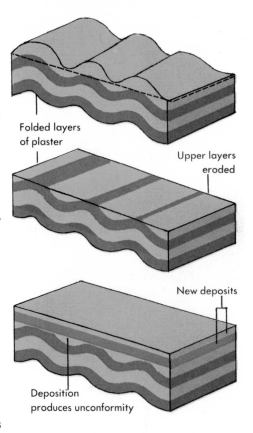

Folded layers of plaster

Upper layers eroded

New deposits

Deposition produces unconformity

An unconformity

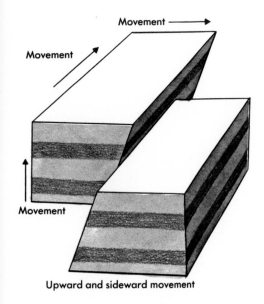

Movement ⟶

Movement ↗

Movement ↑

Movement

Upward and sideward movement

Faults. Sometimes, during the making of folds in moist plaster, the layers break and slide past each other. Geologists call a similar break in rock layers a *fault*.

Fill a cigar box mold with layers of colored plaster. When the block is hard, remove it from the mold and saw it into two parts along a sloping line, as shown at the left. Press against the ends of the two parts until one part slips over the other. The movement of a block may be up or down. Movement may also be sideward if the force is sideward.

The photograph below shows a fault in bedded rock. Try to match up the beds on either side of the fault. Which side is higher? Can you tell which of the two blocks moved? Explain your answer.

Compare the fault in this photograph with the fault in your model. In what ways are they alike? In what ways are they different?

Did this fault in the earth's crust result from forces pushing together or pulling apart? Has erosion taken place since the fault was produced? How do you know?

Earthquakes. The top photo on page 157 shows a fault that appeared in Nevada in 1954. At the same time, there was a violent earthquake.

The movement of rocks along a fault is believed to be the cause of many earthquakes. The rocks slip, catch, and slip again, possibly bending and springing back at times. These motions set up vibrations that spread outward from the fault line. The larger vibrations are felt as earthquakes; the smaller ones may go unnoticed.

A fault

Most movements along a fault are believed to be small, measuring only a few centimeters. Some large movements have been measured. An earthquake in Alaska in 1899 was produced by an upward slip of over 12 meters. The San Francisco earthquake of 1906 was caused by a sideward slip of about 6 meters. The 1964 earthquake in Alaska was caused by a sideward slip of about 11 meters.

Slickensides. Rocks along a fault are sometimes polished as smooth as glass by the friction caused when the two blocks rub against each other. The polished surfaces are called *slickensides*.

Slickensides are one clue to the presence of a fault. Note the slickensides in the photograph at the left. What was the direction of the rock motion?

Slickensides

SUMMARY QUESTIONS

1. Describe an anticline and a syncline.
2. Why does folding of rock layers result in metamorphism?
3. Explain why the tops of folded mountain ranges are rarely seen.
4. What is a geosyncline?
5. What is an unconformity? Using a model, explain how an unconformity develops.
6. What is a fault? How do faults occur?
7. How is it possible to date the occurrence of earth movements?

157

Mauna Loa, a shield volcano

Fluid lava volcano

VOLCANIC ACTION

A volcanic eruption occurs when the pressure of molten rock and gases forces a break in the rocks of the earth's crust. Some eruptions are violent, shattering the overlying rocks, and hurling fragments high into the air. Many eruptions, however, are relatively quiet; lava pours up through the break and spreads out around the opening, where it cools. The history of a volcano can often be inferred from the nature and structure of the deposits produced.

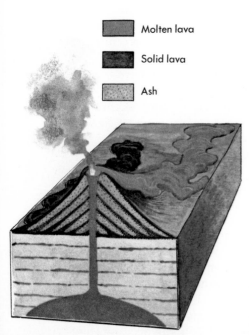

Molten lava

Solid lava

Ash

Formation of a composite cone

Forms of Volcanoes. Lava has a low angle of rest before it has completely solidified. A series of quiet lava flows from a single volcano gradually builds up a mound that is much broader than it is high. The structure is called a shield volcano because of its similarity to an ancient Greek shield. Shield volcanoes are found in the Hawaiian Islands.

Angular fragments produced by a volcanic explosion have a high angle of rest. They fall back around the opening in a steep-walled pile called a cinder cone volcano. Such cones erode rapidly.

Gases may bubble violently up through lava and hurl globs of the molten material into the air. These cool so that many are in a semiliquid state upon landing. The result is a steep-sided, strong-walled cone called a spatter cone volcano. Spatter cones are rarely very large, because conditions suitable for their formation do not last for long periods.

Most large, well-known volcanoes are the products of a long series of eruptions, usually an alternation of explosions and lava flows. The result is a layered structure like that shown in the

Spatter cones   A composite cone

diagram on the opposite page. If all eruptions occur
at a central crater, the cone may be regular in
shape. However, the central crater often becomes
plugged, and eruptions burst out through the sides
of the cone. Note the picture of Mt. Shasta on
page 162.

Active and Inactive Volcanoes. The Western
Hemisphere possesses a number of active volcanoes
in Hawaii, Mexico, Central America, and along the
Andes in South America. Mt. Katmai in Alaska
exploded violently in 1912, and California's Lassen
Peak erupted on a smaller scale in 1914. See the
map on the next page.

Both North and South America also possess
thousands of volcanic cones which show no sign of
activity. Some of these may be truly extinct; that is,
they will never be active again. Others may be
tightly plugged with cold lava. However, they may
someday explode because of steam forming inside,
or they may break open if lava is again forced
toward the surface. Such volcanoes are said to be
dormant.

Geologists try to predict volcanic activity by
measuring and recording earthquakes. The eruption
of a Hawaiian volcano was predicted by measuring
the tilt of the earth around the volcano. The upper
part began bulging upward and tilting outward.
However, there are no simple methods of
determining whether an inactive volcano is extinct
or dormant.

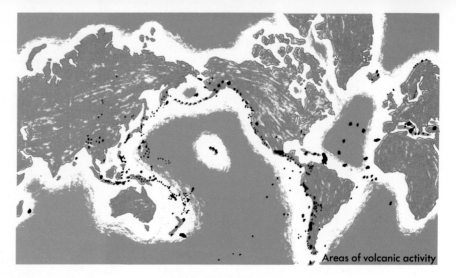
Areas of volcanic activity

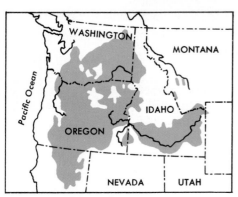
Areas of lava flows in the
northwestern United States

Devil's Tower, Wyoming, is 260
meters high

Distribution of Volcanoes. During the past 400 years, some 500 volcanoes have erupted. They have killed hundreds of thousands of people. Huge quantities of carbon dioxide, water vapor, and ash have been poured into the atmosphere. Volcanoes have also provided some of the world's richest soils.

Study the map at the top of this page. How are earthquakes and volcanoes related? What landforms are common in earthquake and volcanic regions?

Lava Flows. An important factor in determining the shape of a volcano is the way the lava flows. The lava may flow over the ground at a speed of more than 30 kilometers per hour, or it may be thick and move at a snail's pace.

There are volcanoes from which the magma issues from long fissures or cracks. The resulting floods of lava produce a plateau. Usually a thin, flowing lava produces a flat landscape.

In general, there are two kinds of lava. One is dark, dense, and contains very little silica. This kind hardens into basalt. The other is lighter in color, less dense, and much richer in silica. This hardens into rhyolite. A basaltic magma is usually hotter and the lava flows faster than a rhyolitic lava.

Craters and Calderas. The pit at the top of a volcanic cone is commonly called a *crater*. A crater is usually produced by a small explosion during the last eruption of the volcano. Earlier craters are often destroyed or filled in with the products of later eruptions. However, traces of earlier craters are visible in many volcanoes.

A crater may also be formed if the lava in the mouth of a volcano recedes, perhaps because a lower exit exists. When the lava in the botton of such a crater remains molten it is often called a *firepit*.

Occasionally, a major explosion blows away the larger part of a volcanic cone, leaving a huge cavity near the top of the volcano. In other instances, the lava under a volcano recedes, so that the volcano's top collapses. Either type of pit is properly called a *caldera*.

Inactive craters often fill with water and form small lakes that are true crater lakes. These are not apt to occur in cinder cones because the loose materials of the cone are very porous.

Calderas also fill with water if conditions are suitable. Crater Lake in Oregon is, properly speaking, a caldera lake. The lake and the caldera are described more fully on the next three pages.

Erosion of Volcanoes. Volcanic cones tend to erode rapidly. Even active and growing volcanoes are trenched with gullies. The loose materials of a cone are easily dislodged by wind and running water. Certain types of solidified lava weather rapidly.

On the other hand, some forms of solidified lava may resist erosion, and may remain after other parts of the cone have been removed. Sometimes the throat of a volcano is plugged with a resistant lava which remains as a tower after the volcanic cone itself is eroded away. The Devil's Tower in Wyoming is thought to be a volcanic remnant.

Periods of Volcanic Activity. The age of volcanic eruptions can sometimes be determined by fossils in sedimentary rocks above and below the volcanic deposits. Such evidence shows that there has been more activity at certain times than at others. In western North America, volcanoes were especially active about 250 million years ago, again about 150 million years ago, and from 50 million years ago almost to the present.

Large areas in the western United States and Canada are covered with lava flows and ash deposits that look very fresh. However, only Mt. Katmai in Alaska and Lassen Peak in California have erupted during recent times.

Volcanic activity sometimes went on for a long period over a large area. At the right is a diagram of a cliff in Yellowstone Park. Volcanic materials

making up this cliff buried a series of forests, one on top of another. The trunks of many trees were preserved, as shown in the photo on page 161. How many forests were buried? Estimate the age of the trees. Estimate the time that the volcanic activity continued.

Volcanic Mountains. Mt. Shasta (on the left) and Shastina are shown above. These mountains were once active volcanoes. Note the lava flow in the foreground. Did this lava come from the peaks? Look for evidence to support your answer.

Note the ridge crossing line *AB* on the map on the next page. What is the difference in elevation between *A* (or *B*) and the top of the ridge along line *AB*? Propose a possible cause of this ridge. Trace this ridge up the slope of Mt. Shasta. What is the highest elevation at which this ridge can be detected? Locate similar regions.

Trace the 9600-foot, 10 800-foot, and 12 000-foot contour lines with your finger. Describe the shape of Mt. Shasta.

Note the glaciers near the top of Mt. Shasta. On which side of the mountain are glaciers common? On which side are there few glaciers?

Compare the lowest elevation of Whitney, Bolan, Hotlum Glaciers on the northern slope with Konwakiton Glacier on the southern slope. Why do these glaciers melt at different elevations?

Note Mud Falls Creek and Clear Creek on the southeast slopes of Mt. Shasta. Explain how these creeks got their names.

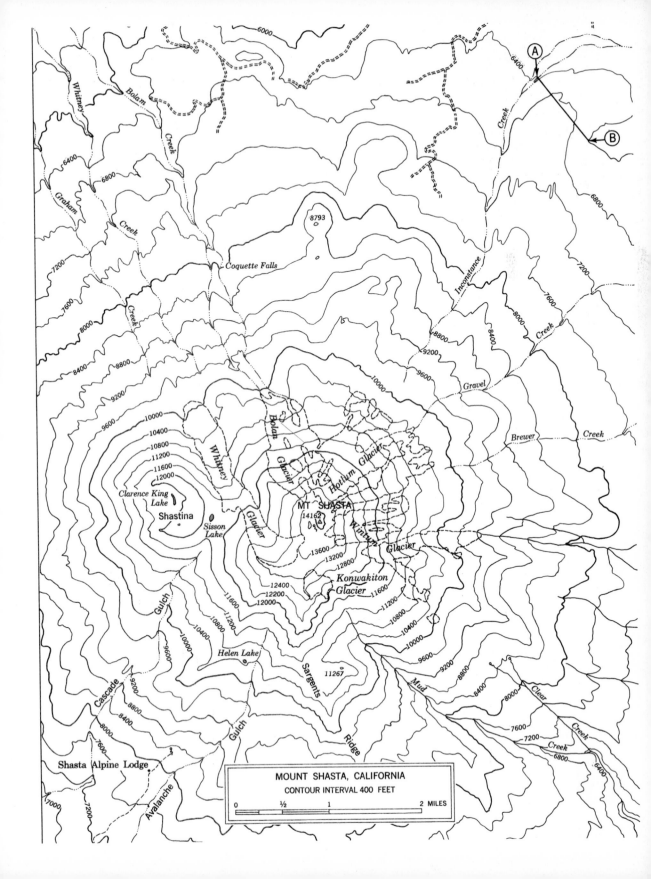

Crater Lake. Crater Lake, shown above, is located at the top of Mt. Mazama in Oregon. Note the cliff called Llao Rock to the right of the island in Crater Lake. Use the map to determine the elevation of Llao Rock above Crater Lake. Where is the lowest point on the cliff around Crater Lake?

Make a profile of the crater in Mt. Mazama from point A to B. Note that the depth contours are given in feet below lake level (6176 feet). Add a second line to your profile to show the level of water in Crater Lake.

Crater Lake is in the top of an ancient volcano. Geologists believe that Mt. Mazama was once 14,000 feet high. About 17 cubic miles of rock material are now missing from the top of this volcanic mountain. What is the highest point now on the cliffs around Crater Lake? Since no evidence has been found of rock material blown off the top of Mt. Mazama, geologists think that the top collapsed.

Locate two small cones within the crater, other than Wizard Island. Suggest how these cones developed. Why did they probably develop after the crater formed?

Kerr Valley is located along the upper rim of Crater Lake. What is the shape of the cross profile of this valley? Apparently, a glacier once flowed through this valley. What can you infer about the time in history when Mt. Mazama lost its top?

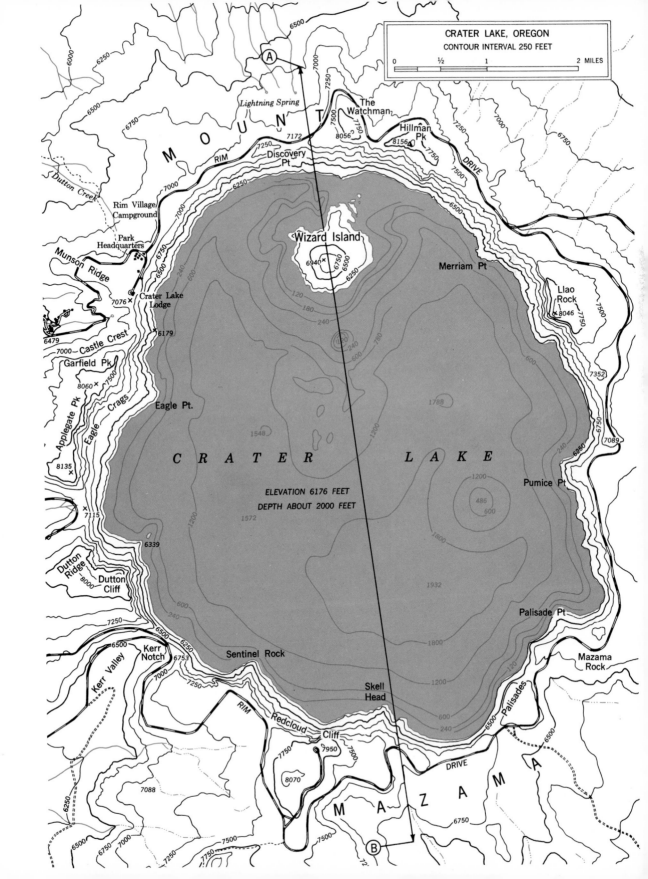

MOUNT

Lightning Spring

The Watchman
7172
8056
Hillman Pk
8156

RIM   Discovery Pt

DRIVE

Rim Village
Campground

Park
Headquarters

Munson Ridge

Crater Lake
Lodge   6179

7076

6479
Castle Crest
Garfield Pk
8060
Applegate Pk
Eagle Crags
8135
7115
Dutton
Ridge
Dutton
Cliff
8000

Wizard Island
6940

240
600
120
180
240
240
780
600
240
660

Merriam Pt

Llao
Rock
8046

7352

7089

Eagle Pt.

1788

1548

CRATER   LAKE

1200

ELEVATION 6176 FEET

DEPTH ABOUT 2000 FEET

1572

1200

1800

Pumice Pt

1200
486
600

Palisade Pt

1932

1800

Mazama
Rock

6339

600
240

Sentinel Rock

6753

Kerr
Notch

Kerr Valley

7250
6500
6250
7000
7250

1200

Skell
Head

1200

600
240

Palisades

Redcloud
Cliff
7950
7750
8070
7088

DRIVE

M   A   Z   A   M   A

A

B

GLACIATION

The climate on many mountaintops is so cold that more snow falls during the winter months than melts during the short summer. The snow piles up deeper and deeper, and the bottom layers are compressed into ice. This ice may then be squeezed downward along the mountain slopes in slow-moving streams called glaciers. The photograph above shows some of the 20 or so glaciers on Mt. Rainier in Washington. How many do you see?

Nisqually Glacier

Mountain Glaciers Today. The photograph at the left shows Nisqually Glacier on Mt. Rainer. Snow falls throughout the year at the top of Nisqually Glacier. Note the snow field near the top.

The weight of snow in the snow field squeezes and compresses the snow below into ice, and pushes it very slowly down the valley. Rocks freeze into the glacier as it moves slowly downward. These rocks scratch the floor and walls of the valley, wearing away more rock material.

As the glacier moves into lower elevations, melting occurs. Rocks and dirt appear on the glacier's surface as the ice surrounding them melts away. Suggest how a layer of rock and dirt might slow the melting of the glacier.

Glacial Erosion. The slow but powerful movement of ice down a mountainside produces erosion that can be seen long after the ice has melted. Loose rocks become frozen in the ice and are dragged along the bedrock, sometimes polishing it and sometimes scratching it. These rocks too are worn and frequently scratched as the glacier moves. Loose rocks are deposited at the end of a glacier where the ice melts. Two of these rocks are shown at the left. Explain their shapes.

Glacial Cirques. Water from a glacier may flow into cracks in the bedrock and freeze. Then pieces of the bedrock may be pulled from the mountainside as the glacier moves. This type of erosion produces a curved, steeped-walled excavation in a mountainside at the head of a glacier. Such an excavation is called a *glacial cirque* because it resembles a type of outdoor theater that Romans once called a "circus."

The photograph above shows a cirque formed by a glacier that used to be larger than it is now. How would you recognize a cirque in a mountain that no longer has a glacier? What inference could you make from the cirque?

Glacial Troughs. A glacier erodes the sides of a valley as well as the bottom. Thus, the valley is given a rounded bottom, and its cross section has the shape of the letter U. What is the shape of the cross section of a valley which has been eroded by a swift stream?

Study the photograph at the left. Why do geologists believe that a glacier moved through this valley? What evidence is there that ice did not cover the mountains on either side of the valley? A valley through which a glacier has passed is called a *glacial trough*. What inferences can be made from the discovery of a glacial trough in a region that has no glaciers?

167

Moraines. Glaciers carry boulders, rocks, and sediments with them as they flow. This debris is plucked out of the sides and bottom of the valley through which the glaciers flow. Rocks, pebbles, and fine particles from the valley walls tumble down on the glaciers.

Unlike a stream, which carries only small particles in suspension, a glacier will carry a mixture of debris ranging in size from large boulders to fine clay. Whenever a part of the glacier melts, the debris is deposited all in a jumble. How do glacial deposits compare to stream deposits?

Notice in the photograph above how ridges of rock and gravel have piled up on the sides and at the front of the glacier. Such ridges of unsorted debris are called *moraines*. Those ridges forming along the sides of the glacier are called *lateral moraines*. The ridge which marks the limit of the glaciers advance is known as a *terminal moraine*. Often a series of moraines are found behind the terminal moraine. How can you explain these moraines?

Outwash. Water flowing downhill from a melted glacier carries sediments. The running water sorts out the sediments by size. Depending on the volume of flowing water and its rate of flow, the water will carry or deposit the sediments. The sand and gravel deposited by the melted glacial water is called *outwash*.

Sometimes the glacier will advance over the outwash, and upon retreating will leave large blocks of ice behind. When these blocks melt, they leave craters in the outwash called *kettles*.

Glaciated Valleys. This map shows the valley of a stream called Lake Fork. Determine the slope of the stream from Virginia Lake to the swamp. Compare the slope of Lake Fork with the slopes of streams studied earlier. Which of these streams have slopes like the slope of Lake Fork?

Construct a cross profile of Lake Fork Valley from point *A* to *B*. How does the cross profile of Lake Fork Valley differ from the cross profiles of streams studied earlier? What type of erosion produces valleys of this cross profile?

The ridge of land around the eastern half of Turquoise Lake is composed of rounded rocks, pebbles, and sand. These materials are not distinctly bedded. What is the name of land ridges such as this? How do they form? Where could you find bedrock similar to the loose rock in this ridge?

Locate two pits in this ridge. What are they called? How did they form?

Describe the shape of the land at the several beginnings of Lake Fork. What are these landforms called? How did they form?

Tracking Glaciers. Glacial cirques show where some glaciers started. What does a glacial trough show? How might you determine the area once covered by a glacier?

Glaciers sometimes push underlying loose materials into long, cigar-shaped ridges called *drumlins*. How might the direction of glacial movement be inferred from these drumlins?

Direction of ice movement can sometimes be determined from scratches in bedrock. How might the point of a sharp rock change as it is dragged across bedrock? What would be the effect on the shape of the scratch? How can the changes in a scratch be used to determine the movement of ice?

Two sets of scratches in different directions may show that two glaciers crossed the region. How can a geologist decide which scratches were made by the earlier glacier, and which by the later one?

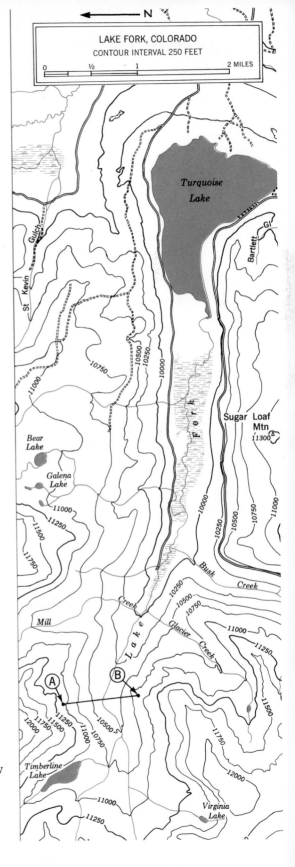

LAKE FORK, COLORADO
CONTOUR INTERVAL 250 FEET

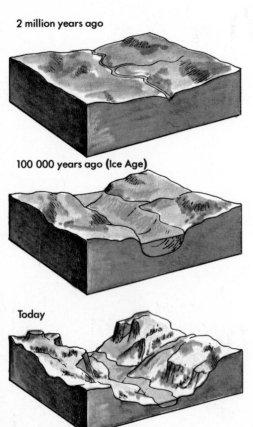

2 million years ago

100 000 years ago (Ice Age)

Today

Hanging Valleys. The photograph above shows part of the region mapped on the next page. The photograph was taken from the edge of the road somewhere between Yosemite (pronounced Yo-'sem-ity) Village in the northeast corner and Chinquapin Ranger Station in the southwest corner. Use the map to locate the exact spot where the picture was taken. (Hint: Was the photographer standing on the valley floor, part way out of the valley, or completely out of the valley?)

Use the map to determine the height from the top of the rock cliff at the left to the valley floor. Estimate the height of the waterfalls on the right side of the valley.

Measure the straight-line distance from Camp Curry (*A*) to Glacier Point (*B*) in the northeast. Estimate the road distance between these two points. Suggest a reason for the difference.

Describe the profile from El Capitan to Cathedral Rocks. Does this profile resemble one of a glaciated valley? What is the name of the creek which produces the waterfall?

Before the Ice Age, this flowing stream was probably a tributary to the main river in the valley. At that time, the creek probably flowed gradually downhill into Merced River. Glacial erosion of Merced River Valley left the tributary streams thousands of feet above the main valley floor. Today, this area is famous for its spectacular waterfalls from such hanging valleys.

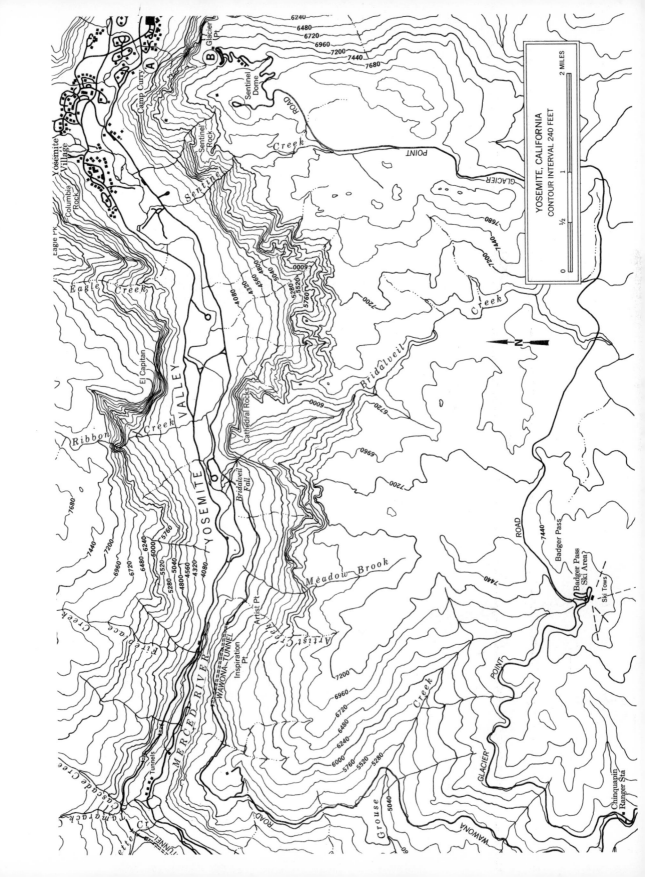

YOSEMITE, CALIFORNIA
CONTOUR INTERVAL 240 FEET

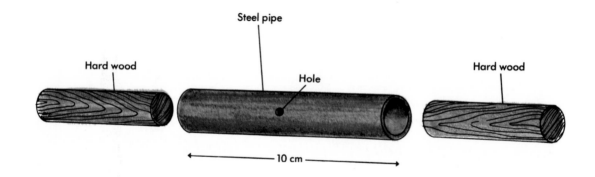

Hard wood   Steel pipe   Hole   Hard wood

← 10 cm →

Crushed ice

Flowing ice

Ice Flow. Geologists are not certain how ice deep within a glacier is able to flow. Some suggest that ice is plastic and can flow if a great force acts on it for a long time. Others suggest that high pressure melts the ice so that it flows to a region of less pressure, and then refreezes.

Ice flow can be demonstrated even if it is not understood. Bore a small hole at the midpoint of a 10-cm length of steel pipe. Whittle two pistons from hardwood (such as a broom handle) and sandpaper them to make a tight fit in the pipe.

Pack the pipe with crushed ice, put the pistons in place, and squeeze them with a vise. At first, the pistons will do little more than squeeze air from the crushed ice, but by adding more ice a solid mass will finally be formed. Then more pressure will squeeze the ice out through the hole.

Striations. Scratches such as those photographed below are found along glacial valley walls and on the bedrock of the valley floor. These scratches are called *striations*.

Notice that the striations in the rock are nearly all parallel. How can you explain their parallelism? If the bedrock is in the same position as it was

when scratched, what other information can be gained by studying the striations?

Evidences of Continental Glaciers. Draw a profile of line *AB* on the Whitewater, Wisconsin map. Describe the shape of this landform. Trace this landform from one edge of the map to the opposite edge.

This ridge is composed of loose rock and gravel which appear to have been dumped at this site. The rocks are rounded and scratched as though they were dragged here. Many different types of rocks are found in this ridge. They are identical to the bedrock found five to fifty miles north. From this evidence, how do you think this ridge was formed? What is this type of ridge called?

The tops of these ridges are often covered with low cone-shaped hills called *kames* and shallow depressions called *kettles*. Kettles become kettle ponds if they fill with water. Find some kames and kettles on this map.

The *X* on the map on the next two pages shows the location of the map below. The dark gray regions on this map show other ridges similar to the one at Whitewater, Wisconsin. What do these ridges suggest about a possible ice age in the past?

The map also shows the direction of striations on outcrops of bedrock. How do these scratches support the idea of an ice age?

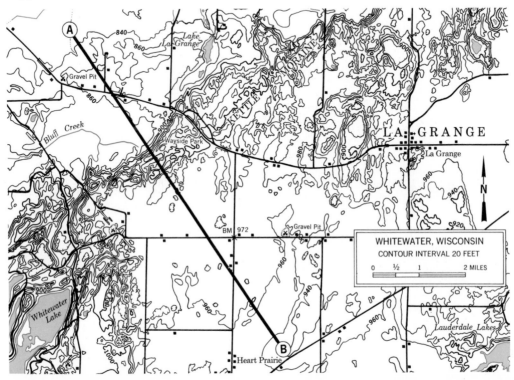

173

Locate regions of the northern United States where cigar-shaped hills called drumlins and narrow, winding ridges called eskers are found. Eskers are believed to be beds of streams which carried rocks and melting ice under the glacier.

Students have made the interesting observation that the glacial deposits seem to stop at the Missouri and Ohio rivers. What explanations can you propose for this observation?

The Great Ice Age. Nearly all of Canada and Alaska, together with a few northern states, show many signs of recent glaciation. The high mountains of the West once contained more and larger glaciers than can be found there today.

Geologists infer from this evidence that the earth has passed through a cold period that produced enormous piles of snow in certain parts of Canada. This snow changed to ice, and flowed outward as a huge glacier or ice sheet.

Radioactive dating of plant remains suggests that the cold period began about one million years ago. The last of the ice sheet melted from the United States about 10 000 years ago, but glaciers still remain in the western mountains. Greenland is still covered with an ice sheet. The cold period may not yet be over.

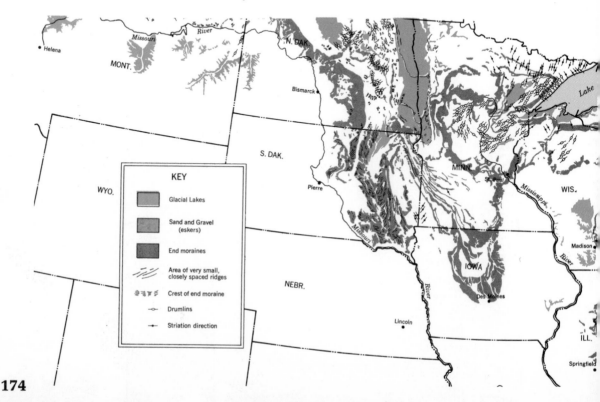

KEY

Glacial Lakes

Sand and Gravel (eskers)

End moraines

Area of very small, closely spaced ridges

Crest of end moraine

Drumlins

Striation direction

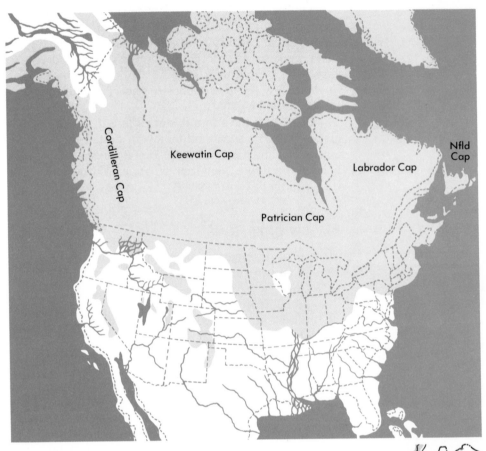

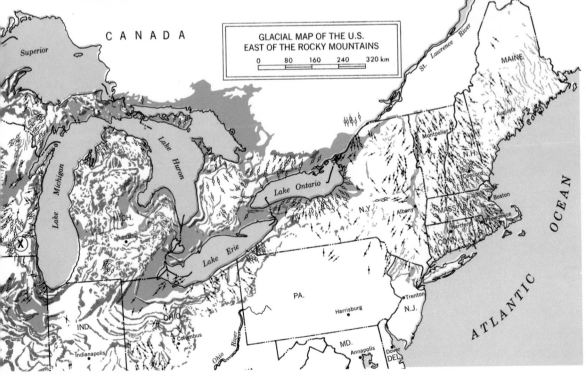

GLACIAL MAP OF THE U.S.
EAST OF THE ROCKY MOUNTAINS

0   80   160   240   320 km

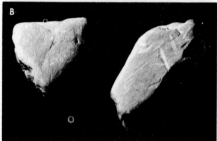

The map on the previous page shows the location of the terminal moraines deposited when the ice sheet had made its greatest advance. The same map also shows the location of the four ice caps from which the ice sheet is believed to have flowed. How could scratches in bedrock be used to locate these ice caps?

Geologists have found four different sets of glacial deposits separated by layers containing fossils of plants and animals. Therefore, they believe that there were four separate cold periods, each producing an ice sheet.

REVIEW QUESTIONS

1. What is an earthquake?
2. What can cause an earthquake?
3. Under what conditions were the beds at A probably deposited?
4. Which of the beds at A are probably the oldest?
5. What changes in conditions seem to have taken place as the beds at A were being formed?
6. What processes gave the rocks at B their shapes and markings?
7. What can be inferred if rocks like those at B are found in a bank of loose rock materials?
8. What is the name of the rock structure shown at C?
9. Why are rocks in structures like those at C often metamorphic?
10. What happened to the rocks at D?
11. Name three states that were covered with ice during the recent Ice Age?
12. What is a geosyncline and how is it thought to be formed?
13. How can fossils be used to locate rock layers of the same age?
14. What is radioactive dating? Give an example.

THOUGHT QUESTIONS

1. What evidence do we have to support the theory that there was a great ice age?
2. Why is it difficult for geologists to be certain about the past history of the earth's surface?
3. How can faulting produce a mountain range?

4

Life in the Past

The true significance of fossils has been realized for less than 200 years. Many people once believed that fossil bones belonged to individuals drowned by the Great Flood described in the Bible. Other fossils have been explained as unsatisfactory forms of life discarded by the Creator while making the world.

Our knowledge of fossils has grown greatly during the last century. Thousands of extinct species have been identified. Nevertheless, only a small part of the fossil-bearing rocks have been explored. There are still countless numbers of fossils which will add to our knowledge of ancient life when they are uncovered.

 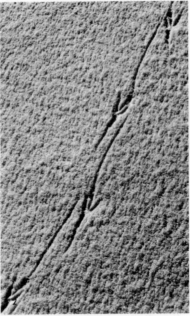 

CLUES TO THE PAST

Complete remains of long dead plants and animals are very scarce. Usually, only bones, shells, wood, or other hard parts are preserved. Such fossils provide incomplete and uncertain clues as to the nature of ancient life.

Geologists compare fossils with living organisms in order to make guesses, or inferences, about the nature and habits of extinct plants and animals. What type of training should a geologist have for making inferences about life in the past?

Tracks as Clues. The first picture above shows that an animal crossed a snow-covered field. People who have studied common animals can tell what animal made the tracks and which way it was going. They can also make inferences about the speed and size of the animal.

Examine the photograph. Which prints were made by the hind feet and which were made by the front feet? Which way was the animal going? What gait was it using? Check your inferences with those given later in this chapter.

Two other sets of tracks are shown above. Set 2 is freshly made. Set 3 was found in a slab of ancient sandstone. What inferences can you make about these tracks? Which inferences are probably more trustworthy? Why? Check your inferences with those given later in this chapter.

Using Experimental Evidence. Dampen clay soil until it shows a clear footprint when you step on it. Measure the depth of the print. Mark off a sheet of paper into squares 2 centimeters on a side, cut it to the size of the print, and determine the area.

With what force did you step on the soil? What was the pressure on each square centimeter of soil?

Ask a child and an adult to make footprints in the same soil. Use measurements of depth and area to estimate the weight of each person. How close are your estimates? How might you improve them?

Estimate the area and depth of the dinosaur track in the photograph. Assume that the animal stepped in soil like that used in your experiment. Infer the weight of the dinosaur. Would it be different if the animal walked on two legs or four legs? What assumptions might make your calculation too large? What assumptions might make it too small?

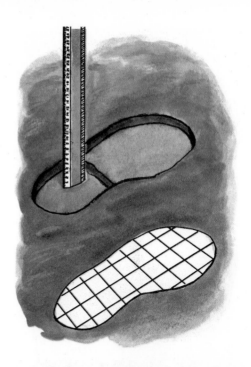

Dinosaur tracks in Colorado Jurassic rock

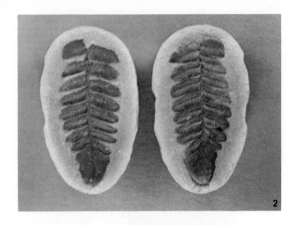

Inferences from Incomplete Remains. A number of fossils are shown on these two pages. None of those shown are the complete remains of an original organism. Study the pictures and look for clues that can help you make inferences about these plants and animals that have been dead for many years. Compare your inferences with those made later in this chapter.

Study fossils in a school collection and fossils that you collect for yourself. Write down your inferences about them. At the end of this chapter, review your inferences. Make changes in agreement with your added knowledge of fossils.

Courtesy of Field Museum of Natural History

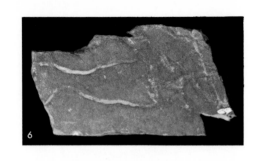

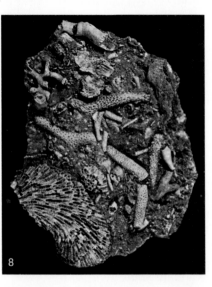

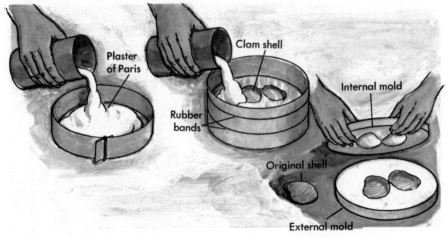

Plaster of Paris · Clam shell · Rubber bands · Internal mold · Original shell · External mold

HOW FOSSILS ARE FORMED

A fossil is any trace of a plant or animal that has lived somewhere in the distant past. A piece of wood, a shell, or a bone is a fossil. So too is an animal track or the burrow of a long-dead worm.

Very few actual remains of plants and animals are preserved as fossils. Chemical changes usually take place, and the original material is replaced with new substances. Sometimes the original material dissolves completely, leaving only prints in the sediments that surrounded the remains.

Molds of Shells. Prints of shells and bones are called *molds*. A mold of a shell has the shape of the shell in reverse. Thus, a mold shows a groove where a shell has a ridge, and a ridge where a shell has a groove.

Make a clam shell mold in plaster of Paris. Prepare a thick mixture of the plaster and water. Press the clam shell into the mixture until it is level with the surface. Let the plaster harden. Then coat the plaster with Vaseline, and pour another layer of plaster on top. On the following day, separate the layers and remove the shell. Compare the markings on the mold with those on the shell.

The mold of the outside of a shell is called an *external mold*. The mold of the inside of a shell is called an *internal mold*. The photo at the left shows both internal and external molds. *A* and *B* are both molds of the same shell. Which is an external mold and which is an internal mold? What is *C*? *D* and *E* are also molds of the same shell. Which is internal and which is external?

Casts of Shells. Ground water seeping through a rock may dissolve shells within the rock. Empty spaces are left between the molds. Fit together the plaster molds you have made. Describe the space remaining between the molds. What would you see if you made a break across the molds? Note the space between F and G in the photograph on the opposite page.

After ground water has dissolved a shell, it may deposit other minerals in the space. The new minerals form a cast. In what ways is a cast like the original shell? In what ways is a cast different from the original shell?

Original Remains. The photograph at the right below shows the remains of a relative of elephants found in frozen arctic soil. Note that skin, hair, and flesh remain on parts of the body. Why did they not decay? How might the animal have become buried before decaying or being eaten?

The bodies of insects are sometimes found in pieces of the mineral called amber. Geologists believe that amber was a sticky gum given off by the bark of cone-bearing trees. Explain how insects might have been trapped in this gum. How did the gum prevent decay?

Why are original remains like these especially useful to geologists?

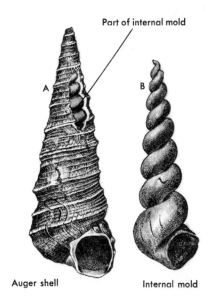

Part of internal mold

A   B

Auger shell   Internal mold

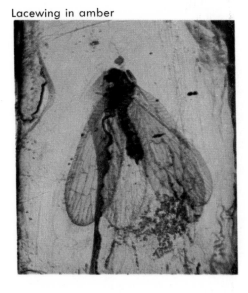

Lacewing in amber

Part of baby woolly mammoth found in Alaska

Carbonization. Plants and animals buried quickly in clay or sand may not decay if oxygen and bacteria are absent. However, other types of chemical changes usually take place. These changes produce gas and oil, leaving part of the carbon. Coal is formed in this manner.

The carbon that remains in the rocks sometimes shows clearly the nature of the buried plants and animals. Sometimes internal organs can be seen.

Petrifaction. Petrify means turn to stone. The process of petrifaction is not well understood.

To the best of our knowledge, ground water sometimes dissolves the materials of which an organism is made, replacing the molecules with equal amounts of some minerals. Thus the fossil is almost exactly like the original organism, even when viewed through a microscope.

Shells, bones, teeth and wood are the parts most often petrified. The picture below shows a petrified tree trunk. The wood has been replaced by silica. Notice how well the appearance of the wood has been preserved. Even the growth rings can be seen. The shapes of cells that made up the wood can also be seen with the aid of a microscope.

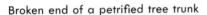
Broken end of a petrified tree trunk

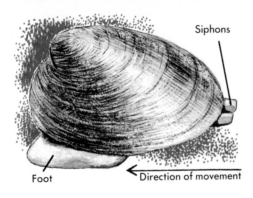

Siphons

Foot ← Direction of movement

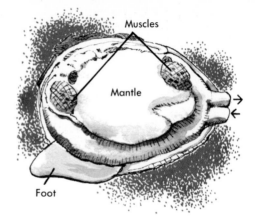

Muscles

Mantle

Foot

INTERPRETING FOSSILS

Many animals of the past have left only their hard, outer coverings as fossils. No one knows exactly what these animals were like. Satisfactory inferences about fossils are possible only if there are living relatives. The closer the relationship, the more reliable the inferences that can be made.

Fossils show that relatives of clams have been numerous for millions of years. Therefore, a knowledge of living clams is very helpful in understanding some of the most common fossils.

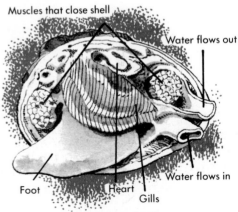

Muscles that close shell

Water flows out

Foot   Heart   Gills   Water flows in

Structure of a clam

A Clam's Exterior. Drop a clam into warm water until its shells open. Push a chip of rock between the shells to keep them open.

Look for the tough ligament that serves as a hinge for the shells. This ligament also acts as a spring which opens the shells.

Look between the shells for a thin membrane that comes to the edge of each shell. This membrane is called the *mantle.*

The oldest part of a shell is the point near the hinge. The shell grows, as glands along the edges of the mantle give off calcium carbonate and other chemicals. Note that a shell does not grow equally in all directions. In which direction is growth most rapid?

Ridges on a shell also show that growth is not regular. Each ridge was formed during a period of slow growth. Large ridges probably represent long periods of little growth, such as during winters. The age of a shell can be estimated from these large ridges.

Interior of a Clam. Slip a thin knife blade between the shells, keeping it close to one shell. Cut the muscles that hold the shells together.

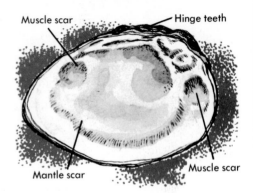

Muscle scar   Hinge teeth

Mantle scar   Muscle scar

Scallop   Mussel

Fossil pelecypods

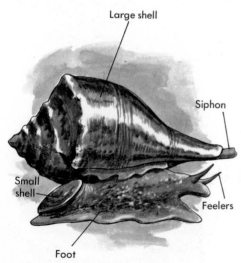

Large shell

Siphon

Small shell

Feelers

Foot

Fold back the mantle from the body of the clam. Find the muscular foot which is thrust out between the shells when the clam crawls.

Notice that the mantle is joined at the rear of the clam to form two tubes called *siphons*. Water is drawn in through the lower siphon, bringing in oxygen and tiny particles of food. Water is forced out through the upper siphon, carrying away wastes.

Remove the body of the clam from the shell. Notice how the muscles which close the shells were attached. Notice also where the mantle was attached to the shell.

Boil the two shells for a few minutes and remove all the tissues. Study the markings on the interior of the shells. Which scars show where muscles were attached? Which show where the mantle was attached? Describe the scar where the siphons were located.

Fit the two shells together. What keeps the shells from being twisted out of line?

Shells of Related Animals. Study shells of animals related to clams, including oysters, scallops, mussels, and fossils like those at the left. Compare each with your clam shells.

Try to imagine the animals in their shells. What was the natural position of the animals? How did they move about, open and close their shells, and obtain food? Try to explain the shell markings.

Mollusks. Clams belong to a group of animals called *mollusks*. Most mollusks have shells.

One division of the mollusks includes clams, oysters, mussels, and scallops. These animals are called *pelecypods* (*pelecy*–hatchet, and *pod*–foot) because the foot of some species reminds biologists of ancient battle axes.

Gastropods. Another division of mollusks includes snails. These animals crawl on a broad foot, as shown here. They are called *gastropods* (*gastro*–stomach, and *pod*–foot). Why do you think this name was given to these animals?

Generally, gastropods have one large, coiled shell. The shell is made up chiefly of calcium carbonate given off by glands in the mantle. One side of the mantle grows much faster than the other. How does this affect the shape of the shell?

The gastropod in the diagram has a second shell. Where is this second shell when the animal is crawling? Where is it when the animal is inside?

186

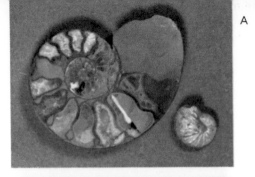

A

Living nautilus

B

C

D

Watch a snail crawl up the side of an aquarium. Note the movement of the foot. Find the mouth, feelers, and eyes. Does your specimen have a second shell? Does it have a siphon? What happens if you disturb the animal?

Drop a snail into boiling water to kill it. Saw open the shell with a fine blade. Does the snail's body go to the end of the shell? Find the muscles that pull the body inside.

Cephalopods. A third group of mollusks includes octopuses and squids. These animals are called *cephalopods* (*cephalo*–head, and *pod*–foot) because they move about by means of tentacles attached to their heads.

Living cephalopods have well developed eyes and tentacles. How does an octopus use its tentacles? How many tentacles does it have?

Only one living species of cephalopod has a shell. This is the nautilus shown above.

A nautilus shell is divided into many chambers, but the animal's body occupies only the outer chamber. Each of the other chambers represents a location of the animal's body when the nautilus was smaller. The nautilus produces a new partition each time it moves its body farther toward the end of the shell. How many times did the nautilus move in making the shell shown in B?

Fossil cephalopods can be identified by the partitions and chambers of their shells, as in A. How else are the shells different from those of other mollusks?

The restorations at C and D show some cephalopods that lived several million years ago. Note that one type has a straight shell and one has a coiled shell. Which parts of these models are probably correct? Which parts are based upon inferences? How might the models have appeared if they were based upon different inferences?

187

Brachiopods. Brachiopods (pronounced 'brack-i-o-pods) were among the most common animals in many ancient seas. Some rocks are made up almost completely of brachiopod shells. Brachiopods are not nearly as numerous today as they were in the past.

Brachiopods are not closely related to clams even though both types of animals have two shells. The diagrams at the left show two important differences between the shells: (1) The two shells of a pelecypod are much alike, but those of a brachiopod are unlike. (2) A brachiopod shell grows equally on both sides of a center line, but a pelecypod shell does not. Try to identify brachiopod and pelecypod shells in a collection of fossils.

The upper right drawing on page 189 shows the natural position of a living brachiopod. The brachiopod is attached to a rock by a fleshy stalk which passes through a hole near the hinge line of the shells. Compare the position of this brachiopod with the natural position of a clam.

A brachiopod has two armlike organs within the shells, as shown at the top of the next page. Biologists once thought that these organs served for moving about, and so named the animals brachiopods (brachio–arm, and pod–foot). However, brachiopods remain fixed in one place most of their lives.

The armlike organs of a brachiopod are used to obtain food and oxygen. They are covered with microscopic fingers called cilia, which wave a current of water into the shells. Oxygen and food particles are carried in with the water.

A brachiopod opens and closes its shells by means of muscles. One set of muscles opens the

Pelecypod shells

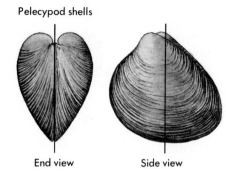

End view   Side view

Brachiopod shells

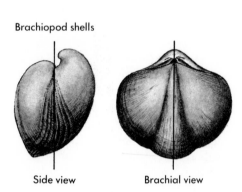

Side view   Brachial view

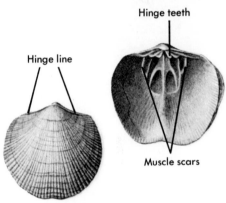

shells; another set closes them. Scars on empty shells show where muscles were attached.

Fossil brachiopods are found in many shapes and sizes. Examine as many kinds as you can. Hold a specimen in its natural position. Find the hinge line. Look for the hole through which the stalk passed. If there is no hole, this species may have cemented itself fast to a rock.

Notice the ridges on the outside of the shells. These ridges show changes in speed of growth. Where is the oldest part of the shell?

Notice also the large ridges and grooves on some specimens. These shells are stronger than smooth shells. It is believed that animals with such shells lived in rough water along wave-swept coasts.

Look for muscle scars on the insides of brachiopod shells or their internal molds. Well-preserved specimens may also show supports for the arms.

Corals. Corals were as important in many ancient seas as they are today. They often formed huge reefs which may have become islands, like certain Pacific islands today.

Modern corals are found in colonies made up of a great many tiny animals. In the past, there were both colonial corals and solitary corals.

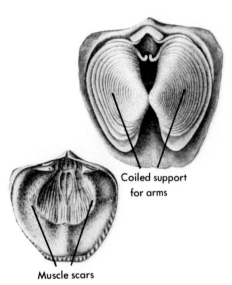

Solitary coral fossil

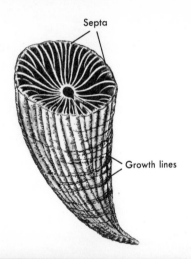

Septa

Growth lines

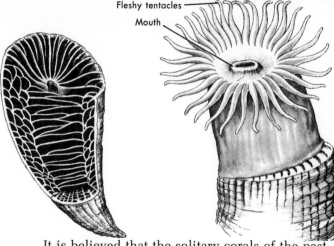

Fleshy tentacles

Mouth

Sea anemone

It is believed that the solitary corals of the past were much like the sea anemone at right above. A sea anemone is a saclike animal, open at the top. Around the opening, which serves as a mouth, are many small tentacles. These tentacles can grasp small animals, sting them, and place them in the mouth. But unlike corals, sea anemones have no hard parts.

Fossils show that a solitary coral built a cone of calcium carbonate around its body. The cone was strengthened inside with plates running up and down. As the animal grew, it added to the cone at the top. Explain the shape of the coral at the top left.

As the cone grew taller, the animal had to pull itself upward from time to time. Each time it did so, it produced a supporting floor under its body. The diagram shows the location of some of these floors. If possible, break open some solitary corals and find the parts described.

Three specimens of colonial corals are shown below. The type at the left lives today; the types shown in the two photographs below are extinct.

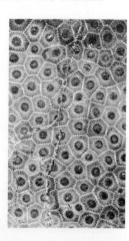

Restoration of ancient marine life

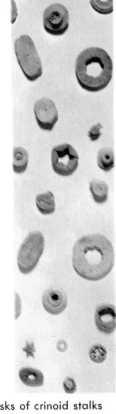

Disks of crinoid stalks
A fossil crinoid

Large reefs have been made by such ancient colonial corals.

What you see are the skeletons of corals made of calcium carbonate. The holes on the surface are the cones within which the many tiny coral animals once lived. The growth of the cones can be traced if a specimen is broken open.

The restoration above shows a number of extinct animals that might have lived in an ancient sea. Upon what evidence would such an inference be based?

Which models are of solitary corals? Which are of colonial corals? What other animals can you identify? Which parts are probably accurate, and which are doubtful?

Crinoids. These animals look so much like plants that they have been named *crinoids* (*crino*–lily, and *oid*–like). They were once much more numerous than today.

A crinoid's body is at the top of a stalk. Its mouth is surrounded by several arms bearing cilia. The cilia wave a current of water along grooves in the arms toward the mouth. Particles of food are swept along with the water.

The body of a crinoid is surrounded with plates of calcium carbonate. The stalk is made up of disks of the same material. These disks and plates usually fall apart when the soft parts decay. Hence, complete fossil crinoids are rarely found.

191

Fossil trilobites

Restoration of early crinoids

A fossil trilobite

Trilobites. These animals were given their names because their bodies have three lengthwise divisions called lobes (*tri*–three, and *lobos*–lobe). Trilobites were once common but are now extinct.

A *trilobite* had an external skeleton divided into many segments like those of crayfish, centipedes, and insects. It is believed that trilobites shed their skeletons several times while growing, just as insects do. These skeletons usually fell apart and are rarely found complete.

Some fossil trilobites show limbs which may have been used for walking. A few show limbs that may have been used for swimming. Most trilobites had eyes; the eyes of some were compound, like the eyes of many insects. Some trilobites had feelers on the head and a few had feelers on the rear as well.

The restoration below shows two kinds of trilobites. What animal is capturing a trilobite? What other animals can you identify?

Restoration of ancient marine life

Restoration of mammoths

RECONSTRUCTING ANCIENT ANIMALS

Bones of mammoths have been found in many parts of the Northern Hemisphere. Some people claimed they were the skeletons of giants. Others thought they were the bones of huge moles, because they were always found buried in soil. Early attempts to assemble skeletons produced some amazing creatures, especially when the bones of two animals were mixed!

Finally, a few complete skeletons were discovered, showing them to be similar to present-day elephants. Specimens found frozen in arctic soil gave information about the hair and skin. The reconstruction shown above is probably accurate, but as late as 1900 the tusks were shown pointing outward. New discoveries may make other changes necessary.

The Skeleton. The reconstruction of an animal often begins with a pile of bones. A geologist must first sort out these bones, keeping only the ones belonging to the animal being studied. If some bones seem to be missing, the skeletons of related animals are studied to infer the sizes and shapes of the missing bones.

The geologist then mounts each bone in what is believed to be its natural position. Why is a knowledge of living relatives of the animal a great help at this time? Why does a geologist record the arrangement of the bones before removing them from the rocks?

193

The Fleshy Parts. A geologist must make many inferences while reconstructing the muscles and other soft parts. If one can decide from the skeleton whether the animal flew, swam, hopped, ran, or waddled, one can decide which muscles had to be large. Scars on the bones show where the muscles were attached. Knowledge of living animals is again of great help.

A reconstruction is not made on the original skeleton. Instead, a framework of rods and wire is used to support clay or other modeling materials which can be carved to the desired shape.

Scales, Hair, and Feathers. A geologist can only guess at the covering to put on many reconstructions. A few prints of feathers and scaly skin have been discovered. A little hair has been found in caves and frozen soil. Generally, a geologist must depend upon knowledge of living animals when deciding whether to give the model a bare skin or to cover it with hair, scales, or feathers. The color chosen must be a reasonable one.

The restorations in museums are based partly on facts and partly on inferences. When looking at these restorations, keep in mind which portions are probably accurate and which portions may not be accurate.

Obtaining Skeletons. Skin a small animal, such as a mouse or chicken, and remove the internal organs. Cook the animal until the meat is soft. (A pressure cooker speeds the process.) Next separate the bones from the meat. Dig the brain from the skull with a stiff wire. Boil the bones again, this time in strong soap, to remove the grease.

Geologists recovering fossils in Bone Cabin Quarry, Wyoming

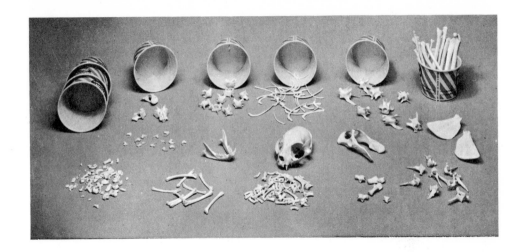

If possible, obtain skeletons from several kinds of animals. Mark the bones of each animal with colored paint to help in sorting them in case they become mixed accidentally.

Assembling a Skeleton. It is helpful to assemble a skeleton yourself to better understand how a geologist works. First sort out the loose bones, putting similar ones together. Then place the bones in what seems to be their proper positions in the skeleton, fitting them as closely together as possible. Do not try to mount the skeleton. Lay the bones flat as shown below. If you cannot find places for some bones, place them to one side.

After you have assembled a skeleton, notice how many loose bones you can identify.

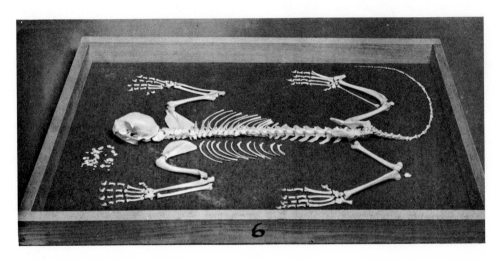

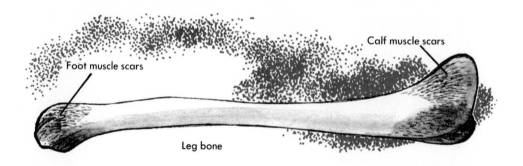

Calf muscle scars

Foot muscle scars

Leg bone

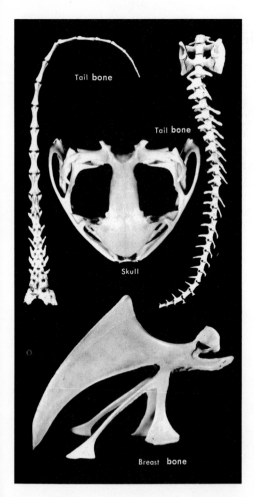

Tail bone

Tail bone

Skull

Breast bone

Inferences from Bones. Scars on bones show where muscles were attached. Usually, the larger the muscle, the larger the scar. Note the scars on the leg bone above. This animal had big calf muscles. What kind of animal needs big calf muscles? What type does not need big calf muscles?

Study the picture of the breast bone and compare this with your own breast bone. Big chest muscles were attached to both sides of the center keel. What type of animal needs big chest muscles?

The larger bones of many animals are hollow. What is the advantage of a hollow bone? What types of animals might have hollow bones?

Most of the bones of the neck, back, and tail have spines and other projections. Muscles are attached to these projections. Two sets of tail bones are shown. Which belonged to a strong, muscular tail? Which were part of a weak tail? How might an animal use a strong tail? What animals have need for such tails?

Where are the eye sockets in the skull shown here? In what position would you expect this animal to hold its body most of the time? What are some animals commonly seen in this position? Compare your inferences with those given later in this chapter.

Make a study of bones of different animals. Note the size and arrangement of foot and leg bones, then make inferences about the habits of these animals.

Early Amphibians. The skeleton on the next page belonged to an amphibian which has been extinct for many millions of years. Note the shape of the skull. Study the structure of the legs. What inferences can you make from these bones?

Compare the ribs, backbone, and tail bones with those of a cat. What inferences can you make about this amphibian from its bones?

Courtesy of Field Museum of Natural History

Skeleton of extinct amphibian

The drawing below shows a restoration based on one geologist's inferences. What parts of the animal are probably correct? What parts may be inaccurate?

What inferences can you make about the habits of this ancient amphibian? Base your thinking on both the skeleton and on your knowledge of the habits of living amphibians.

Eryops

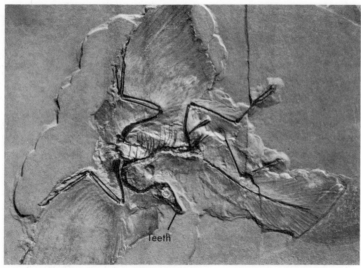

Fossil remains of earliest known bird

The Earliest Known Bird. This fossil of a bird is the earliest one known to geologists. How do they know that this prehistoric animal was a bird?

In what ways are the front legs of this fossil like the wings of a living bird? In what ways are they different? Compare the tail of the fossil with that of a living bird. What inferences can you make about the flying ability of this ancient bird?

Note the teeth in the skull of this fossil. What living birds have teeth?

The restoration shows an animal that looks more like a reptile with feathers than a living bird. Geologists believe that this ancient bird was closely related to certain reptiles.

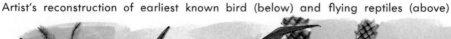

Artist's reconstruction of earliest known bird (below) and flying reptiles (above)

Inferences from Teeth. Our own jaws contain four types of teeth: (1) cutting teeth (incisors); (2) tearing teeth (canines); (3) crushing teeth (biscuspids); and (4) grinding teeth (molars). These types are about equally developed. We could probably eat a mixed vegetable and meat diet if we were wild animals.

Which type of teeth in the skull at *A* are most highly developed? Which teeth are so small as to be almost useless? Describe the molars. For what purpose might the molars be used? What inferences can you make about the food of this animal?

The skull at *B* lacks one type of tooth. Which is lacking? Which type is very highly developed? How might this type of tooth be used? Describe the molars. What is the advantage of this type of molar? What food does this animal probably eat?

Study the skull at *C* and the diagram of one of its molars. How might this animal use its front teeth? For what type of food are its molars well fitted? What inferences can you make about the food of this animal?

Inferences about food habits are not always dependable when based on teeth alone. Name some animals that have teeth fitted for eating meat but which eat as much plant material as meat. Can you name some animals with teeth fitted for eating plant material but which eat meat?

What type of food is probably eaten by the fish whose skull is shown on the next page? How may the fish use these teeth?

Compare the bills of the bird skulls on the next page. What are some possible differences in the food of these birds?

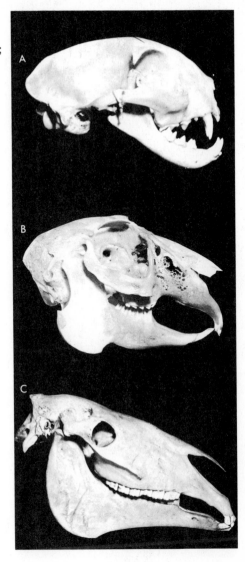

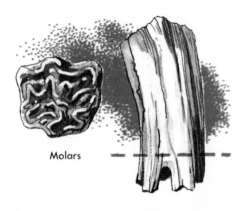

Molars

CHECKING INFERENCES

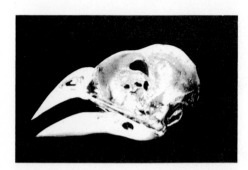

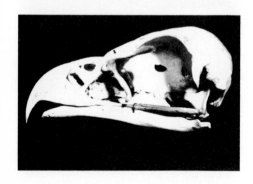

Tracks as Clues. (1) The larger prints at the top of each group were made by the hind feet. The animal was hopping away from the observer. (2) The tracks in the snow were made by a walking bird. (3) The fossil tracks were made by a reptile that walked on two legs. (See page 178.)

Inferences from Incomplete Remains. (1) A giant animal lived in the region where these bones were found. (2) This fossil shows that ferns lived at some time in the past. (3) Empty snail and clam shells were probably washed into a place where they were later covered by sediments. (4 and 4a) The segmented bodies suggest that these animals were relatives of insects and crayfish. (5) These are fossil reptile eggs. Bones of unhatched dinosaurs have been found in some of them. (6) These look like worm burrows made in mud that became rock long ago. If they were made by worms, no trace of their soft bodies has been found. (7) Count the legs. It was probably a relative of the spiders. (8) The lower right specimen was a coral, small colonial animals that lived together. (9) It is a starfish. They have been numerous for millions of years. (10) This specimen was once labeled as the skull of a man drowned in the Great Flood. Today, it is believed to be the skull of a giant salamander. (See page 180.)

Inferences from Bones. (1) Animals that walk and run need much larger calf muscles than do animals that swim, fly, or crawl. This is the leg bone of a dog. (2) Animals that fly need strong chest muscles. This is the breastbone of a chicken, which does not fly well. Hollow bones are light. Flying animals would need hollow bones. (3) The right set of tail bones belonged to an alligator which used its tail for swimming. The set of bones at left

200

Artist's reconstruction of reptile skeleton shown below

Skeletons of extinct reptiles

belonged to a cat. (4) The eyes of this animal were at the top of its head. Probably it kept its head low and its body in a crouching or crawling position. It is the skull of a frog. (See page 196.)

Habits of Dinosaurs. The dinosaur restoration suggests that different dinosaurs had different feeding habits. The skulls provide the evidence for this inference. Study the skulls. What type of food do you think each animal ate? Why do you think so? Explain why your inference might be wrong.

Note the position of the dinosaurs in the restoration. Why is it unlikely that the skeletons were in this position when found? What type of evidence may have suggested that the animals stood differently?

The large dinosaur in the foreground, called Tyrannosaurus rex, is shaped somewhat like a kangaroo. Why do you suppose it is shown walking rather than hopping? Note the way this dinosaur holds its tail. Explain this position in terms of a dinosaur's need to balance itself.

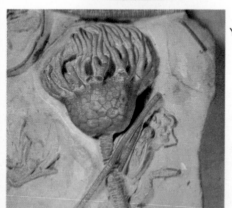

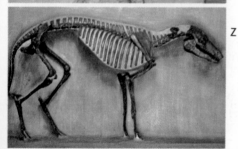

REVIEW QUESTIONS

1. To what group of animals does each of the lettered models in this reconstruction belong?
2. Which of the groups represented became extinct without leaving any present-day descendents?
3. How does the animal at *A* get its food?
4. How does the animal at *B* get its food?
5. What are some living relatives of the animal at *D*?
6. How can a brachiopod shell be told from a pelecypod shell?
7. What are the objects shown at *X*?
8. Describe four ways by which plant and animal remains can be preserved as fossils.
9. What inferences can a geologist make from the teeth of a fossil?
10. How does a geologist decide upon the type of covering to use for a model of an animal for which only the skeleton is known?
11. What type of organism is shown fossilized at *Y*?

THOUGHT QUESTIONS

1. Why is a fossil found in a low layer of sedimentary rock considered to be older than a fossil in a higher layer?
2. Why do geologists show plants in restorations as being green?
3. What inferences can you make about this fox-sized skeleton at *Z*?

5

Investigating on Your Own

"I have been wondering, Mr. Weaver. Why does the ocean always look so clean even though rivers and waves are always adding mud to it? Even the Gulf of Mexico looks clean."

"I guess I haven't thought about it, Sally, but you are right. Ocean water does usually look clear."

"Maybe the salt in ocean water makes mud settle out fast, Mr. Weaver."

"That is possible, George. Does anyone have another idea?"

"Aren't there other things dissolved in ocean water? Maybe one of these minerals makes the mud drop out."

"I think the oceans are just so big that the muddy water doesn't make much difference."

Proposing Hypotheses. Three pupils suggested reasons why ocean water is usually clear. A suggested explanation, not yet tested, is called a *hypothesis.* What three hypotheses were proposed by the class? What other hypotheses can you propose?

Testing Hypotheses. A hypothesis is only a guess. It must be tested before it can be accepted or discarded. Many reasonable hypotheses have been proven wrong, and some that seemed unreasonable were later found to be correct.

Scientists test hypotheses with experiments whenever possible. Conclusions drawn from experiments are based upon facts that can be rechecked.

Above are shown the materials that can be used to check the suggestion that dissolved salt speeds up the rate at which mud settles from water. Plan an experiment using these materials.

Try the experiment. What are your observations? What conclusions do you draw from the experiment?

The Variables of an Experiment. An experiment is used to discover how a change in one condition may affect another condition. For example, the student here is trying to find out if the rate of evaporation is affected by dissolved salt.

Any condition that changes during an experiment is called a *variable.* In the experiment at left, the amount of dissolved salt is one variable. What is another? The student can change the amount of dissolved salt as desired. Therefore, this variable is called the *independent variable.*

The student wishes to find out if the evaporation rate depends upon the dissolved salt.

Problem	Independent Variable	Dependent Variable
Effect of pebble shape on movement by water	Shape of pebbles; flat, round, angular	Distance moved down a trough by flowing water
Effect of particle size on soil drainage	Size of soil particles	?
Effect of water on flow of clay	?	Angle of slope down which clay flows
Effect of wind speed on drying soil	?	?

Hence, the evaporation rate is called the *dependent variable*.

Every properly planned experiment has one independent variable (which the experimenter changes), and one dependent variable (which may depend upon the independent variable). The table on page 204 lists several problems that may be solved by experiments. Identify the independent and the dependent variables in each problem.

Planning an Experiment. Two pupils learned that pendulums are used in studying the earth's shape. They planned experiments with a pendulum.

One pupil suggested that the rate of a pendulum's swing depends upon its weight. The other suggested that the rate depends upon a pendulum's length. What other hypotheses can you propose?

The pupils decided to test the first hypothesis. Their first step was to select a method for changing the weight of the pendulum, the independent variable. They decided to add metal washers, one at a time, as shown in the diagram.

Next, they had to find a way to measure changes in the rate of the pendulum's swing, the dependent variable. They decided to count the number of complete swings the pendulum made in 30 seconds.

Finally, the pupils had to think about other possible variables. Suppose that the pendulum was started from a different height each time.

The pupils listed other variables: the pendulum's length, the type of string, the way the string would be fastened, and the starting height as possible variables which should be kept from changing. What other variables can you think of? How might you keep each of these variables from changing? Try the experiment.

Because the pupils tried to control other variables which might affect the results, the experiment is called a *controlled experiment*. Plan and carry out a controlled experiment to find out whether or not a pendulum's rate depends upon the length of the pendulum.

Testing by Prediction. Many hypotheses cannot be tested by experimentation because conditions cannot be controlled. A geologist may believe that ice flows faster in the center of a glacier than along the edges. But no one can develop a controlled experiment with a glacier because you cannot

Mass	Rate: Swings per minute
2 Washers	
3 Washers	
4 Washers	
5 Washers	
6 Washers	

control such variables as thickness, composition, and pressure. You would have to find some other method to test such a hypothesis.

Suppose you start with the hypothesis that some parts of a glacier flow faster than others. You might predict that a straight row of stakes across the glacier should become curved within a few months. You could then visit a glacier, drive in a row of stakes, and wait to see if your prediction is correct.

If the row of stakes remains in a straight line, your prediction is wrong, and you will probably discard the hypothesis. If the row becomes curved as predicted, you will have obtained evidence supporting the hypothesis.

The Problem of Proof. A geologist should not be satisfied with a single set of observations. The flow of ice might be different in other glaciers, and the pattern of flow might change from time to time.

The hypothesis should be tested in as many other situations as possible. Each correct prediction increases confidence in the hypothesis. A wrong prediction shows that the hypothesis is faulty and should be changed or discarded.

How many correct predictions would you consider necessary to prove any hypothesis? Why is it more difficult to prove a hypothesis correct than to prove it incorrect? Why do scientists prefer to test hypotheses with experiments whenever possible?

Testing Earth Science Hypotheses. At the present time, controlled experiments with the weather, the stars, and the earth's interior are not possible. Direct observations of plants and animals which lived millions of years ago are also impossible. For these reasons, geologists, astronomers, and meterologists depend upon predictions followed by observations to test their hypotheses. Biologists, physicists, and chemists however, can more easily test their hypotheses.

Astronomers depend almost entirely upon predictions for testing hypotheses. For example, after an astronomer has measured the position of a comet on three or more nights, he or she can calculate its probable path, and then predict the comet's position at any future time. If the comet is seen in the predicted locations, the calculations are correct, or nearly so. If the comet is not seen where predicted, the path must be recalculated and tested again.

The table below lists several hypotheses that cannot be tested satisfactorily by experimentation. Make a prediction for each of the hypotheses illustrated by the first line of the table. Describe ways to check each prediction.

Discuss each hypothesis in terms of the number of correct predictions you would need before the hypothesis gained your confidence. What would be the effect of an incorrect prediction in each case?

Propose some other hypotheses that must be tested by the method of prediction. Test one of the hypotheses.

Hypothesis	Prediction
Atmospheric pressure decreases as altitude increases.	A barometer will read less on top of a tall building than at the bottom.
An east wind brings rain.	?
Granite makes longer lasting cemetery monuments than marble.	?
The swiftest part of a stream is in the middle.	?
Leaves cool more rapidly than the air around them on clear nights.	?
The moon is larger when it first rises than later in the night.	?

Theories. About 300 years ago, Isaac Newton suggested that any two objects pull or exert a force on each other. This force is now called *gravity*. Thus, Newton's hypothesis became known as a theory. Today, it is often taught as fact, proven beyond all doubt. Nevertheless, a few scientists point to certain predictions based on Newton's theory which have been found to be incorrect. These scientists have developed other hypotheses to explain the behavior of objects in space, out of the earth's atmosphere.

A Discarded Theory. At one time, many people believed the earth to be a mass of molten rock covered with a thin crust which formed as the surface cooled and hardened. Volcanoes were explained as openings in the crust, and mountains as the wrinkles produced as the earth cooled and contracted.

Certain geologists pointed out that lava should be the same everywhere if all volcanoes connect to the same mass of molten rock. Other geologists pointed out that the moon should cause large tides in molten rock similar to tides in the oceans. Geologists find that lava varies widely throughout the earth. They also do not find evidence of tidal effects in lava flows. Therefore, these predictions were found to be inaccurate, and the theory was discarded.

A Popular Theory. During the earth's four billion year history, the continents have not eroded completely. To explain this observation, a hypothesis was developed describing continents as blocks of light rock floating in a heavier plastic rock. This hypothesis states that: (1) as part of a continent is unloaded through erosion, it rises

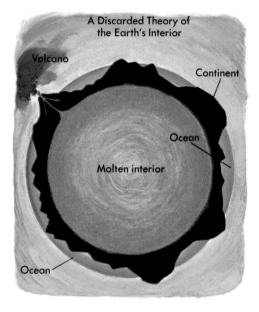

A Discarded Theory of the Earth's Interior

Volcano

Continent

Ocean

Molten interior

Ocean

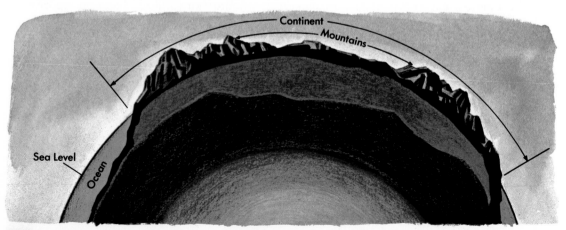

Continent

Mountains

Sea Level

Ocean

slowly; (2) as part of a continent becomes loaded with sediments, it sinks slightly.

This hypothesis explains the difference in thickness of the earth's crust beneath mountains and oceans. It has become a widely accepted theory which is used as a basis for other theories.

Proof of this theory could be obtained by observing the rocks underneath the continents. For this reason, geologists are developing methods of drilling deep holes through the edges of continents. If the rocks underneath are heavy and plastic, as predicted, the theory will probably be considered as proven. But if the rocks are different, a number of theories will have to be discarded.

Rival Theories. Fossils in Europe and North America show many similarities. So do fossils found in other now distant places. One hypothesis explains this relationship by land bridges between continents along which animals could travel, as North and South America are connected today. Another hypothesis suggests that the continents were once joined as shown here, but broke apart and drifted away from each other. This is called the *continental drift theory*.

Where might there have been some land bridges in the past? Where were land bridges unlikely?

Lay a thin paper on a globe and trace the continents. Cut out the outlines and push them together as suggested by the continental drift theory.

If scientists have two equally good theories, they usually accept the simpler one. Therefore, the land bridge theory was popular for many years, although there were no signs that some of the continents could have been joined by bridges. Recently, scientists have made observations which support the continental drift theory.

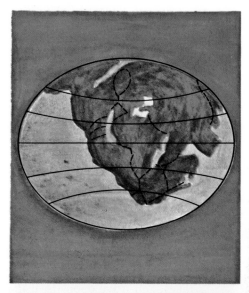

THEORIES IN EARTH SCIENCE. The majority of the broad explanations in earth science are not yet proven. Most of today's theories are accepted only because they provide accurate predictions. As scientists develop ways for making direct observations of places out of range of our present instruments, many accepted hypotheses and theories will be changed or discarded.

EXPERIMENTAL RESEARCH

1. Make a study of a pond or small lake. Look for evidence that it is filling up with plant remains and sediments. Look for evidence that the outlet is wearing down. Make a map showing the present shoreline, probable past shorelines, and possible future shorelines.

2. Study the speed in wide and narrow parts of the same stream. Measure a distance of 5 meters or more along the bank of each part selected. Time the motion of a stick moving through this distance. Make a graph that shows the relation of width to rate of speed.

3. Visit a shore at high and low tides. If the shore is rocky, make diagrams of sea cliffs, wave-cut terraces, and wave deposits. If the shore is sandy, make a study of the shape of the sandbars.

4. Find out how the shape of a stream bed affects the rate of flow. Get two troughs of different widths. Place each trough at the same slant, and siphon water into each at a constant rate. Measure the water speed by timing the movement of a piece of cork.

5. Make a stream table like the one below. Cover the stream table with fine sand, and place some stones in the sand. Start the water flowing and observe the effect of the water on the sand.

6. By positioning the stones and changing the flow of the water, describe how each of the following features may be formed: a sandbar, a bend in the river, ripple marks, a waterfall, rapids, undercutting, a delta.

7. Almost fill a two-liter cardboard milk carton with water and freeze it. Remove the ice from the carton and place the ice on a thick layer of sand in

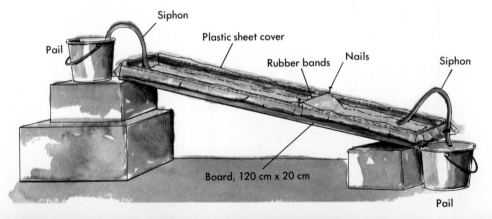

Siphon

Plastic sheet cover

Pail

Rubber bands   Nails

Siphon

Board, 120 cm x 20 cm

Pail

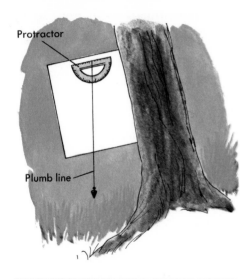

Protractor

Plumb line

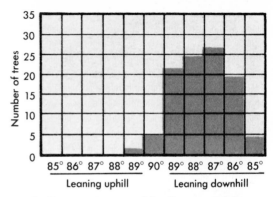

the upper end of your stream table. Pour a thick layer of sand on top of the ice. Observe the features made in the stream table as the ice melts.

8. Build a dam using a rubber band stretched across the stream table under the plastic sheet. How can you control the flow of water downstream? When the inflowing water increases, how can you prevent flooding downstream?

9. Investigate the creep of soil on steep hillsides. Measure the slant of all the trees in an area using a protractor and plumb line. Use the measurements to make a graph called a histogram, as shown above. Make a study of trees growing on level ground for comparison.

10. Investigate the erosion of a stream bank. Drive a series of stakes at equal distances from the edge of the bank. Hammer each stake in level with the ground. Return and measure the distances regularly. Make a map that shows the erosion occurring between visits.

11. Make a study of a sand dune. Watch the sand being blown around the dune. Measure the angles of the slopes of the dunes. Compare the angles with the angle of rest for the same kind of sand.

12. Collect soils from hillsides, road cuts, and the like. Measure the angle of rest for each soil type when dry, wet, loose, and packed together. Compare the angles obtained with the angles of the slopes where the soils were collected.

13. Put paint on several stones on a talus slope. Measure the distance of each painted stone from the base of the cliff. Return regularly and measure the distances again to determine the rate of movement, if any.

211

Sedimentary rocks

Fault

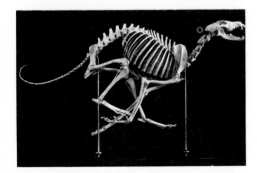

14. Visit a cliff of sedimentary rocks. Study the types of sediments and the way they were deposited. Collect fossils and other indications of past events. Develop inferences about conditions in the region at the time the sediments were being deposited.

15. Investigate the geological history of your region. Use bulletins published by national or state geological commissions, by local colleges or universities, and by local museums. Prepare a report summarizing the history.

16. Cast two blocks of plaster 2 cm thick. Let one block represent bedrock. Break up the other block and let the pieces represent loose rocks being dragged over the bedrock by a glacier. Study the scratches in the bedrock and the changes in the shapes of the small pieces.

17. Show how a glacial kettle may have formed in a moraine by burying chunks of ice in a tray of sand and watching the changes as the ice melts. Then pour water into the tray to show how the kettles may become kettle ponds.

18. Reconstruct the skeleton of some animal, such as a rabbit. Thread the backbone on a heavy wire and bend it to the proper curves. Drill tiny holes at the joints and fasten the bones together with fine wire. Drill the toes lengthwise and thread them on stiff wire. Attach the ribs with quick-setting cement.

19. Make a collection of fossils. Label each, giving the name of the group to which it belongs, where it was found, and any inferences you can make from the rock in which it was found.

20. Prepare an exhibit showing different ways plants and animals have been fossilized.

21. Make a chart to show how different kinds of volcanoes are formed.

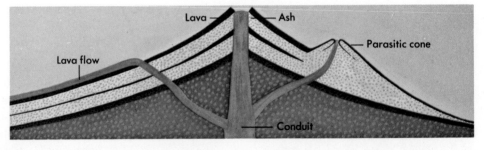

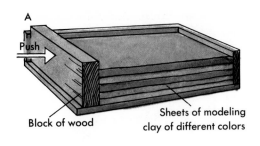

A

Push

Block of wood

Sheets of modeling
clay of different colors

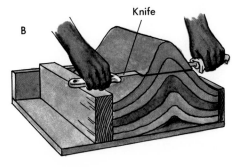

B

Knife

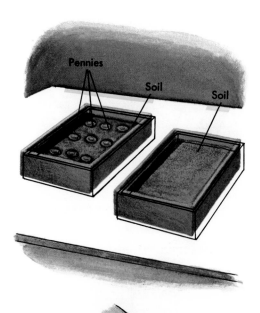

Pennies

Soil

Soil

22. Develop and outline the procedure you would follow to compute the volume of water flowing in a river near your home. Describe your procedure for computing the load of the river. Trace the source and mouth of your river.

23. Select three pieces of modeling clay, each a different color. Form each piece into a flat sheet about one centimeter thick. Find a rectangular container and a block of wood that is as wide and as high as the container, but loose enough to slide within the container. Grease the insides of the container with Vaseline. Put the flat sheets of clay in the bottom, one on top of the other. Slowly push the block against the layers until they start to fold. Once you have observed the folding, use a knife to slice off a layer from the top as if erosion had taken place. Carve out a river canyon. Write a report of the geological processes represented by your model landscape.

24. Investigate to find out how much water expands upon freezing. Pour water into a plastic cylinder, mark the level of water, and place it in a freezer. How much did the ice expand? By what percent of the volume of the water did the ice expand?

25. Investigate the effects of raindrops on bare soil. Put a layer of soil in two flat boxes. Place several pennies on the surface of one, and put the boxes outside during a rain storm. After the rain has ended, examine the surface of each box.

26. Study the effects of waves on a sandy beach. Pile sand in one end of a deep tray. As illustrated below, partly fill the tray with water, placing a sponge in front of the "beach." Move the sponge back and forth to produce gentle waves. Describe effects of the waves on the beach.

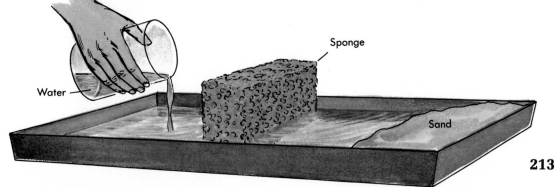

Sponge

Water

Sand

Nebular theory   Tidal theory   Dust cloud theory

ADDITIONAL INVESTIGATIONS

1. Read about some of the hypotheses that have been proposed to explain the origin of the earth. Three of these hypotheses are illustrated above. Try to find out why some of the hypotheses have been discarded. Which hypotheses are popular at present? Prepare an illustrated report of one of these hypotheses.

2. Scientists are trying to develop low-cost methods of producing fresh water from salt water. Test the hypothesis that dissolved salt can be removed from ocean water by freezing.

3. Look up the theory of isostasy in a geology book. Make a model to illustrate the theory. Use wood blocks in water to show that continents may sink slightly when loaded with sediments, and rise slightly when worn away by erosion.

4. Plan a method for testing the hypothesis that rain packs bare soil. Carry out your experiment, and report the findings in class.

5. Invent at least three hypotheses to explain the presence of fossil trees in Antarctica, and present the hypotheses to your class. Suggest how the hypotheses might be tested.

6. Investigate the theory of continental drift, and prepare charts to illustrate it.

7. Use a geology book to find out about the different kinds of geologic faults. Make plaster models of the common types.

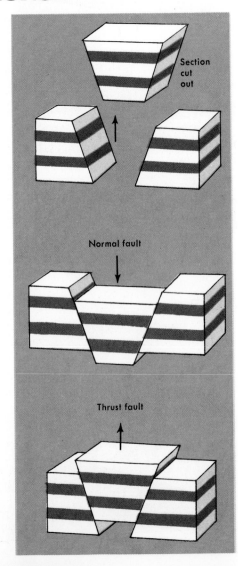

Section cut out

Normal fault

Thrust fault

8. Prepare a report on the history of geology. Describe how geologists like James Hutton changed our ideas about the earth.

9. Study the pictures in geology books of different kinds of folds. Make plaster models of these folds.

10. Collect and display glacial erratics, labeling each with the kind of rock. Display samples of the local bedrock also.

11. Procure bulletins describing the geological history of some national park. Summarize this history for your classmates, illustrating your talk with pictures or slides.

12. From a book on historical geology, copy a series of maps of ancient seas and shorelines to show changes in North America during past eras.

13. Make a geological time chart on a sheet of wrapping paper at least 4 meters long. Divide the chart into eras, and divide the eras into periods, all according to scale. Make drawings of the typical plants and animals living in each period.

14. Make a restoration of one of the fossils you have collected, using clay, paper, and other materials as needed.

15. Read about the different kinds of fossil horses that have been found. Write a report and illustrate it with pictures, diagrams, and maps.

16. Study the fossils and rocks of your region and make inferences about conditions that existed during the periods of time represented.

17. Study a U.S. Geological Survey topographic map of your area, and prepare a relief map from it. Trace every fifth contour line onto a piece of cardboard or wallboard. Cut out the pieces and glue them together as shown here. Cover with a coat of plaster of Paris to smooth the hills and valleys.

Millions of		Epochs	ERAS
years ago	years duration	PERIODS	
.025	.025	Recent	CENOZOIC
2	2	Pleistocene	
	11	Pliocene	
	12	Miocene	
	11	Oligocene	
	22	Eocene	
63	5	Paleocene	
135	72	CRETACEOUS	MESOZOIC
181	46	JURASSIC	
230	49	TRIASSIC	
280	50	PERMIAN	PALEOZOIC
310	30	PENNSYLVANIAN	
345	35	MISSISSIPPIAN	
405	60	DEVONIAN	
425	20	SILURIAN	
500	75	ORDOVICIAN	
600	100	CAMBRIAN	

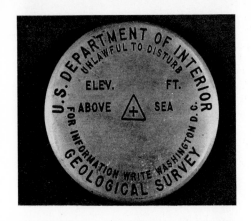

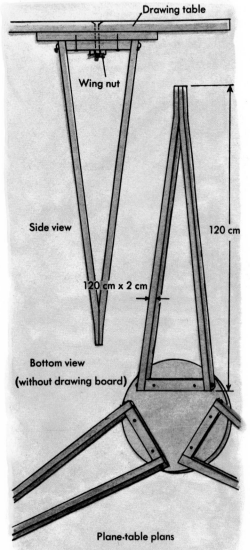

Drawing table

Wing nut

Side view

120 cm

120 cm x 2 cm

Bottom view
(without drawing board)

Plane-table plans

Paint the relief map green. Then add the streams to the map with blue paint, and paint in some of the major roads in red.

18. Locate a high point on your topographic map marked with a small triangle and the elevation. This indicates the location of a bench mark. Visit the exact spot so marked on your map and find the bench mark. Bring to class the information on the bench mark, but **do not** disturb it in any way.

19. Make a plane table from pieces of wood, 2 cm thick. Cut a 15 cm circle of wood for the top of the plane table. Nail or screw 3 pieces of 2 cm by 8 cm wood to the underside of the top for leg supports. Cut 6 pieces 2 cm wide and 120 cm long for 3 legs. Nail together at one end two of these long pieces of wood. Screw the other ends of these pieces into the sides of the leg supports as shown at the left. When the table is finished, drill and countersink a hole through a drawing board and through the table top. Mount the drawing board to the table top with a flat-headed bolt and a wing nut. Then use the plane table to aid in surveying an area.

20. Arrange a visit with a surveyor. Find out what training the job requires. Ask the surveyor what kinds of projects they are asked to do, and how they go about accomplishing them.

21. Visit a team of surveyors at work. Prepare a report about their methods and equipment.

22. Find maps in the library that were made hundreds of years ago, before mapmakers were able to locate points accurately. Compare some of these maps with modern maps and photographs taken from space.

23. Geologists believe that Mt. Mazama was originally about the size of Mt. Shasta. Use the Mt. Shasta map to estimate how many cubic kilometers of rock would disappear if a crater developed at the same elevation as the one in Mt. Mazama. (The formula for the volume of a cone is $V = \frac{1}{3} \pi r^2 h$.)

24. Construct a profile from A to B on the Donaldsonville, Louisiana map. (Note the two contours closest to the river are both 20-foot contours.) Suggest a possible reason why natural levees are never steeply sloped. See page 134.

Geologic Time

Every year a thin layer of sediment is deposited in the Mississippi delta. Well-drillers tell us these sediments are thousands of meters thick. These deposits must have been forming for a very long time. The sediments making up the walls of the Grand Canyon in the photograph above were probably deposited at the same rate. As you look from the top to the bottom of the canyon, you see the results of geological events which took place over an enormous span of time.

The time of a geological event may be described in two ways: (1) by placing the event before and after other events, and (2) by calculating how much time has passed since the event took place. In the first case, a geological event is described in terms of relative time. In the second case, the event is described in terms of absolute time. Suppose you describe your age by saying you are older than your sister but younger than your brother. In what terms are you describing your age? In what terms would you be describing your age if you say, "I am 15 years old"?

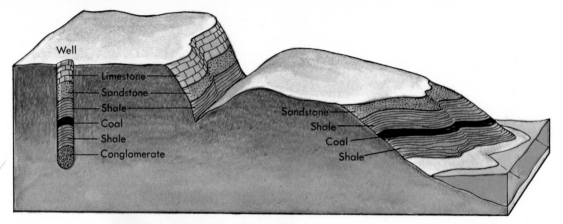

Well
— Limestone
— Sandstone
— Shale
— Coal
— Shale
— Conglomerate

Sandstone
Shale
Coal
Shale

RELATIVE TIME

In a previous chapter you were introduced to the principle of superposition. The principle states that in a series of sedimentary rocks the topmost layer is the most recent, while the bottom-most layer is the oldest. Using this principle you can quickly determine the relative age of any one layer to any other in a cliff.

A difficulty lies in the fact that unknown hundreds of thousands of meters of sediments have been deposited throughout the earth's history. There is no one cliff where every sedimentary layer which has ever been deposited can be found. The layers in one place may be older, younger, or the same age as those in another place. The problem can be solved if we develop a relative time scale to which any sedimentary layer can be compared.

Correlating Sedimentary Rocks. It is impossible to find one place where the sedimentary rock represents all of geologic history. We must therefore piece together the sequence of layers from one place to another. We call this process *correlation*.

Study the illustration above of an exposed sea cliff. The topmost layer is sandstone. Beneath the sandstone is shale, then a layer of coal, and more shale extending down to the level of the beach.

Inland, in an eroded canyon, there are exposed sedimentary layers. How are the layers in the face of the sea cliff related to the layers in the canyon? Why is there no limestone on top of the sea cliff?

Further inland from the canyon, a well is being dug. The digging has gone through a layer of limestone, sandstone, shale, and has entered a coal seam. As digging continues the drill reaches more shale, then it passes through a layer of conglomerate. How are these various rock layers in the well related to the canyon and sea cliff layers? What kind of a rock layer would you expect to find under the beach?

William Smith

Georges Cuvier

Correlating Fossils. Fossils have long been recognized as the remains and impressions of animals living in the distant past. However, the thinking and philosophies that existed in times past discouraged investigation of them. Fossils were not seriously studied until the 1800's. At that time fossils played an important role in two very significant developments in the study of earth history.

William Smith, an English mine surveyor and canal builder, studied in great detail stream banks, sea cliffs, quarries, and canal beds. He must have been a sharp observer to notice that different rock layers contained distinct groups of fossils. Using this information he was able to identify the same rock layers in different regions. His work carried him throughout England and Wales. After 24 years of keeping careful notes, he was able to construct the first useful geologic map of a large area.

During the time of Smith's studies the machine age was making increasing demands for coal and iron. New methods for finding these resources were needed. By identifying rock layers, it is possible to find mineral deposits of iron and coal more readily. Geological mapping by correlating fossils was extended into France, Germany, Italy, and Russia. Such maps made it possible to compare the sequences of rock layers in widely separated areas. The sequences of fossils from deep layers to shallower layers were in the same order in places even 2000 kilometers apart.

Knowledge of the fossils enabled geologists to predict accurately that coal would be found beneath an area south of London. This spot was many kilometers from the nearest outcrop of coal.

While Smith was at work in England, a Frenchman, Georges Cuvier, observed that not only do the fossils differ from one bed to another, but the changes in fossils appear to follow definite patterns. The fossils in the deepest and oldest beds are least like the plants and animals living today. Fossils found in the younger beds above look more and more like present day forms of life.

Tens of thousands of geologists have probed the earth's surface on every continent. Their observations continue to support the principles proposed by Smith and Cuvier. When all their observations are pieced together, a second important conclusion can be made. The fossils found in rock layers differ in obvious ways. A fossil from any layer can be placed properly in a relative time sequence, even if the layer from which it came is unknown.

STANDARD GEOLOGIC COLUMN

ERA	PERIOD	AGE (in Millions of Years)	LIFE FORMS				
CENOZOIC	Cenozoic	Pleistocene 2 / Pliocene 11 / Miocene 25 / Oligocene 36 / Eocene 58 / Paleocene 63	Flowering Plants		Mammals		Man / Grasses / Horses first appear
MESOZOIC	Cretaceous / Jurassic / Triassic	230	Conifers, Cycads			Reptiles	Extinction of dinosaurs / Birds appear / Dinosaurs appear
PALEOZOIC	Permian / Pennsylvanian / Mississippian / Devonian / Silurian / Ordovician / Cambrian	500	Ferns / Marine Plants	Marine Invertebrates	Fish	Amphibians	Coal-forming swamps / First vertebrates appear (fish) / First abundant fossils (Marine Invertebrates)
PRECAMBRIAN							Scanty fossil record / Primitive marine plants / One-celled organisms

Standard Geologic Column. From studies of rock layers the world over, geologists have drawn up a chart which lists geological time sequences in historical order. Included in the chart are groups of fossils identified with each span of time.

Rock layers formed during a specific span of time are called by the name of that time span. For example, rocks formed during the Cambrian period are called Cambrian rocks. Thus, all Cambrian rocks the world over were formed during the same period of geologic history.

ABSOLUTE TIME

For everday purposes we choose two recurring events to keep track of time. One is the regular rising and setting of the sun by which we determine the length of a day. The other is the seasons. What recurring units mark each day? Each year? In geology, the problem is to determine how many of these yearly events have occurred in the dim past when no one was around to keep track of them.

In a previous chapter of this unit you considered some natural timekeepers, including tree rings and varves. These timekeepers tell us the length of time during which the tree was growing, or the span of time when sediments were being deposited. They do not actually tell us how many years ago these events occurred.

For centuries, humans have been curious about the exact age of past geological events. Only recently have we found a reliable way to measure very long spans of time. The discovery of radioactivity in 1896 by Henri Becquerel and the additional discoveries of Marie and Pierre Curie have made these measurements possible. Vast spans of time such as the age of the earth can be determined now.

Radioactive dating is a complicated process. Using a simpler situation will help to explain it. If you know the size of a log and the rate at which it burns, you can tell how long it has been burning by measuring the amount of ashes left.

Measuring Time Using Radioactivity. Most of the elements found in nature are mixtures of several isotopes. Isotopes are atoms which are chemically identical, but contain different numbers of nuclear particles. Some isotopes are radioactive. They change into other elements as the process of radioactivity continues. This process is called radioactive decay.

The chart in the margin lists some radioactive isotopes. Why are some radioactive isotopes more useful for measuring the age of recent events?

Isotope	Material	Half-Life
Carbon-14	Wood, peat, charcoal, bone, shells	5.7×10^3
Potassium-40	Mica, some whole rocks	1.3×10^9
Rubidium-87	Mica, feldspar	4.7×10^{10}
Uranium-235	Many whole rocks	7.13×10^8

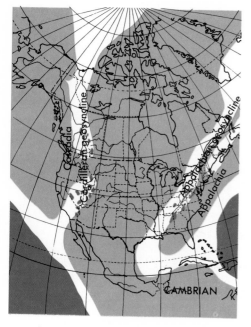

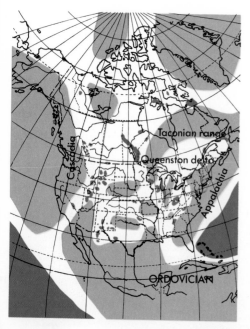

Using Carbon-14. One of the isotopes of carbon, carbon-14, is radioactive. Carbon-14 comprises only about one million millionth as much as other isotopes of carbon in living organisms. One-half of an amount of carbon-14 will decay in about 5600 years. This span of time is called the *half-life* of carbon-14. It is possible to detect this decay with sensitive instruments.

Carbon-14 is constantly being produced in the upper atmosphere. Atoms of nitrogen struck by cosmic rays may be transformed into carbon-14. As far as anyone can tell, carbon-14 has been produced at a constant rate above the earth for at least 50,000 years before the first atomic bomb was exploded.

The newly produced carbon-14 soon is evenly mixed with other isotopes of carbon in the air. Carbon-14 is taken up by all living plants and animals. Living things contain a constant proportion of carbon-14 as long as they are alive. When a living thing dies it no longer takes in carbon. The carbon-14 in its system continues to decay.

Geologists can calculate the age of a dead tree. They determine how much carbon is carbon-14. Then they calculate back to the time when the radioactivity from carbon-14 was the same as we find in living things today.

Events in the Earth's History. Geological events seem to repeat themselves regularly. Mountain building has been accompanied by continental uplift, and the general retreat of the oceans. At the same time, volcanoes have poured vast amounts of molten rock over the surface. Changes in climate have caused ice caps to form at the poles. The snow falling on newly formed mountains have become flowing glaciers. Running water has constantly worn down the land and has moved large amounts of sediments into lower lands. Such a sequence of events took place in a time period of approximately 250 million years. During this time many new types of animals and plants have replaced extinct species.

Since the formation of the earth's crust, there have been at least three such episodes. One occurred about 500 million years ago at the beginning of the Cambrian period. Another, about 250 million years ago, took place between the Carboniferous and Permian periods. The third apparently occurred during the Pleistocene age. The earth is just emerging from the Pleistocene age.

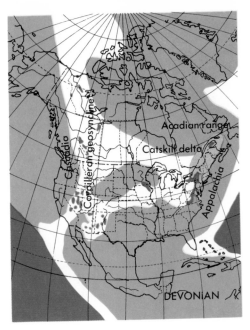

There may have been other such episodes in the long time span before the Cambrian period. The earth's history was erased by the geological events of the Precambrian era. Geologists are very uncertain about the history of the earth's surface in Precambrian times.

Geologic Maps. Geologists have drawn maps showing the locations of seas and coastlines during the earth's history. Maps for the Paleozoic era are shown here.

Study these maps, and describe in a general way what happened to North America during the Paleozoic era. What happened to your region?

Salt deposits of the middle Paleozoic era are found in New York, Pennsylvania, Ohio, and Michigan. What have geologists shown in that region on the middle Paleozoic map? Why?

Notice the map on which geologists have drawn the Catskill Delta. Where did the material for this delta come from? Are there any high mountains in this region today? What may have happened to the mountains?

At what period in the earth's history were the Appalachian Mountains formed? These mountains are not very high or rugged today. What might have happened to them?

Study the maps for the Mesozoic era shown here. Which parts of North America were beneath sea level during this era? What happened to your region?

	Deep seas
	Shallow seas
	Land
	Deltas

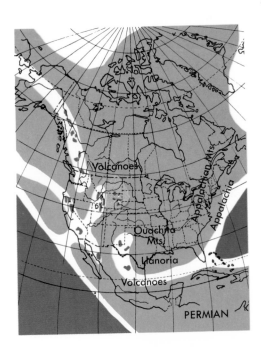

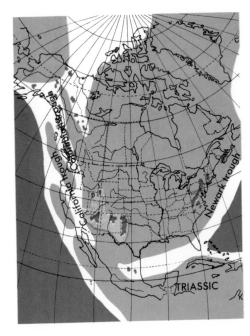

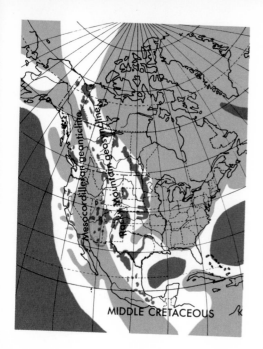

MIDDLE CRETACEOUS

During the Cenozoic era the shorelines were much as they are now. Notice that there are no high mountains shown for western North America during the Paleozoic and Mesozoic eras. It is believed that the Rocky Mountains were folded and pushed upward near the end of the Mesozoic era and again in the Cenozoic era. The Cascades seem to have been formed late in the Cenozoic era. Explain why these western mountains are more rugged than the eastern mountains.

Why would you expect most fossils of eastern North America to be of Paleozoic age? Where might you expect to find fossils of land animals and plants of late Paleozoic age? Where might you expect to find fossils of land plants and animals of Mesozoic age?

REVIEW QUESTIONS

1. What is the difference between absolute time and relative time?
2. How is it possible to correlate one layer of rock with another miles away?
3. Of what practical value are geologic maps?
4. How are the fossils found in rock layers near the surface different from those found in deep layers?
5. What is the standard geologic column?
6. How is absolute time measured?
7. What are some geologic events which repeat themselves regularly?

THOUGHT QUESTIONS

1. How can fossils be used to correlate rock layers?
2. How do geologists know that some forms of animals lived only during certain periods of earth's history?
3. How is the standard geologic column useful?
4. What assumptions are made for dating by carbon-14?

ASTRONOMY

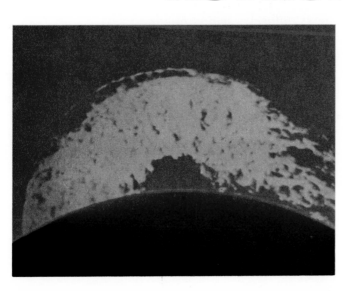

Watching the Sky

People in ancient times watched the sky closely, seeing much that puzzled them. For centuries, these early sky watchers thought that some magic or their gods caused the movements of objects in the sky. Gradually, they realized that motions of most objects in the sky are regular, and can be explained by mechanical models.

1

The earliest models were simple because so little was known about the sky. New observations were continually being made, however, and models had to be changed to fit them. The process of change has not ended. New information about the sky is constantly being obtained, and we may expect today's ideas and today's models to be replaced in the future.

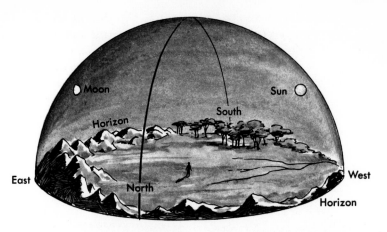

A MODEL OF SKY MOTIONS

We view the sky as though it is a large bowl over our heads. The sun, moon, and stars appear to be on the inside of this bowl.

For thousands of years sky watchers believed that the bowl slowly turned each day. Even today, astronomers use this idea to describe the motion sky objects appear to take each day.

In the following sections you will observe how objects in the sky appear to move. From your observations, you will build a model of the motion of sky objects.

Carry out one of the sun observations described below. Then go on to the section called Interpreting Sun Observations.

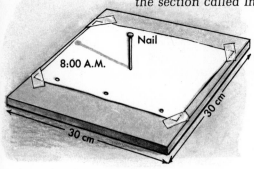

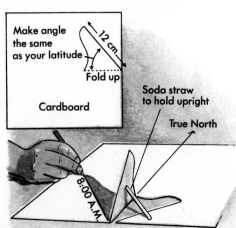

Shadow Stick Observations. Pound a long nail through a piece of plywood or stiff cardboard. Tape a sheet of paper to the board, as shown at the left. Set this device in a place where the sun shines all day.

Trace the position of the nail's shadow on the paper. Label the time of this shadow next to the shadow's position. Repeat your observations at intervals throughout the day.

Sundial Observations. Construct a sundial from a piece of cardboard, as shown at the left. Set the sundial in a sunny place, with the upright pointing north. Mark the position of the upright's shadow. Label the time of this shadow. Repeat your observations at intervals throughout the day.

Observing Direction and Elevation of the Sun. Make a circle about 30 cm across on a sheet of cardboard. Use a protractor and a ruler to mark off ten-degree spaces around the circle.

Lay the cardboard flat in sunlight with the zero point toward the north. Stand a 2 cm × 10 cm × 15 cm board on edge along a radius of the circle. Turn

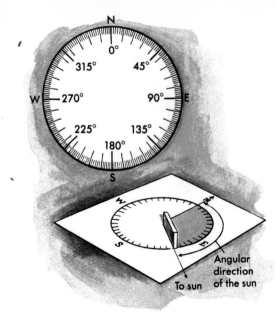

Angular direction of the sun

To sun

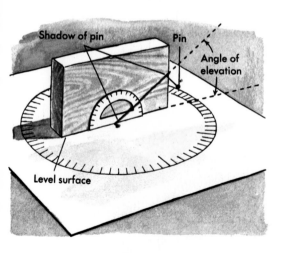

Shadow of pin

Pin

Angle of elevation

Level surface

the board until its shadow is as thin as possible. Record the angle between the board and due north. This is the angular direction of the sun.

Push in a pin halfway up one edge of the board used above. Stand the board on edge on a level surface in sunlight. Turn the board until the shadow of the pin falls across the board. Use a protractor as shown below to measure the angle between the shadow and the bottom edge of the board. This is the angle of elevation of the sun above the horizon.

Use this instrument to measure the direction and elevation of the sun at times throughout the day. Record your observations in a table like the one below at the left.

Interpreting Sun Observations. Point to the place in the sky which corresponds to the sun's position as recorded in your first observations. Repeat this for each observation. Now show the path the sun makes each day as it crosses the sky.

Examine the sun data you collected. At what times of day was the sun low in the sky, making long shadows? At what time of day was the sun highest, making the shortest shadows? In what direction does this shadow point?

The time of day when the sun casts the shortest shadow is called solar noon. How does solar noon compare with noon on your clock? Before time zones were invented (about 100 years ago), each town kept its own time. Each town determined noon by when the sun was highest in the sky. As a result, the clocks in towns a few miles east or west of each other showed different times.

Plot graphs of the angular direction and angle of elevation of the sun against the time of day. How do these observations help you interpret the sun's daily motion? What is the direction of the sun when it is at its highest elevation? What is the westward speed of the sun in degrees per hour?

SUN'S MOTION DURING OCT. 8

Time	Angular Direction	Angle of Elevation
8:00		
9:00		
10:00		
11:00		
12:00		

Masking tape circle for sun

Model horizon

Direction of rotation

Colored water

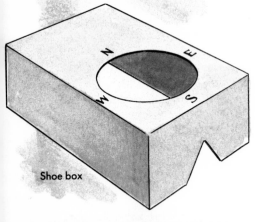

Shoe box

N E S W

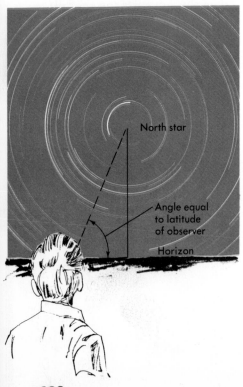

North star

Angle equal to latitude of observer

Horizon

Making a Sky Model. Partially fill a round-bottomed flask with colored water. Stopper and turn the flask upside down, as shown at the left. Add or take away water so that exactly half the round part of the flask is empty.

The upper part of the flask represents the sky. In this model, you would be a tiny dot at the center of the surface of colored water. The horizon is represented by the line where the surface of the water meets the glass.

Cut out a 1-cm circle from masking tape to represent the sun. Tape this sun onto the model sky. Use this model to demonstrate (1) how the sun sets, (2) how the sun rises, and (3) the sun directly overhead.

On the bottom of a shoe box draw a circle with a diameter 1 cm larger than the diameter of the round-bottomed flask. Cut out the circle. Label N, E, S, and W on the bottom of the box, as shown at the left.

The Sun's Daily Motion. Place the flask inside the shoe box. The water level inside the flask should be level with the bottom of the shoe box. Now turn the flask so that the model sun rises in the direction you observed. Turn the sky so that the sun is in the noontime position you observed. Make the model sun set in the direction which you observed. Use the model to show how the sun crosses the sky as you observed. Where is the sun during the night?

Testing the Sky Model. Our knowledge of astronomy began with observations of the sky. After making many observations, astronomers try to describe what is happening by making models and developing ideas to describe how the motions occur. Such ideas in science are called theories. Generally it is easy to construct a model to fit data you have observed. If the model and theory are good ones, they should be able to predict new observations. If the new observations do not agree with the predictions, then the theory and its model must be changed to fit the new observations. In this way, new theories are developed with better models.

The photograph at the left provides you with a new observation about sky motions. This observation was made by pointing a camera on a clear night toward the northern night sky. The lens of the camera was left open for 8 hours while the

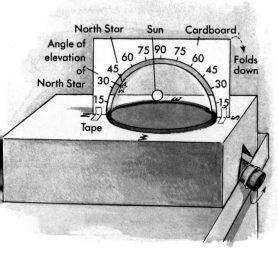

Angle of elevation of North Star

North Star Sun Cardboard

Folds down

60 75 90 75 60
45 45
30 30
15 15

Tape

Sky equator

NOONTIME ANGLE OF ELEVATION OF THE SUN

Date	Angle	Degrees from Sky Equator
9/21		
10/22		
11/21		
12/20		
1/22		

sky "turned." As a result, each star made a trail across the film. The star near the center of the trails is the North Star.

Turn your model of the sky through a complete day. Does your model fit the observation shown at the bottom of page 230?

Changing the Model to Fit New Data. Set up your sky model as follows to fit the new observation about star motions. Add a cardboard angle measuring device as shown here. Then tape a model North Star on the middle of the bottom of the flask. Tape the model sun onto the sky as shown here. Tilt the model sky so that the North Star is at an angle of elevation which is the same as your latitude.

Now slowly turn the flask as shown by the blue arrow. Where does the sun rise? Set? What is the sun's angle of elevation at noon? How do these observations of the model sun compare with your observations of the real sun? If necessary, move the model sun so that its motion is more like the real sun. Would stars near the North Star move in the way shown in the photograph? Why is this model better than your earlier one?

The Sky Equator. Put a rubber band around the flask as shown at the left. This imaginary line around the sky is called the sky equator. Where is your model sun in relation to the sky equator?

Seasonal Changes in the Sun's Path. Select a window through which the sun shines at noon. Tape a strip of paper across a beam of sunlight, as shown below. Mark the position of the edge of the light beam at noon each day for a week. Explain the changes in the position of the light beam.

Record the position of the light beam at noon for a school year. Record the angle of elevation of the sun at noon and its relative position to the sky equator. Keep a monthly record of your observations. When is the sun's path closer to the horizon? When does it cross the sky equator?

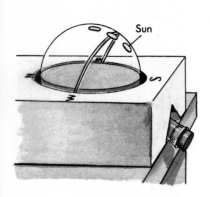

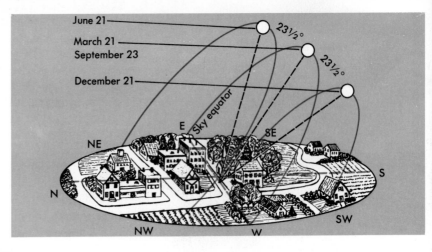

The diagram above shows three different paths of the sun on the days listed. Note that the position of sunrise and sunset shifts, as does the position of the sun at noon.

Where would the sun's path be on April 21st? On May 21st? On January 21st? During which months of the year is the sun north of the sky equator? When is the sun south of the sky equator?

A Model for Seasonal Changes. Tape the model sun onto the sky equator. This represents the sun on March 21st or September 21st. Tilt the flask to represent the sky in your latitude.

Rotate the flask. Where does the model sun rise on September 21st? Where does it set?

Move the model sun to a position which is 23½ degrees above the sky equator. What date is now represented? Slowly turn the flask through a day. Where does the sun rise and set on this date?

Move the model sun to 23½ degrees below the sky equator. What date is now represented? Slowly turn the flask through a day on this date. Describe the daily motion of the sun.

Tape three model suns in a row as shown at the upper left. Rotate the flask. Which sun rises first? Which sets first? Compare the length of time each sun can be seen in the sky. How does the length of daylight and the angle of the sun affect our temperature throughout the year?

Effect of Latitude. The two photographs at the left were both taken at noon. Change the tilt of your model sky until conditions represent those shown in the photographs. Discuss the date and latitude at which each of these pictures was taken.

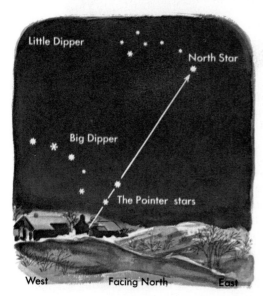

Little Dipper

North Star

Big Dipper

The Pointer stars

West   Facing North   East

The Nighttime Sky. A study of the nighttime sky with all its stars provides the best understanding of positions and motions of objects in the sky. Pick a clear, moonless night. Plan and make at least three observations one hour apart. Continue the hourly observations longer, if possible.

Finding the North Star. The North Star is one of the most useful stars for finding directions and locations on the earth. Unfortunately, it is not a very bright star. Also, the constellation in which it is found is not always easy to pick out.

Look first for the Big Dipper, a constellation that contains seven bright stars. The Big Dipper is on your right when you face the place where the sun went down. Follow an imaginary line through the two stars at the end of the bowl of the Big Dipper. This line points to the North Star.

Finding True North. The point in the sky which is directly over the earth's North Pole is called the *north sky pole*. The North Star is almost exactly on the north sky pole. Thus, this star can be used to locate true north. True north is the point on the horizon exactly below the north sky pole (or North Star).

Sight over a magnetic compass. Is there a difference between magnetic north and true north? If so, how great is the difference?

The earth's magnetic north pole does not lie exactly upon its geographic north pole. Therefore, compasses in some parts of the world point east of true north; compasses in other places point west of true north. In only one narrow strip of the United States do compasses point toward true north.

Little Dipper

Big Dipper

Measuring the Night Sky. The pointer stars of the Big Dipper are separated by about five degrees. Use this value to estimate the number of degrees between the Big Dipper and the North Star. Estimate the number of degrees between the North Star and the constellation Cassiopeia. Estimate the angle of elevation of the North Star.

Vertical and Horizontal. A weight hanging from a thread points toward the center of the earth. The line that is formed by the thread is called a *vertical line*. What angle does a vertical line make with a horizontal line?

Use a thread with a weight at the end to find two stars that lie along a vertical line on the sky globe. Find the true north on the horizon, the point that is directly beneath the North Star.

Brightness of Sky Objects. Sky objects are classified into *magnitude* numbers according to how bright they appear. The brightest stars have a magnitude of 1, less bright stars have a magnitude of 2, and so on. The fainter the sky object, the larger the magnitude number. A person can see fifth magnitude stars on clear nights with the unaided eye.

The stars on these charts are shown by symbols that tell their magnitude. A key for these symbols is given below. What is the brightness of the North Star? What is the brightness of the stars making up the Big Dipper? Does Cassiopeia or Draco have the brighter stars? What sky chart shows the greatest number of bright stars?

Star Magnitude	
✹	1
✱	2
∗	3
●	4
·	5

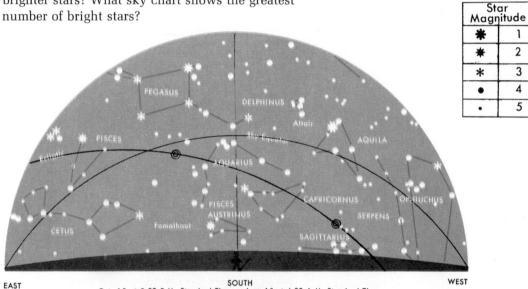

EAST   SOUTH   WEST
Oct. 15 at 8:30 P.M. Standard Time or June 15 at 4:30 A.M. Standard Time

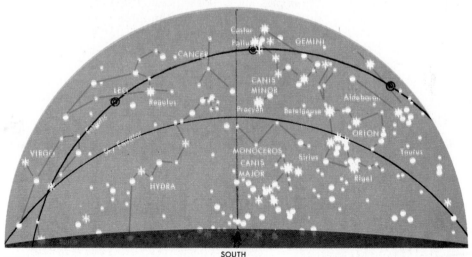

EAST   SOUTH   WEST
Oct. 16 at 5:30 A.M. Standard Time or March 1 at 8:30 P.M. Standard Time

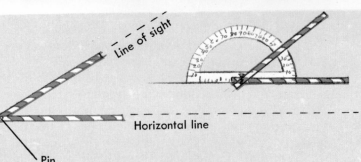

Line of sight

Horizontal line

Pin

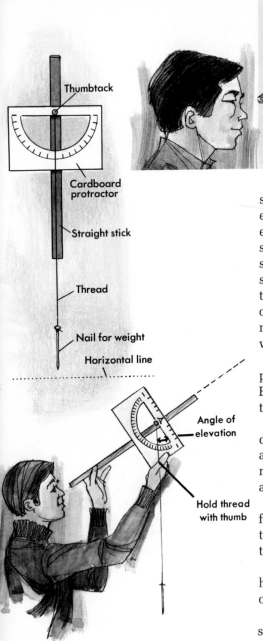

Thumbtack

Cardboard protractor

Straight stick

Thread

Nail for weight

Horizontal line

Angle of elevation

Hold thread with thumb

Angle of Elevation

Star	8 PM	9 PM	10 PM
North star			
Pointer star			
Star in east			
Star in south			
Star in west			
Star overhead			

Measuring Angles of Elevation. Some type of sighting device is needed to measure angles of elevation of stars. Push a straight pin through the ends of two soda straws, as shown above. Hold one straw level. Move the other straw until you can sight along it at an object in the sky. Then lay the straws on a protractor to measure the angle between them. Find the angle of elevation of the moon and other objects in the sky. (**CAUTION:** Do not try to measure the angle of elevation of the sun in this way; direct sunlight can damage the eyes.)

Set the sighting device at 45°, and locate a point on the sky globe that is 45° above the horizon. Estimate other points on the sky globe, and check them with the sighting device.

An Angle-Measuring Device. The sighting device at the left measures angles of elevation more accurately than the soda straws above. This device measures elevation using a vertical line rather than a horizontal line.

Center a protractor carefully on a stick, and fasten it in place with tape or thumbtacks. A thumbtack should pass through the center mark of the protractor to support the thread and weight.

Use this sighting stick to locate the true horizon. Also find a star that is almost directly overhead.

Changes in Star Elevation. Use the sighting stick to measure the angle of elevation of the North Star. Sight along the stick at the star while the weight hangs free. Then press the thread against the protractor so that it cannot move while you read the angle.

Measure the angle of elevation of several other stars. Choose a star in the east, one in the south, one in the west, and another as nearly overhead as possible. Make a record of the angles.

Repeat the measurements one hour later, two

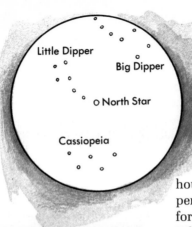

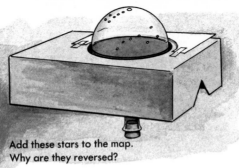

Add these stars to the map.
Why are they reversed?

Minneapolis   8:30 P.M.   December 12

New Orleans   8:30 P.M.   December 12

hours later, and after as many additional one-hour periods as you can. Study the changes in elevation for each star. Which star changed its position least? Which star changed its position most? Do these changes fit your model of sky motions?

Circumpolar Constellations. Cassiopeia and the Big and Little Dippers are called *circumpolar constellations* because they travel around the sky pole without setting. Mark these constellations on the sky model using a glass marking pencil or gummed dots.

Tilt the flask to the proper angle for your region and rotate it slowly. How do the movements of the dots compare with your observations of the real constellations?

Where are the circumpolar constellations during the daytime? If you could see them, in what positions would they be at different times throughout the day?

Effect of Latitude. The view of the sky changes as the latitude changes, that is, as you move toward or away from the equator. Tilt the sky model until the North Star is directly overhead. Where is the sky equator? From what place on earth does the sky appear in this position?

Tilt the model until the sky equator is directly overhead. What is the angle of elevation of the North Star? What location does this position represent?

Rotate the sky model in several different positions. Note the motions of the circumpolar constellations in each position. Explain the two pictures here.

What is the angle of elevation of the North Star at the North Pole? At the equator? At your location? Compare these figures with the latitudes of the same regions as shown by a map. How can a ship's navigator determine the ship's latitude?

The midnight sun looking north

REVIEW QUESTIONS

1. What are some methods of making records of the sun's daily path across the sky?
2. What other evidence is presented in this chapter to show how the sky moves?
3. When is the sun directly overhead in your area? Explain your answer.
4. What does star magnitude mean?
5. How can you locate the North Star at night?
6. Define angle of elevation, sky pole, and sky equator.
7. What is the latitude of your location?
8. Describe the daily motion of the sun, moon, and stars in the sky.
9. Describe the yearly changes in the sun's path across the sky.
10. Why is June 21st a national holiday in countries such as Iceland, Norway, Sweden, and Finland?
11. Explain the difference between magnetic north and true north.

THOUGHT QUESTIONS

1. From what direction is the sun shining on the moon craters in the photograph at the left?
2. Answer the same question for the moon photograph at the beginning of Chapter 3.
3. Why do star trail photographs like the one below not show complete circles? Is there any place on earth where complete circles could be photographed? Where?
4. Where is the Big Dipper in the daytime?
5. A student took the star trail photograph below, claiming that the trail through the picture was a meteor. What other explanation can you give?
6. Why can you see more stars some nights than others?
7. Who gets more sun each year, a person on the equator or a person at the North Pole? (Hint: Use your sky model.)

The Earth in Space

Astronomers made important observations about the earth thousands of years ago. Many correct **conclusions** were made about the shape, motions, and other **features** of the earth. Unfortunately, the scientific method of making conclusions based on observations was not widely used before the 17th century. Many ideas which we now accept about the earth were not accepted before that time.

Today we observe the earth and draw conclusions based on our observations. The practice of making conclusions based on observations is an important part of modern science. As better observational methods are developed, some of our present ideas about the earth are likely to change.

1:03 A.M.   12:03 A.M.   11:03 P.M.

THE EARTH'S SHAPE AND MOTIONS

For hundreds of years students have been taught that the earth is round, that it spins causing day and night, and that it travels once each year around the sun. However, the eyes of these students see an earth that appears to be flat, and their other senses detect no evidence that the earth is moving. Most people can offer no proof or evidence that the earth is not flat and standing still. People who accept new ideas without understanding the evidence for the idea are not reasoning or thinking scientifically.

There are many observations about the earth which cannot be explained in terms of a flat earth which is standing still. After you make some of these observations, you will be able to decide scientifically about the earth's shape and motions.

The Earth's Shadow. The photographs above were taken one night during a partial eclipse of the moon. In an eclipse of the moon, the moon passes through the earth's shadow. Study the photographs. What do you see? How do you interpret your observations?

Many people think that Columbus was the first person to wonder if the earth was round. However, Aristotle, a Greek philosopher and teacher, proposed this idea about 2400 years ago. Aristotle watched the edge of the earth's shadow on the moon. He suggested that only a round earth would cast such a shadow.

Make a study of shadows cast on a ball by objects of different shapes. Which objects cast round shadows? Does a curved shadow on the moon support the idea that the earth is curved? Does this evidence prove that the earth is a sphere?

Aristotle didn't know about cameras and film. How might he have made a record of an eclipse as shown in the photographs on the opposite page?

Ships at Sea. Note the picture showing a ship sailing over the horizon. What part of the ship is seen last as the ship sails further from shore? How does this observation support the idea that the earth is round? Why does this evidence fail to prove that the earth is a sphere?

Elevation of Stars. Changes in the positions of stars and constellations were first noticed about 2 400 years ago by travelers moving northward and southward. As travelers went north, the constellations in the northern sky appeared higher in the sky. The southern constellations appeared lower in the sky.

Study the drawing below. What happens to the angle of elevation of the North Star as an observer moves northward from C to B to A? What shape of the earth makes possible this change in angle of elevation of the North Star?

Time Zones. When you watch ball games on television, you probably notice that the sun sets at different times in different parts of the country. How does this observation support the idea that the earth is curved in an east-west direction?

Photographs from Rockets. Recently, rockets have taken pictures of the earth from great distances. One such photograph is shown on the first page of this chapter. What does this tell about the shape of the earth?

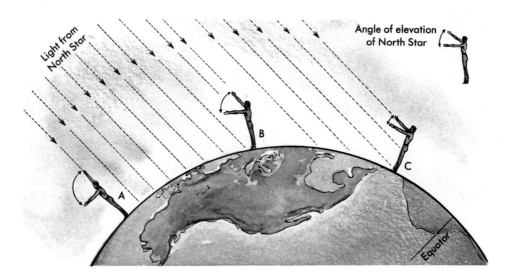

Pendulums. Tie a weight such as several metal washers to one end of a string. Tie the other end of the string to an overhead support, such as the top of a doorway or a lab stand as shown above. Carry out three separate experiments to find out how length, weight, and angle of swing affect the number of swings a pendulum makes in 15 seconds.

Record your observations from each experiment in a data table. Use these data to make three graphs of your observations. What can you conclude about pendulums from these graphs?

Foucault Pendulum. Set up a pendulum on a stool as shown above. Turn the stool slowly as the pendulum swings. Describe the motion of the pendulum with respect to the turning stool.

The photograph shows a *Foucault pendulum* five stories high at the United Nations in New York City. This pendulum does not appear to swing back and forth in the same direction. Instead, the pendulum seems to change its direction very slightly with every swing.

Use the idea of a rotating earth to explain why the direction of a Foucault pendulum seems to change. Does this prove that the earth rotates?

Artillery Fire. Artillery shooters long ago learned that when a long-range gun was fired northward in the Northern Hemisphere, the shell landed east of the target, as shown at the left. A shell fired southward landed west of the target. In both cases, the gun and target behave as though they are moving at different speeds.

Study the diagram on the next page. Note the position of the gun and target shown in red. A shell is fired at this instant. The figures in white show

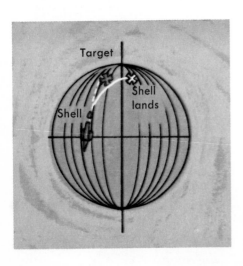

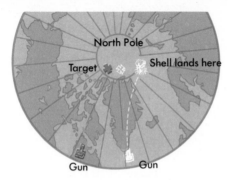

the gun, target, and shell at the moment the shell lands. Compare the distances the gun and shell moved to the right while the shell was in the air. How far did the target move to the right during this time? How does this observation support the idea that the earth turns?

Pendulums and the Earth's Shape. Set up a pendulum about one meter long. Adjust its length so that the weight swings from one side to the other exactly once each second (or over and back in two seconds). Measure the distance from the center of the weight to the top of the pendulum. Record this length in centimeters.

Early clockmakers made use of pendulums to give them accurate time intervals. Many early clockmakers in England developed a reputation for making excellent, accurate clocks. However, when these clocks were shipped to other places in the world, complaints were received. These clocks did not keep accurate time. After many such complaints, the clockmakers decided to consult one of the brightest people in England, Sir Isaac Newton. Newton read the complaints, examined the clocks, and suggested that the world might not be perfectly round.

Years later, scientists measured the length of a pendulum which swings once a second at different places on the earth's surface. The results are recorded in the table here. Note that a pendulum which swings once each second at the poles of the earth will be a half centimeter longer than one which swings at the same rate at the equator.

Pendulums behave as though gravity is greater at the poles of the earth than at the equator. To explain this difference in gravity, scientists have proposed that the earth's surface at the poles is slightly closer to the center of the earth than at the equator. This difference in distance is shown in the diagram at the right.

EFFECT OF LATITUDE ON THE LENGTH OF A PENDULUM WHICH SWINGS ONCE PER SECOND

Latitude	Pendulum Length at Sea Level
0°	99.09 cm
10°	99.11 cm
20°	99.16 cm
30°	99.23 cm
40°	99.31 cm
50°	99.40 cm
60°	99.49 cm
70°	99.56 cm
80°	99.60 cm
90°	99.62 cm

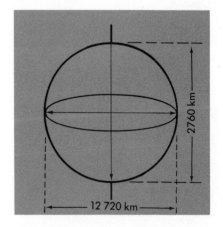

Jupiter
Saturn

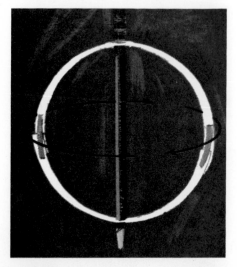

The Shapes of Other Planets. Look at the photographs of the two largest planets, Jupiter and Saturn. Both of these large planets rotate once every ten hours. Thus a day is ten hours long on these planets.

Use a ruler to compare the pole-to-pole diameter with the diameter through the equator of each planet in the photographs. Describe the shape of each.

Studying Shape Changes. Cut out two 20-cm strips of paper 2 cm wide. Tape the strips together, and push a pencil through them as shown below at the left. The holes through the paper should be large enough for the pencil to slide through.

Adjust the strips to form a nearly perfect circle. Measure the polar and equatorial diameters of the circle. Twirl the pencil. Remeasure the two diameters. What change has occurred in the shape of the circle?

How does the shape change as a result of the spinning? How would our earth's shape change if it were spinning?

A Scale Model of the Earth's Shape. Construct a scale model of the earth on a chalkboard or a large sheet of paper. Use this scale: 1 cm = 250 km. First, draw a circle with a diameter of 50.9 cm to correspond to the earth's polar diameter of 12 720 km. Next, calculate how much the equatorial diameter of the scale model should be increased to represent the 12 760 km equatorial diameter. Make this correction on the model.

Satellite studies have found further irregularities in the earth's shape. Sea level at the North Pole is 15 meters higher than it was thought to be, and sea level at the South Pole is 15 meters lower. Some people describe the earth as being pear-shaped on the basis of these findings. What effect would these differences have on the shape of the scale model earth?

Mount Everest is 8700 meters high. The deepest known spot in the ocean is about 11 000 meters deep. How large would these distances appear on the scale model? Many relief maps do not give an accurate picture of the earth's surface. Why is the scale not shown correctly?

Parallax. One type of evidence that our earth moves comes from observations of *parallax*. To learn what parallax is, hold a pencil as shown in the picture on page 245. Sight past the pencil with

244

one eye open. Note where the pencil appears with respect to objects behind it. Now sight with the other eye open. Note the change in position of the pencil with respect to the objects behind it. The pencil appears to change position because it is viewed from two different positions. This effect is called parallax.

Study the drawing below. Imagine you are on the earth in January looking at star A. Does it appear to be near star B or C? Where does A seem to be when viewed in July?

Early astronomers thought the stars were much closer to the earth than they are. These observers knew about parallax. However, since they could not detect any star parallax with their best instruments, they believed that the earth did not move.

Repeat the parallax activity described in the first paragraph, but move the pencil downward as you view it first with one eye and then with the other. Compare your observations with those recorded in the photograph at the left.

This illustration was made from three photographs taken six months apart. The three negatives were placed on top of each other with distant stars I and II lined up. The pictures showed that star A_1 moved to A_2 in six months, and then A_3 in six more months. Astronomers know that many stars are moving. The white line shows the direction of motion of star A. Propose an explanation for the shift of this star from one side of the white line to the other at six-month intervals. How does this photograph support the idea that our earth travels around the sun once a year, as shown below?

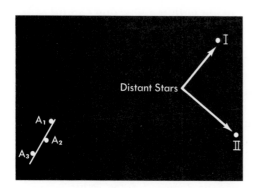

Distant Stars

SUMMARY QUESTIONS

1. What evidence do you have that the earth is round?
2. What evidence do you have that the earth spins?
3. What evidence do you have that the earth travels in an orbit around the sun?

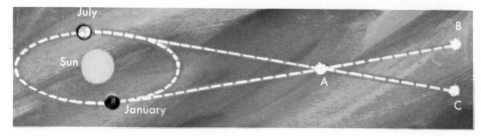

July
Sun
January

245

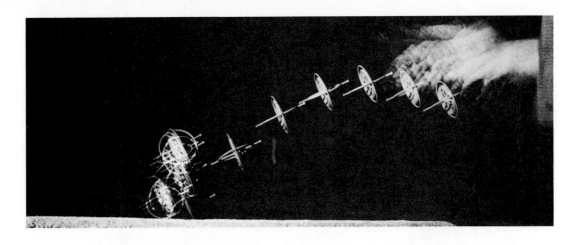

MOTIONS OF GYROSCOPES
AND OF THE EARTH

The earth spins once every 24 hours as it revolves around the sun. It is not possible to experiment with the spinning earth, but we can experiment with a spinning gyroscope. If we assume that all spinning bodies behave the same, the observations made with a gyroscope should be useful in understanding the earth's spin.

Stability of Gyroscopes. Gently toss a gyroscope which is not spinning to a friend a few meters away. Repeat the toss with a spinning gyroscope. What difference do you notice in the stability of the gyroscope in space?

Was the gyroscope spinning in the multiflash photograph above? Explain your answer.

Seasons. The diagram below shows the earth's position in its orbit at four different dates. Why does the earth's axis (an imaginary line through the

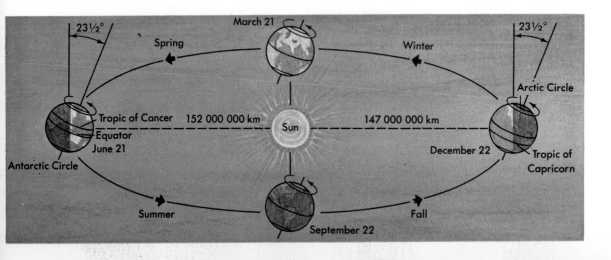

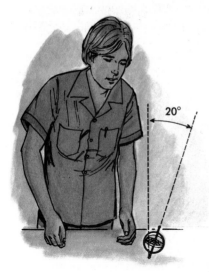

20°

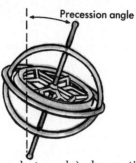

Precession angle

earth from pole to pole) always tilt in the same
direction?

Note the position of the earth on June 21. On
this date, the North Pole is tipped toward the sun.
Where on the earth would the sun appear to be
directly overhead at noon? Which half of the earth
receives more sunlight and heat—the Northern
Hemisphere or the Southern Hemisphere? Compare
the number of daylight hours on this date at the
Arctic Circle, the equator, and the Antarctic Circle.

Precession of Gyroscopes. Spin a gyroscope on
a tabletop. Notice the circular path taken by the top
of the gyroscope shaft. This movement is called
precession.

Use your finger to block the precessing shaft of
the gyroscope. What effect does precession have on
the stability of a spinning gyroscope? Use your
finger to speed up the precession of a spinning
gyroscope. What happens to the amount of
precession?

Evidence of the Earth's Precession. The two
circumpolar star trail photographs below were taken
ten years apart. Note the star trail located closest to
the center of the arcs. What change do you notice in
the position of this star with respect to the sky pole?

247

The stars near the north sky pole appear to move very little during each night. For this reason, these stars had special importance to early man. Star maps 5000 years old have been found showing the position of the north sky pole at that time. The star map here shows the path the north sky pole has taken through the sky during the past 5000 years. Note the predicted position of the north sky pole 5000 years from now. How long will the earth take to precess once, if precession continues at the present rate?

A Model of the Earth's Precession. Spin the gyroscope in the same direction that the earth spins. Hold the gyroscope shaft so that the shaft forms an angle of 23½° with a vertical line. Note the point on the ceiling to which the gyroscope points. Allow the gyroscope to precess. Note the imaginary arc on the ceiling to which the precessing shaft points.

Hold over the spinning gyroscope the star map showing the precession of the sky pole. Does the gyroscope precess in the same direction as the earth, or does the gyroscope precess in the opposite direction?

List some differences between the earth and a gyroscope which might account for differences in the behavior of a gyroscope compared to the earth.

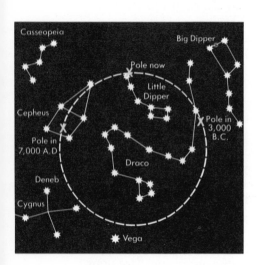

1. What is an analogy? Give an example.
2. Why are analogies not always good ways of explaining how something behaves?
3. What evidence have you seen that the world is flat? What evidence have you seen that the world is round? How do you explain to yourself the two different sets of observations?
4. What evidence do you have that the earth does not spin? What evidence was presented in this chapter which supports the idea that the earth does spin? How do you explain to yourself the two different sets of observations.
5. What evidence was presented in this chapter to support the idea that the earth travels around the sun once each year?
6. Why does the earth always point in the same direction as it moves in its yearly trip around the sun?
7. What is precession?
8. What is a Foucault pendulum?
9. What is parallax?
10. In what two ways have pendulums helped us in our study of the earth's shape and motions?
11. How does the rate of a pendulum's swing depend upon its length, weight, and height of swing?

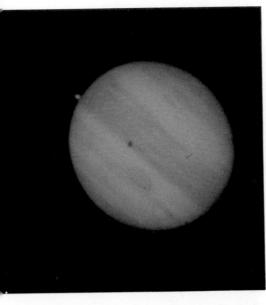

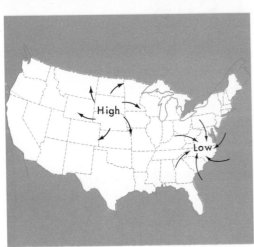

THOUGHT QUESTIONS

1. What causes the black spot near the center of Jupiter in the photograph above?
2. Why is it colder in January than in July, although the earth is closer to the sun in January?
3. Weather maps such as the one shown here usually show winds blowing out of high pressure areas turning clockwise. Winds blowing into low pressure areas turn counterclockwise. Why? (Hint: Winds turn the other way in the Southern Hemisphere.)

3

Our Neighbors: The Moon and Sun

One never-ending search in astronomy has been to build better and better instruments to study sky objects that are far away. However, instruments which astronomers build to learn more about distant sky objects also help them to know more about objects which are relatively close. Much of what we know about stars has been gained by studying the one star which is quite close, the sun. Much of what we know about the moons of other planets has been gained by studying our own moon. Which planet is the easiest to study?

Plato

Alps

Sea
of
Showers

Archimedes

Apennines

Sea
of
Serenity

Sea
of
Crises

Ocean
of
Storms

Copernicus

Sea
of
Tranquility

Sea
of
Fertility

Sea
of
Clouds

Tycho

THE MOON

Since earliest times, people have wondered about the moon. Sometimes this sky object is seen in the daytime, more often at night.
Sometimes it appears as a round disk or circle, other times as a thin
crescent. Some people believe they can see a face on the moon.

In the past the moon has been considered a special object by poets
and songwriters. Perhaps all this will change now. Because humans have
walked on the moon, it may no longer symbolize something mysterious
that we can never understand.

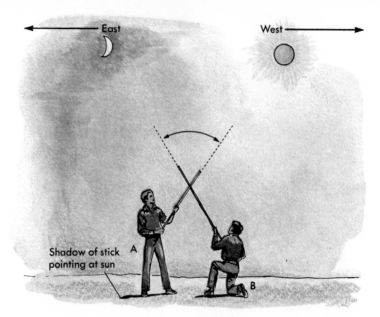

East

West

Shadow of stick A
pointing at sun

B

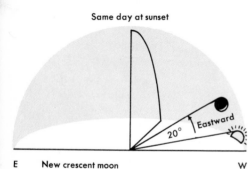

Same day at sunset

20° Eastward

E   New crescent moon   W

5 days later at sunset

E   First quarter moon   W

Daily Changes in the Moon. The drawing above shows one useful way of accurately determining the moon's position when both sun and moon are visible. The student at A holds a stick so that it casts the smallest possible shadow. In this position, the stick points directly at the sun. (**CAUTION:** You should never look directly at the sun.)

The student at B sights along a stick at the moon as this stick rests against the one pointing at the sun. Then the angle between the two sticks is measured with a protractor. This measurement is the angular distance between the sun and moon.

Measure the angular distance from the sun to the moon. Record this distance, the time, and the moon's shape on the science room chalkboard. Compare your data with the data collected at other times during the day. Describe the path of the moon during the day. Discuss the data collected, and decide whether the moon's shape and distance from the sun change during the day.

Add a model moon to your sky model of two chapters earlier. Turn the model sky as you did before. Does the model moon behave in the same way that the real moon behaves?

Monthly Changes in the Moon. Look for the moon in the sky every clear day for a month. Measure the angular distance from the sun to the moon and record it on a data table. Also record the direction from the sun to the moon and the moon's shape.

SHAPE AND POSITION OF MOON

Date	Shape	Distance from Sun (degrees)	Direction from Sun
10/17	🌙	20°	East

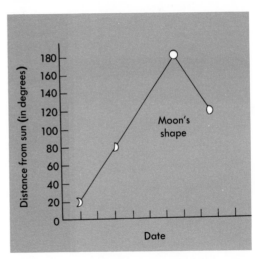

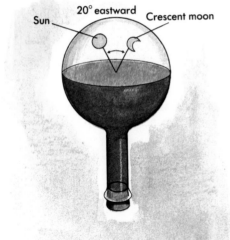

20° eastward — Sun — Crescent moon

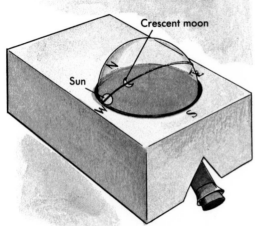

Crescent moon — Sun

Study the data you have collected. Describe the changes in appearance of the moon. Between what dates did the moon appear to increase in size? When did it decrease? How many days pass before changes in the moon's shape are repeated?

Graph your data as shown above. Show on the graph the moon's shape as well as the angular distance from the sun to the moon for each date you recorded. Note the angular distance between the sun and moon for several pairs of points on the graph. Calculate the angular distance the moon moves toward or away from the sun each day. At this rate, how many days will the moon take to travel 360° (to return to the same position in the sky)?

Position and Shape of the Moon. Note the drawings of the moon and setting sun on the previous page. The moon is 20° east of the sun in this drawing. In the left diagram here, a moon of the same shape has been drawn on the sky model. If you were at the center of the flask, the crescent moon would point in the same direction as the crescent moon on the previous page. Add your other moon observations to the sky model.

Set up your model sky as before. Where in the model sky is the new crescent moon at sunset? At what time will this moon set? At what time of day in the model sky does the full moon rise and set? Compare your observations of the sky model with your observations of the real sky. Is the model a good one? Does it behave like the real sky does?

New Crescent

First Quarter

New Gibbous

Full Moon

Phases of the Moon. The different shapes of the moon which you observed during the month are called *phases*. Note the phases of the moon shown at the left. Other phases of the moon are called old gibbous, third quarter, and old crescent. These phases appear similar to the top three photographs at the left except that the left side is lighted. When none of the brightly lit surface of the moon can be seen from the earth, the phase is called new moon. Describe the positions of the sun, moon and earth at this time.

Phase Changes Similar to the Moon. Set up a projector, a white ball, and a stool for the activity shown below. Remove the lens from the projector. Imagine that the projector is the sun, the ball is the moon, and the stool is the earth.

Sit on the imaginary earth. Notice that the region you view from the earth is shaped like a dome. Let the region to your left represent east, overhead is north, to your right is west, and down is south, as shown below.

Slowly turn yourself and the imaginary earth counterclockwise, as shown by the arrow. In what direction does the sun appear to rise? To set? In what direction does the moon appear to rise? To set?

Note the shape of the imaginary moon when it is viewed in the position shown below. Keep the imaginary earth turning slowly, and at the same time move the moon slowly around the earth. Which way must the moon move to change phases in the same manner as the real moon?

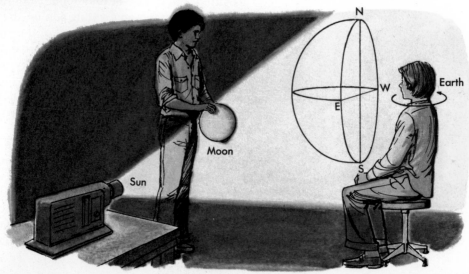

Sun

Moon

Earth

Earthshine makes the dark side of the moon visible.

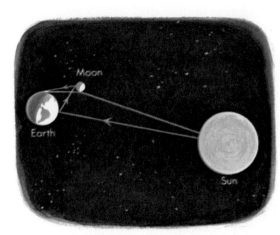

Earthshine. Examine your observations of the phases of the moon. Can the darkened part of the moon be seen in some of the phases? During which phases? At what time of day?

The faint light which makes the dark part of the moon visible is called *earthshine*. Earthshine can be seen just after sunset when the moon appears in the new crescent phase.

Study the diagram of the sun, earth, and moon below. The white line shows how sunlight reflects from the moon to the earth. What phase is produced? The blue line shows how earthshine occurs. Light from the sun reflects from the earth to the dark side of the moon and back to the earth.

Make a diagram of the sun, earth, and moon when the moon is in the gibbous phase. Use the drawing to help explain why earthshine is not noticeable at this time.

The Moon's Features. The moon is famous for its many craters. Most astronomers agree that these craters were produced when meteors hit the moon's surface. The craters can be seen easily with binoculars or an inexpensive telescope. For best viewing, do not observe at full moon. Look on any other evening. Craters appear clearest near the terminator (the region separating the dark and light portions of the moon).

The first astronomers who looked at the moon through a telescope thought the darker regions were seas. Therefore, these regions were given such names as the Sea of Tranquility and Ocean of Storms. These names remain today, although astronauts have confirmed later observations that these regions are dust plains. Note the shape of the Sea of Tranquility and the Sea of Crises on the moon photograph at the beginning of the chapter. Propose a possible explanation for how these areas were formed.

The moon has a diameter of 3480 kilometers. Use this information to estimate the diameter of the largest crater in the photograph. What is the diameter of the smallest crater that can be seen in this photograph?

The earth's atmosphere burns up many meteors which might otherwise hit the earth's surface and produce craters. The few meteor craters on the earth's surface are slowly being erased by the action of wind, rain, and plant growth. Why does the moon have more craters than the earth?

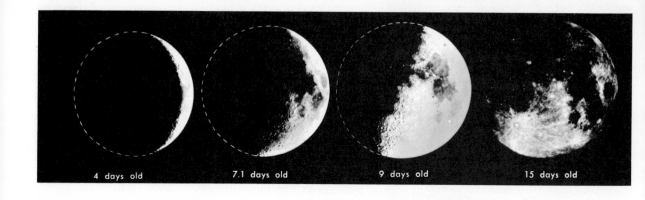

4 days old   7.1 days old   9 days old   15 days old

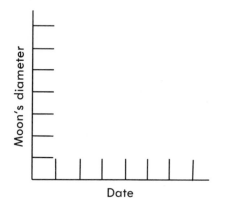

Moon's diameter

Date

Apparent Size of the Moon. Use a ruler to measure the pole-to-pole diameter of the moon in each photograph above. Record each diameter on a graph like the one here. Propose a hypothesis to explain the apparent change in diameter.

The moon travels in an elliptical (oval) orbit around the earth. Sometimes the moon is about 356 000 kilometers from the earth. The point where the moon is closest to the earth is called *perigee*. Label perigee on the graph.

When the moon is at its greatest distance from the earth, it is about 407 000 kilometers away. This point in the orbit is called *apogee*. Label apogee on the graph.

The Shadow of a Model Moon. As the moon travels in its orbit, sunlight is blocked from the space directly behind the moon. As a result, a shadow is produced. Form a shadow with a ball model of the moon, as shown at the left. Note the two distinct shadow regions that are on the paper. The black region [B] is called the *umbra*. The gray region [A] is called the *penumbra*.

Move the paper closer to the ball. What happens to the size of the umbra? What happens to the size of the penumbra?

Use a sharp pencil to poke three small holes in the paper. Poke one hole in region *A*, one in region *B*, and one in region *C*. Look through the hole at *B*. The bulb's light as viewed from *B* is said to be eclipsed (blocked out). How does the bulb's light appear to an observer at *A* and at *C*?

Look through the hole at *B* as you move the paper away from the ball. What happens to the total eclipse (totally blocked out bulb)? Propose an explanation using the terms umbra and penumbra.

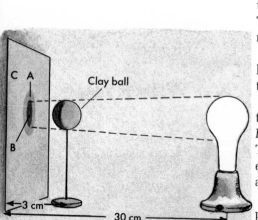

C A

Clay ball

B

←3 cm→

←—— 30 cm ——→

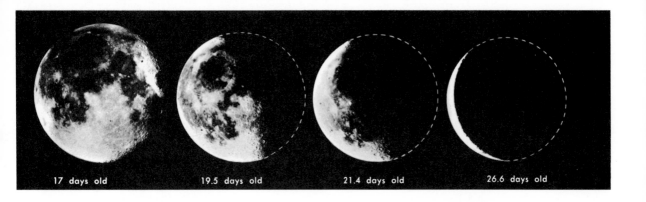

17 days old   19.5 days old   21.4 days old   26.6 days old

Eclipses. Use a long, thin nail to attach a Ping Pong ball to one end of a long stick, as shown at the right. Let this 4-cm ball represent a scale model of the earth. The moon's diameter is one-fourth that of the earth. What would be the diameter of a model moon on this scale?

The distance from the earth to the moon averages about 30 earth diameters. How far from the model earth should the model moon be placed to show the distance to the correct scale? Put a nail in the stick at this distance from the model earth. Around the nail, mold a clay model moon of correct size.

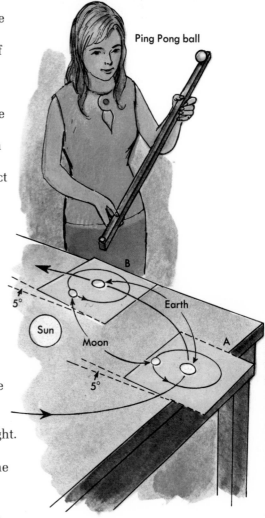

Take the model earth-moon system to a sunny place. Line up the sun, earth, and moon so that the moon passes through the earth's shadow. Identify the umbra and penumbra regions of the model earth's shadow. How would the moon look viewed from the dark side of the earth during an eclipse?

Line up the sun, moon, and earth so that the moon's shadow falls on the earth. Compare the moon's shadow with the earth's shadow in the previous activity.

Locate the umbra and penumbra regions. Describe how the sun would look during an eclipse as viewed from different places on the earth.

Astronomers observe that the moon's orbit is tipped 5° from the earth's orbit, as shown at the right. On the scale model, this means that the moon is sometimes 10 cm above or below a line between the sun and earth. Set up this arrangement with the model. Will an eclipse occur? Explain.

Study the drawing at the right. Will an eclipse occur at *A*? At *B*? Three months after *B*? Six months after *B*?

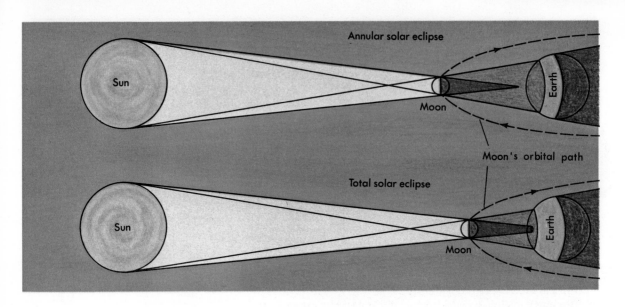

Annular solar eclipse

Sun

Moon

Earth

Moon's orbital path

Total solar eclipse

Sun

Moon

Earth

Lunar Eclipses. A *total lunar eclipse* occurs when all of the moon passes through the earth's umbra. A total lunar eclipse may last as long as 3½ hours from start to finish. The moon may be entirely in the umbra for as long as 1⅔ hours.

If only part of the moon passes through the earth's umbra, the eclipse is called a *partial lunar eclipse*. Partial and total lunar eclipses can be seen from any place on the dark side of the earth. Lunar eclipses occur about twice a year, 6 months apart. What must be the phase of the moon during a lunar eclipse?

Annular Solar Eclipses. Two different types of solar eclipses are possible because of the elliptical shape of the moon's orbit. These are pictured above. The upper diagram shows a solar eclipse occurring when the moon is near apogee. The moon, as viewed from the earth, appears somewhat smaller than the sun. Thus, the moon's disk does not completely cover the sun's disk, leaving a ring of sunlight around the moon. This type of eclipse, shown at the left, is called an *annular solar eclipse* (*Annulus* means ring.)

Total Solar Eclipses. Solar eclipses may also occur when the moon is near perigee in its orbit. At such times, the umbra region of the moon's shadow reaches the earth. The disk of the moon appears larger than the disk of the sun, and all direct light from the sun is blocked. Such an eclipse is called a *total solar eclipse.*

Annular eclipse

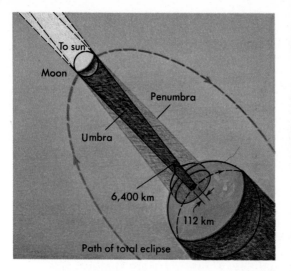

Total solar eclipse with corona

Total solar eclipses, unlike total lunar eclipses, can only be seen from a relatively narrow band across the earth. The band along which the eclipse is total is never more than 110 km wide. The total eclipse never lasts longer than 7½ minutes. The region of the earth from which a partial solar eclipse can be observed may be as wide as 6400 km.

A strange darkness falls over a region during a total solar eclipse. Bright stars and planets can be seen in the sky during the daytime. Some flowers begin to close, and many animals behave as though evening were beginning. The glowing outer atmosphere of the sun, called the *corona*, is an impressive sight in the sky. (**CAUTION:** Never look directly at the sun, even during an eclipse.) The corona is difficult to see except during a total solar eclipse because of the sun's brightness.

The map shows some of the paths of total solar eclipses over a period of many years. Compare two paths 18 years apart. What pattern do you notice in these paths?

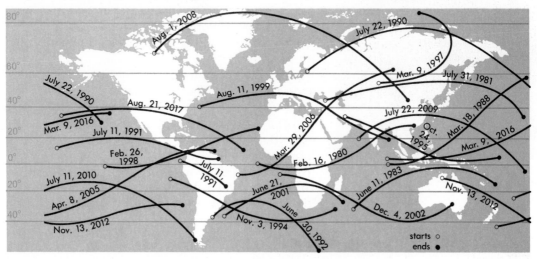

High and Low Tides. Tides are another occurrence caused by the interaction of the earth, moon, and sun. The photographs shown here were taken along the Atlantic Coast at the Bay of Fundy. The first photograph was taken at *low tide*. The second photograph was taken 6 hours 12½ minutes later at *high tide*. High and low tides generally occur twice a day at regular intervals. Tides usually repeat about 50 minutes later each day.

Tidal Range. *Tidal range* is the difference in the levels of the sea at high and low tides. Tidal range along seacoasts averages about one meter. Some places, such as the Bay of Fundy, have tidal ranges as great as 18 meters.

Although tidal ranges on the coasts of continents average one meter, the tidal ranges at ocean islands are usually much less. Many scientists feel that the tidal range in mid ocean is only a few centimeters. They believe that the mid-ocean tidal range has a piling-up effect in the shallow water near the shores of continents. Unfortunately, no one has yet devised a method of measuring tidal range in the middle of the oceans.

Usually tidal ranges are greatest at new moon and at full moon. These tides are called *spring tides*. Tidal range is least when the moon is in the first and third quarters. These tides are called *neap tides*. Does the drawing at the left show a spring tide or neap tide?

Center of Mass. In order to understand tides, you need to understand how large masses behave. The following activities will aid you in this understanding.

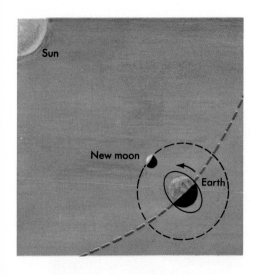

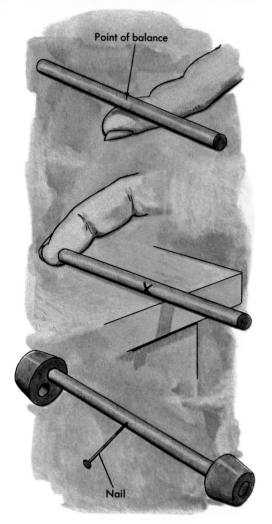

Point of balance

Nail

Balance a wooden rod, such as a pencil, on a finger. Mark the point of balance on the rod.

Push one end of the rod slowly off the edge of a table. At what position is the rod's point of balance when it falls off the table?

Put a large one-hole stopper on one end of the rod and a small one-hole stopper on the other end. Locate the balance point. Again note the position of the balance point as the rod and stopper are pushed off the edge of the table.

We can describe the behavior of the wooden rod by considering that the mass of the rod is centered at one point. This point is called the *center of mass*. Where is the center of mass when the rod is balanced? Where is the center of mass at the instant the rod falls off the edge of the table?

Revolving Objects. Pound a small nail into the center of the wooden rod, as shown at the left. Hold the nail and twirl the device. Note that the rod does not revolve smoothly.

Move the nail to a different position and twirl the rod again. Repeat this process and locate the point around which the object revolves smoothly. Compare the position of the center of mass with the points around which the device revolves smoothly.

Throwing a Revolving Object. Wrap a narrow strip of white adhesive tape around the center of mass of the rod and stoppers. Toss the device from one hand, up in the air, and back down to the other hand. Describe the path which the smaller stopper takes as it travels from one hand to the other. Describe the path which the center of mass takes.

Compare your observations with the multiflash photo below. Note the path of the center of mass.

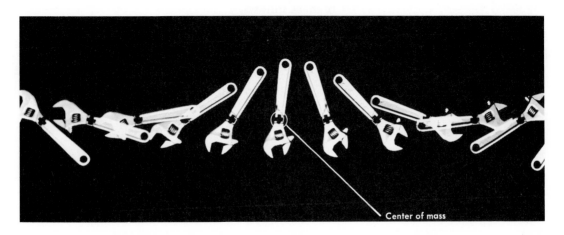

Center of mass

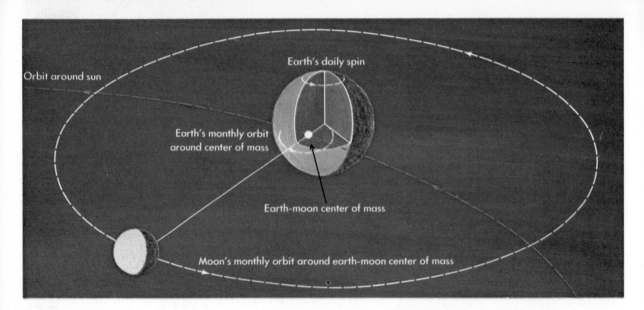

The Earth-Moon System. The picture above shows the earth and moon in orbit around the sun. Note that the center of mass of the earth-moon system travels in a smooth orbit around the sun. The moon and the center of the earth each travel in orbits around their common center of mass once each month.

The center of mass of the earth and moon has been calculated to be 4800 km from the center of the earth in the direction toward the moon. The center of mass is 1600 km below the surface of the earth.

The diagram below shows the positions of the earth and moon at one-week intervals as they revolve around their common center of mass once each month. Note the smooth path which the center of mass follows.

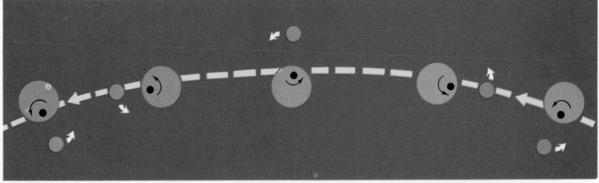

1 Month later   3 Weeks later   2 Weeks later   1 Week later

Explaining Tidal Bulges. The diagram at the right shows the positions of the high and low tides during the new moon. Note the bulges of water (high tides) on the opposite sides of the earth. Which bulge can be explained by the pull of gravity between the water and the sun and moon?

The figure at A on the earth is at low tide. Where will this figure be six hours later on the spinning earth? Twelve hours later?

The moon is believed to exert a greater attractive force on the waters of the earth than does the sun, because the moon is so much closer. Propose an explanation of why tidal range is not as great when the moon is at position B.

The Tidal Bulge Away from the Moon. Make the apparatus shown at the right, using a large two-hole stopper, a small one-hole stopper, a pencil, and a pin. Grease the pencil and push it through one hole of the large stopper. Stick the pin through the small stopper and into the side of the large stopper next to the pencil. Hold the device as shown, and pile sand on the two-hole stopper.

Turn the pencil slowly, and increase the speed until some of the sand flies off the top of the stopper. From where on the stopper did the sand fly off? Explain why the sand leaves the stopper.

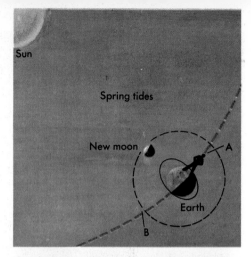

Sun

Spring tides

New moon

A

Earth

B

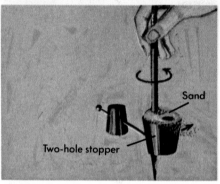

Sand

Two-hole stopper

SUMMARY QUESTIONS

1. What evidence do you have that the moon's orbit is elliptical?
2. What is earthshine?
3. What causes the moon to have phases?

The picture below shows the monthly path taken by the moon and earth as they revolve around their center of mass. Explain why the water bulges out on the side of the earth away from the moon.

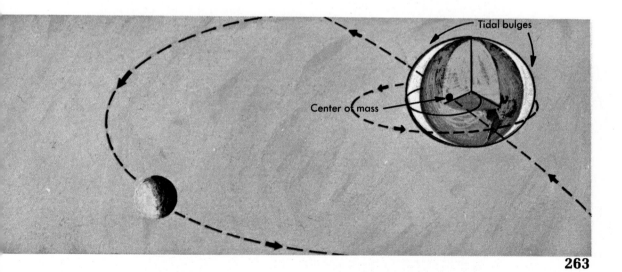

Tidal bulges

Center of mass

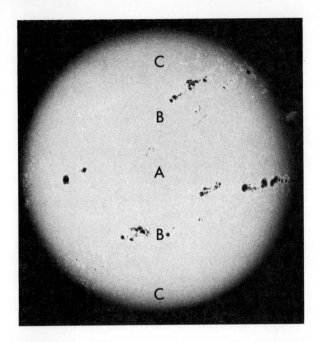

THE SUN

On a clear night about 2000 stars can be seen. If a telescope is used, thousands more become visible. However, even the best telescopes pointed at the brightest stars show us nothing about the details of any star; they are too far away.

There is one exception. This exception is the star at the center of our solar system which we call the sun. Much of what we believe to be true of other stars is known by studying the star which is closest to us, about 150 million kilometers away.

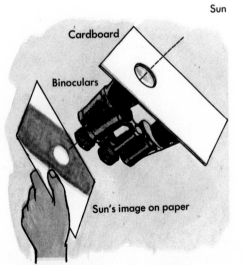

Sun
Cardboard
Binoculars
Sun's image on paper

Observing the Sun's Surface. (CAUTION: Never look directly at the sun, and never look through an instrument pointed at the sun.) Use a piece of cardboard, some white paper, and binoculars to observe the sun's image on a sheet of paper. Make a hole through the cardboard the same size as one of the large lenses. Tape the cardboard to the binoculars, as shown at the left. Point the binoculars at the sun, and hold a sheet of paper in the shadow of the cardboard. Focus the binoculars and move the paper toward and away from the binoculars until a bright circular image of the sun forms on the sheet of paper.

Examine the sun's image. Look for dark spots or groups of spots on the sun's surface, called *sunspots*. Compare your image of the sun's surface with the photograph above.

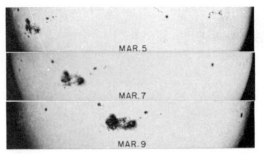

MAR. 5

MAR. 7

MAR. 9

Size of earth

Evidence of the Sun's Rotation. Draw a 6-cm diameter circle on each of several index cards. Form an image of the sun on one card so that the sun's image is the same size as the circle. Mark on the circle the position of a sunspot. Repeat this observation for several days. Propose an explanation for your observations.

The photographs above were taken two days apart. Do these photographs support your explanation above?

Note the regions labeled *A*, *B*, and *C* in the photo on page 264. Astronomers have noted that sunspots in region *A* move such that they return to their original position in 25 days. Sunspots in region *B* take 27 days, and those in region *C* take 33 days. What does this observation tell you about the sun's surface?

Solar Flares and Prominences. The top left photograph shows a huge loop of gas escaping from the surface of the sun. This feature is called a *solar prominence*. Note the small round circle in this photograph. This circle shows how large the earth would appear next to the solar prominence.

The next picture shows a part of the sun facing the earth. The lighter regions are called *solar flares*. A dark line in the middle of a flare is thought to be a solar prominence.

Auroras. People sometimes notice strange flickering colored lights in the night sky. These lights are called *auroras*. In the Northern Hemisphere, auroras appear brightest in the northern sky. The northern aurora is called the *aurora borealis* or northern lights.

Many people have wondered what causes this strange glow. When no simple explanation was found, other events which happened at the same time were studied, in hopes of finding some relationship. One event which occurs at the same time as an aurora is a solar flare.

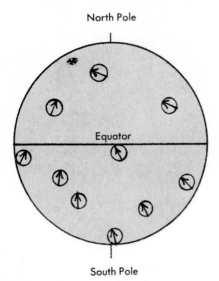

North Pole

Equator

South Pole

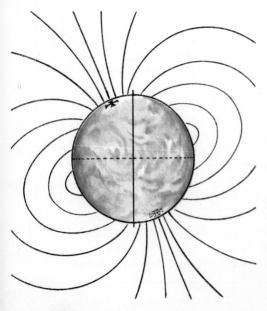

Magnetic Fields. In order to understand the causes of northern lights, you must first understand something about the earth's magnetic field. Find a ba. magnet. Trace the outline of the bar magnet on a sheet of paper. Label the magnetic poles on the paper, and place the magnet under the paper below the outline.

Sprinkle iron filings on the paper. Tap the paper lightly, and note the pattern which forms. Sketch the pattern around the outline of the magnet. The region of influence around the magnet is called a *magnetic field*.

Remove the iron filings from the paper. Replace the paper over the magnet, and place a compass on the paper in the magnetic field. Mark under the compass the direction in which the needle points. Repeat this at several points on the paper.

To which pole does the compass needle point? Compare the behavior of the compass needle with the behavior of the iron filings.

Compass Observations. The globe above shows the direction in which compass needles point at different places on the earth. From this information, describe the magnetic poles and magnetic field of the earth.

The Earth's Magnetic Field. The picture here shows the earth's magnetic field. It combines data from compass readings and from specially built dipping needles. The earth's magnetic field has the same form as the field around other magnets.

Solar Flares and Auroras. Rocket and satellite studies show that large numbers of charged particles, such as electrons and protons, appear around the earth at the time of solar flares. From these observations, scientists think that billions of charged particles are given off by the sun during a solar flare.

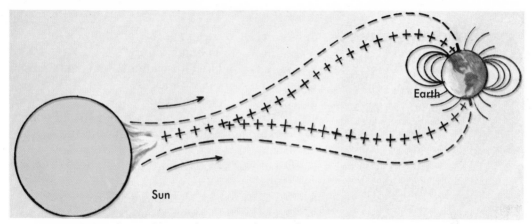

Some of the charged particles enter the earth's outer atmosphere near the magnetic poles, traveling at great speeds. These particles collide with molecules of oxygen and nitrogen. When they collide, green and red light is given off by the molecules which can be seen at night as an aurora.

Some of the aurora light is believed to result from protons and electrons combining to form hydrogen atoms. In this process, light energy is also released.

Van Allen Radiation Belts. Satellite and rocket studies have detected doughnut-shaped regions around the earth which contain fast-moving charged particles. These regions are called *Van Allen radiation belts.*

The radiation belts are much closer to the earth near the earth's poles. Charged particles spiral back and forth within the belts, as pictured below. Near either pole, the trapped particles collide with molecules of oxygen and nitrogen, losing energy at each collision. Eventually the charged particles leak out into the earth's atmosphere.

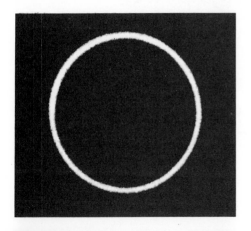

REVIEW QUESTIONS

1. Which photograph at the left shows a total eclipse? An annular eclipse?
2. Why is a solar eclipse annular at some times and total at other times?
3. What is a lunar eclipse?
4. Why doesn't a lunar eclipse and a solar eclipse occur each time the moon goes around the earth?
5. What evidence do you have that the sun spins?
6. What causes craters on the moon?
7. Where is the center of mass of the earth-moon system?
8. Explain the following terms: magnetic field, solar flare, solar prominence, aurora, Van Allen radiation belts.
9. Describe the daily apparent path of the moon across the sky.
10. Describe the monthly path of the moon.
11. What evidence do you have that the moon's orbit is elliptical?
12. What phase of the moon is shown below?
13. In what direction is the sun located in the moon photograph below?
14. What is earthshine?

THOUGHT QUESTIONS

1. Why do spring tides occur in the summer, fall, and winter?
2. Why don't the forks below fall?
3. Stars which change in brightness are called variable stars. What evidence do you have that the sun is a variable star?

Investigating on Your Own

Astronomy is one of the few areas of science in which important discoveries are often made by people with little or no training in astronomy. Such people may be machinists, or medical doctors, or teachers, who study astronomy as a hobby. Many discoveries of new comets, or information about stars which change in brightness, come from these people. Many amateur astronomers first became interested in astronomy when they were your age.

The following pages of projects and independent study problems are planned so that you can work on your own. There are many different types of projects so that you have a wide selection. Report your results to your classmates.

CONSTRUCTION PROJECTS

When people think of astronomy, they often think of construction projects, such as building large telescopes, or sun temples, or planetarium buildings. Certainly our knowledge of astronomy, as well as interesting ways of learning about astronomy, would be very different without these structures. Construct one of the following items. Then use what you have built and share your findings with your classmates.

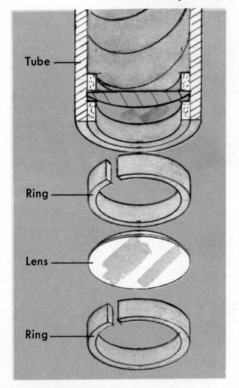

Tube

Ring

Lens

Ring

Thick lens   Thin lens

1. Make a simple two-lens telescope. You will need one thick lens, one thin lens, and two cardboard tubes, one sliding inside the other. The thick lens may be a high-power magnifying lens; the thin lens may be a simple reading glass. Kits containing such lenses are sold by some hobby shops.

Hold the lenses as the girl in the picture is doing. Move the lenses back and forth until you see a clear magnified image of objects far away. Is the image right side up or upside down? Ask someone to measure the distance between the lenses while you are looking at the image.

The outer cardboard tube should be a little shorter than the distance between the lenses. Fasten each lens in place with two rings of cardboard which are glued to the tube, as shown here. Fix the thin lens at one end of the outer tube, and the thick lens at the opposite end of the inner tube.

2. Use a sky chart to locate the Milky Way. Look at part of the Milky Way through your telescope (or binoculars). Count the number of stars you can see, holding the telescope in one position. Use your telescope to look at a part of the sky which is not in the Milky Way. Again count the number of stars you can see. Compare the number of stars.

3. Use a telescope (or binoculars) to study the moon. Make a sketch of details along the line separating the dark and light portions of the moon.

4. Use a telescope (or binoculars) to study the Pleiades, a cluster of stars shown opposite. Look for other star clusters in the sky. Use a sky chart to locate the position of some star clusters in the sky.

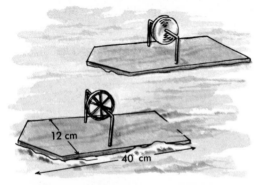

5. Build two similar flat boats. Mount a gyroscope on each boat as shown here. Spin one gyroscope and make waves to rock the boats. How does the spinning gyroscope affect the stability of the boat? Write a report on gyroscopes and stability. Discuss how gyroscopes are used.

6. Most giant telescopes use a concave (curved-inward) mirror in place of the thin lens in two-lens telescopes. Make a simple reflecting telescope using a curved shaving mirror. First measure the distance from the mirror to where the sun's rays focus on a piece of paper. This is the focal length of the mirror. Tape together several large juice cans. Tape the mirror to one end, as shown at the left. Mark the point on the cans which is 7 cm shorter than the focal length of the mirror. Attach a piece of coat hanger wire between opposite sides of the cans at this point. Tape or glue a 2-cm square piece of flat mirror to this wire. Turn the mirror so that light is reflected out a 2-cm diameter hole cut in the can at this point. Look through a good magnifying lens, and focus on a distant building or object. (**CAUTION:** Never look at the sun through any telescope.)

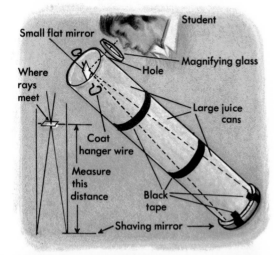

7. Use an optics supply catalog to find the variety and cost of telescopes or telescope mirror grinding kits. Report your findings to the class.

8. Build a classroom planetarium dome. Construct the dome by taping together large cardboard triangles. Paint the underside of the dome flat white. Use a small planetarium projector to project the stars onto the sky, or paint constellations on the sky in the proper positions.

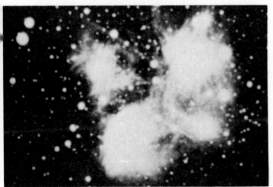

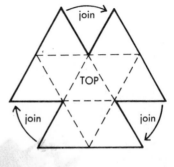

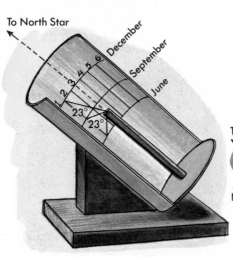

To North Star

December
September
June

23° 23°

9. Make a sundial similar to one of the three sundials shown below. Read the sundial at intervals, and keep records of its accuracy. Prepare a report on the history of the sundial.

10. Make the sundial shown at the left which will tell the time and date. Explain how it works. Keep records of its accuracy.

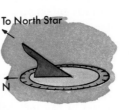

To North Star

N

South

EXPERIMENTS IN ASTRONOMY

One of the best ways to understand what science is really like is to do experiments. After all, that is what scientists spend a good part of their time doing. When scientists carry on experiments they are trying to find out how one variable controls the behavior of a second variable. For example, many astronomers have measured how bright a star is every 15 minutes or half hour. (Variable stars change in brightness.) In this kind of experiment, the astonomer is studying how one variable (time) affects the behavior of a second variable (brightness of a particular star).

In the experiment, the astronomer chooses the times at which to measure star brightness. For this reason, the time is called the independent (or manipulated) variable. The brightness of the star is believed to depend on (or respond to) the time at which it is measured. For this reason, star brightness is called the dependent (or responding) variable. Identify the missing variables in the table below.

Problem	Independent Variable	Dependent Variable
How do the position of sunspots change with time?	Time	
What time exposure takes the best pictures?		Quality of photograph
How does curve of lens affect magnifying power?		

272

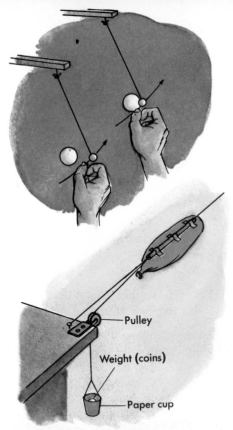

TABLE OF WEIGHTS

penny	3 grams
nickel	5 grams
dime	2½ grams

Pulley

Weight (coins)

Paper cup

Launching
angle

1. Make a study of the motions of a small ball on a string, as shown at the left. Exert a force on the smaller ball using the techniques shown at the left. Note the path the ball takes each time. In which cases will the small ball "orbit" around the larger ball? Describe the forces which act on the small ball, causing it to change direction.

2. Compare amounts of force exerted by different balloon rockets. Set up the apparatus shown here, adding coins to the cup until their weight balances the force exerted by the balloon.

3. Make a study of falling bodies. Drop an empty can and one full of water at the same time. Drop the cans from a second or third story window. Repeat the experiment several times. Do the cans hit the ground at the same time? Explain your findings.

4. Some people say the moon is bigger when it first rises than when it is overhead. Stick two pins upright in the end of a meterstick. Place the pins so that the moon appears to just fit between the pins, as shown below. Then observe the width of the moon when it is overhead. Compare the two diameters. Propose an explanation of your findings.

5. Measure the space between the pins in the last activity. Ask a mathematics teacher to help you calculate the diameter of the moon. The moon is about 384 000 km from the earth.

6. Study the effect the launching angle of a projectile has on the distance the projectile travels. Use a rubber band to launch a small rod. Pull the projectile back the same distance before each launching. Measure the launching angle as shown below. Make a graph of the launching angle and the distance the projectile travels. At what angle does the projectile travel farthest?

7. Borrow a camera which can be set to keep the lens open. Plan an experiment to learn how to take good pictures of the night sky using black and white film. (Hint: If you have a store print your pictures, tell them to make a print of each negative.)

8. Hold a magnifying lens 1 meter from a lamp. Form an image of the lamp on a piece of paper held on the other side of the lens. Experiment to determine how the distance the lamp is from the lens affects the distance from the lens to the image.

LONG TERM PROJECTS AND EXPERIMENTS

Many changes in the sky can only be noticed by observing carefully over many days, weeks, or months. These long term observations, however, offer the beginning astronomer some of the most interesting experiments. Some long term projects are suggested below.

1. Line up the heads of two pins to point directly at a star in the sky. Measure how long the earth takes to make exactly one rotation by noting the time required for the pins to line up again with the star. Explain why the time is not exactly 24 hours.

2. Repeat number 1, but make the observations one or two months apart. Explain your findings.

3. On a clear night, use a compass to determine a line across the sky which is directly south of you. Find the vertical line on the map below which corresponds to this line across your sky. Observe the sky exactly one or two hours later. What vertical line on the map now corresponds exactly to south from you?

4. Use the map below to observe the southern sky. Only half the region shown on the map will be visible at any one time. Look along the ecliptic in the sky for bright sky objects which are planets. Plot the position of any planets on a piece of thin paper placed over the map. Repeat these observations weekly. Keep a record of the movement of the planets among the stars.

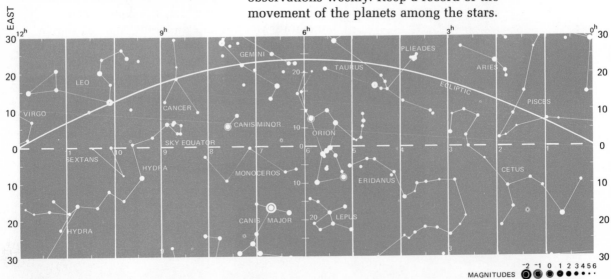

Mark this point

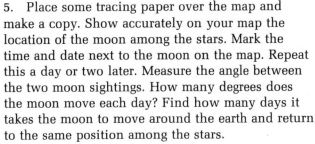

To setting sun

5. Place some tracing paper over the map and make a copy. Show accurately on your map the location of the moon among the stars. Mark the time and date next to the moon on the map. Repeat this a day or two later. Measure the angle between the two moon sightings. How many degrees does the moon move each day? Find how many days it takes the moon to move around the earth and return to the same position among the stars.

6. Repeat number 5 but continue your observations on as many nights as possible throughout the month. Does the moon stay exactly in the plane of the earth's orbit all month? (Is it exactly on the ecliptic all month?) If not, what is the maximum distance (in degrees) from the ecliptic? This shows the tilt of the moon's orbit with respect to the earth's orbit. Review page 256 and explain the importance of this slight tilt.

7. Locate a flagpole or other vertical rod which has a clear view of the southern sky and a level area to the north. Set up a way of keeping track of the shadow of the pole's top at the same time around noon each day for a year. The figure that forms is called an analemma. Look up this term in an encyclopedia or an astronomy book. How does this shape provide evidence that the earth travels in an elliptical orbit?

8. Tape an index card to a west-facing (or east-facing) window which has a clear view of the setting (or rising) sun. Devise a method to record the setting (or rising) position of the sun throughout the year.

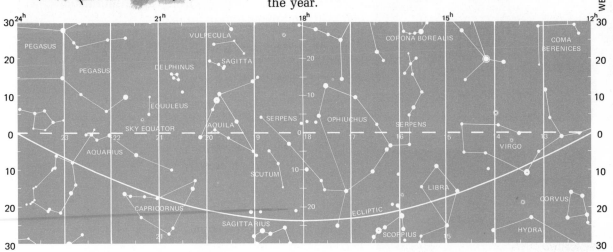

·Horsehead Nebula

Great Spiral Galaxy

Pleiades Star Cluster

Meteor· Trail·..

OTHER INVESTIGATIONS AND PROJECTS

1.   Many additional sky objects can be studied besides those discussed in this book. These sky objects include other galaxies, star clusters, nebulas, and meteors. Examples of these sky objects are pictured on this page. Prepare a report about one or more of these sky objects. Learn where to locate each object in the sky. Then, during an evening sky-viewing session, point out examples of the sky objects you have studied.

2.   Prepare a chart which shows the times during the year of maximum meteor activity. List the dates of these meteor showers, and the constellations in the sky from which the meteors appear to come.

3.   Prepare a table listing the position of each planet in the sky each month during the coming year. Display the table on the science class bulletin board.

4.   Make drawings of some of the constellations. Put the drawings and a short story telling how each constellation got its name on the bulletin board.

5.   Prepare a report about one of the scientists for whom a moon crater is named.

6.   Prepare a table of the manned space flights which have been made to date. Include in the table the date, maximum distance from earth, information obtained, and other important data.

7.   Make a table listing space flights planned for the future. Include probable launch dates and the objects of each flight.

8.   Read about the studies which were conducted during the International Geophysical Year and the International Year of the Quiet Sun. Make a report including some of the information which was discovered in these studies.

9.   Prepare a report on the source of our sun's energy.

10.   Find out what the requirements are for becoming an astronaut.

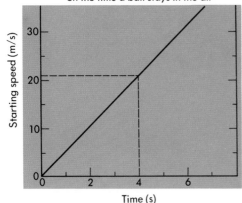

The effect of starting speed
on the time a ball stays in the air

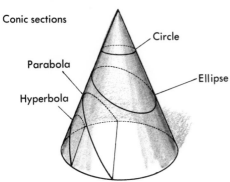

Conic sections

Circle

Parabola

Hyperbola

Ellipse

11. Use pin holes to make constellations in black color slides. Project the constellations with a slide projector.

12. Read about Kepler's laws of planetary motion. Give examples of each law in class.

13. Prepare a report describing the gravitational and magnetic instruments which prospectors use to locate mineral deposits.

14. Find out how Cavendish first determined the mass of the earth. Report your findings to the class.

15. Determine the speed of a ball thrown straight up into the air. Use a stopwatch to measure the time interval between when the ball is thrown in the air and when it lands. Use this time interval and the graph above to determine the speed of the ball as it left the thrower's hand. For example, a ball which was in the air for 4.0 seconds, started with an upward velocity of 21 meters per second.

16. Make a chart showing the orbital and escape velocity for other planets and some of their moons. Explain the reasons for Mars' satellites, Phobos and Deimos, having such low escape velocities.

17. Prepare a short biography of one of the following scientists: Nicolas Copernicus, Galileo Galilei, Johannes Kepler, Tycho Brahe, Isaac Newton, Robert Goddard, Werner von Braun, Albert Einstein, James Van Allen.

18. Models of the paths of objects traveling through space can be made by slicing through a cone at different angles, as shown below. Melt a can of paraffin wax in a pan of boiling water. (**CAUTION:** Paraffin wax vapors are flammable.) Plug the neck of a small funnel with a wad of paper. Add liquid wax to the bowl of the plugged funnel. When the wax hardens, warm the outside of the funnel until the wax cone loosens. Use a coping saw to cut the cone into the sections shown here. The top of each section shows the shape of a different path. Prepare a chart showing the types of objects which travel along each path.

19. Prepare a report about a NASA space flight. As part of your report, make a bulletin board display.

20. Prepare an oral report to your class about career opportunities in astronomy and related fields.

PLANET DATA

Planet	Orbit Radius (km)	$(Radius)^3 = r^3$	Orbit time (days)	$(Time)^2 = t^2$	r^3/t^2
Mercury	5.8×10^7	195×10^{21}	88	7.7×10^3	25×10^{18}
Venus	1.1×10^8		225		
Earth	1.5×10^8		365		
Mars	2.3×10^8		687		
Jupiter	7.8×10^8		4400		
Saturn	1.4×10^9		10 600		
Uranus	2.9×10^9		31 000		
Neptune	4.5×10^9		60 000		
Pluto	5.9×10^9		91 000		

21. Kepler's third law of planetary motion states that a planet's distance from the sun cubed (r^3), divided by the square of the time the planet takes to make one revolution (t^2), is the same for all planets. Check Kepler's third law by completing the table above.

22. Take star trails using high speed Ektachrome film. Use library references to explain the different colored trails. Report your findings to the class.

23. Count the number of stars you can see in the sky. To do this, divide the sky into eight equal sections, then count the stars in one of the sections and multiply by eight.

24. Look up and learn the meaning of the following terms used in astronomy: altitude, azimuth, declination, celestial meridian, A.M., P.M., precession, right ascension.

25. Visit an amateur astronomer. Report your observations to the class.

26. Prepare a display of the constellations of the zodiac. Tell why they are important.

27. Prepare a bulletin board display about the sun temple at Stonehenge, England, and/or other sun temples.

28. Prepare a display of the history of man's progress in getting to the moon and the planets.

29. Prepare a list of the largest telescopes in the world. List their location, size, and whether they are reflectors or refractors.

Position and Time on the Earth's Surface

5

You have probably explained to someone how to walk from one place to another several blocks away. Or, you may have given directions to people in a car so they could get to a stadium or a building. In both cases, you used known routes such as sidewalks and roads to help give directions.

Many travelers must move from one place to another along routes which are not so well marked. These include people who travel by water, by air, or through outer space. These travelers rely on a system of imaginary lines around the earth. Travelers often use these lines to indicate where they are starting from, where they are going, and how well they are progressing in their trip. These imaginary lines are called lines of latitude and longitude.

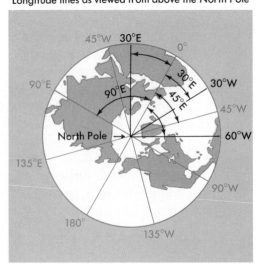

Figure A

LATITUDE, LONGITUDE, AND TIME

Mathematicians say that two pieces of information are necessary to locate a point on a surface. For example, any point on a graph can be located in terms of the distance along a horizontal (x) axis and a distance along a vertical (y) axis. Similarly, points in the sky can be located by measuring the angular distance along the horizon from the north and the angular distance above the horizon. In both cases, two bits of information are needed to accurately locate the point.

Points on the earth's surface can be identified in a similar manner. The two bits of data used to locate points on the earth's surface are called longitude and latitude.

Longitude. Figure A shows lines of longitude labeled 60°W, 30°W, and 30°E. Figure B shows these same lines, in black, as viewed from directly above the North Pole of the earth. Note that the 30°E line forms an angle of 30° eastward with the 0° longitude line. This 0° line of longitude is called the *prime meridian*. What angle does the 60°W line form with the prime meridian?

What is the longitude of London? Of New York City? Moscow? Rome? The western tip of Africa?

Latitude. Latitude is the angular distance from the equator to a point on the earth's surface. Latitude lines connect points of equal latitude. Figure C shows the 45° north latitude line in black. Why is this imaginary line called the 45° north latitude line?

Use Figure A to determine cities in the world which have latitudes of about 45°N. What is the latitude of London? Of New York City?

Study the latitude lines on the globe above. How many degrees of latitude are there between the equator and the poles? How many degrees are there

Figure B

Longitude lines as viewed from above the North Pole

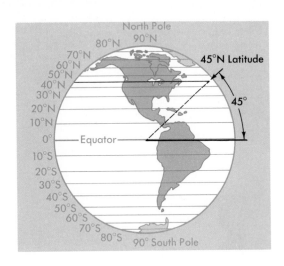

Figure C

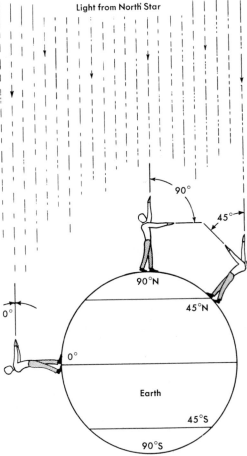

going around the earth through the poles?

The circumference of the earth is approximately 40 000 kilometers. How many kilometers north or south must you travel to cover one degree of latitude?

Each degree of latitude is subdivided into 60 minutes of latitude. How many kilometers north or south must you travel to cover one minute of latitude (¹/₆₀th of a degree)?

Determining Latitude. Latitude at any location can be determined easily by measuring the angular distance from the horizon to the sky pole. In the Northern Hemisphere, the north sky pole's position is very close to the position of the North Star. Navigators measure the angle of elevation of the North Star to determine their latitude.

What is the angle of elevation of the North Star for each observer shown here? What is the observer's latitude? How do Southern Hemisphere navigators determine latitude?

Latitude and Early Map-Makers. Explorers in the 15th and 16th centuries could locate quite accurately the latitude of newly discovered places. But no method was known for accurately determining longitude. Therefore, maps of that period look quite different from maps of today.

Locate several 15th and 16th century maps in your library. Compare the latitude and longitude of places on these maps with the same places on a modern map. Are differences in latitude and longitude consistent with the idea in the previous paragraph?

281

Shadows on a Model Earth. Set a globe in sunlight, as shown above. If someone in outer space could observe the real earth and the model earth at this time, how would the positions of the continents on the real and model earths compare?

Trace a line with your finger that separates the half of the globe which receives sunlight from the half which does not. Which continents on earth are now in darkness?

Hold a pencil so it points outward from the center of the model earth. A pencil in this position points "up" for any position on the model earth. Move the pencil over the sunlit part of the globe until you locate the point where the pencil casts no shadow. What is the time at this location? What is the latitude and longitude? If you were at this location on the real earth, what time would it be there?

Move the pencil to several points where its shadow points due north. What is the time along a line through these points?

Study the shadows on the earth model here. Which nail casts no shadow? On what date was this photograph taken? Describe the conditions of daylight and darkness near each pole on this date.

Nail A casts a shadow due north. It is noon at this location. All the nails to the west of nail A cast shadows to the northwest, indicating that the sun in these locations is in the southeast. Is it morning or afternoon in these regions? Where would you find shadows pointing to the northeast?

Nails B, C, and D are 15° of longitude apart. Describe their shadows.

Sun Shadows and Time. The globe above has four nails in it. The nails and dials act like sundials.

Sundial *A* is 15° longitude west of sundial *B*. What is the difference in time between these two sundials? Sundial *D* is 45° longitude east of sundial *A*. What is the difference in time between these two points?

How many degrees in longitude are equivalent to one hour time difference in both cases above?

Use your previous answer to estimate the time in London when the picture was taken. (London has a longitude of 0°, which is 75° east of sundial *D*.)

Setting Clocks by Sun Time. Less than 100 years ago, each city and town across the country established its own time. Towns usually had the local jeweler establish noon for the town by noting when the sun's shadow pointed directly north.

The signal gun at the left was used to indicate noon. How did it work? In what direction was it placed in order to function correctly?

Suggest a possible reason why time zones replaced local sun time shortly after rail travel began.

Cannon   Fuse

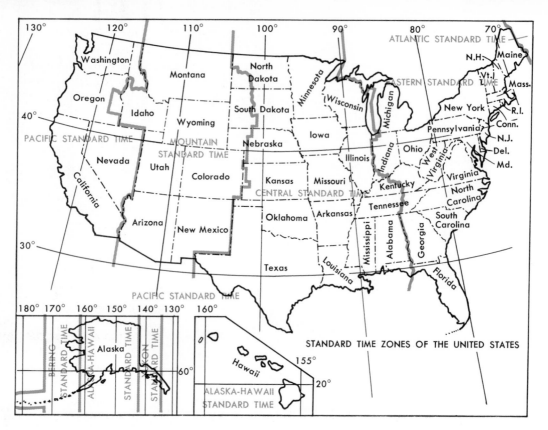

STANDARD TIME ZONES OF THE UNITED STATES

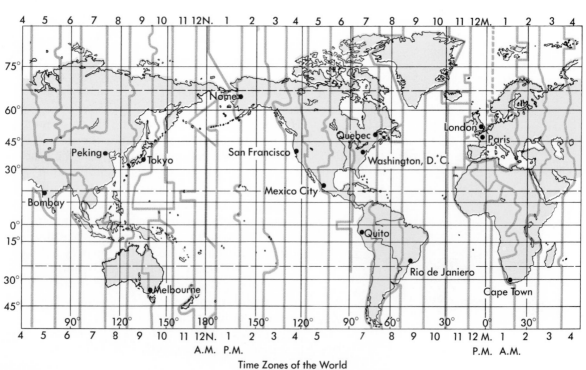

Time Zones of the World

Time Zones in the United States. The longitude of eastern Maine is about 67°W. The longitude of western Washington is about 125°W. How many time zones 15° wide are needed for this region? Compare your answer with the time zone map at the top of page 284.

What time is it now in your time zone? Use the same map to determine the time now in the other time zones of the United States.

Theoretically, each time zone is a section of the earth's surface with a width of 15° longitude. Time zones, however, were established to help us. In order to be of greatest help, time zone boundaries have been adjusted so as to inconvenience the fewest number of people. For this reason, time zone boundaries usually do not pass near large cities.

Time Zones of the World. The time in other places of the world can be estimated by using these two ideas: (1) All time zones are 15° longitude in width; (2) the center of each time zone is along the 0°, 15°, 30°, 45°, etc., longitude lines. For example, Berlin, Germany has a longitude of 13°E. What is the longitude of the center of its time zone?

If it is noon in Berlin, estimate the time in London (longitude 0°). What is the time in Des Moines, Iowa (longitude 94°W)? (Hint: What is the center of the time zone for Des Moines, and how many time zones are there between Berlin and Des Moines?)

The International Date Line. When early explorers returned to Europe from travels around the world, they always found their calendars were one day off. World travelers who went in a westward direction lost a day. Travelers who went in an eastward direction gained a day. Today, these differences can be understood by realizing that traveling 15° in longitude results in the gain or loss of an hour. Thus, traveling 360° in longitude (once around the world) results in the gain or loss of 24 hours or one day.

The International Date Line has been established to correct for changes in dates which arise when people travel around the world. When the date line is passed going westward, the day changes to tomorrow. When the date line is crossed going eastward, the day changes to yesterday.

Locate the International Date Line on a globe. Why is the date line situated in a remote corner of the world?

Going east moves calendar back one day.

International Date Line

Going west moves calendar ahead one day.

The International Date Line

Determining Longitude. In the days of exploring the New World and building the Spanish and English Empires, sailors needed to know their location at sea. They could make accurate estimates of their latitude using the North Star. But no such method was known for determining longitude. Nations established huge rewards for the first person to invent a method of accurately determining longitude.

The device at the left was the first instrument to accurately give longitude. It is called a *chronometer*. It is an accurate clock which is usually set for the time in London, England (Greenwich Time). Suppose the chronometer shows that the time in London (0° longitude) is noon. You measure the position of the sun and determine that it is 2:00 P.M. in your location. How many degrees of longitude are you from 0°? Are you east or west of the prime meridian? What is your longitude? If your latitude is 33°N, locate your position on the globe on page 280.

If your latitude is the same as above but your sun observations indicate that it is 8:00 A.M. at your location, where are you?

If local time is 3:00 A.M., where are you? Suggest a way to find your "sun" time after sunset.

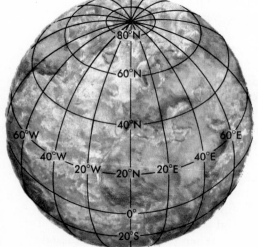

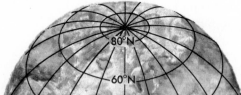

REVIEW QUESTIONS

1. Which lines on the globe at the left are called latitude lines?
2. How is latitude determined?
3. Which lines on the globe are called longitude lines?
4. How is longitude determined?
5. How wide is a time zone, on the average?
6. Why aren't time zones all exactly the same width?
7. What is the prime meridian? The International Date Line?

THOUGHT QUESTIONS

1. Why did the invention of the radio hurt the business of making chronometers?
2. If time measurements were changed to have only 20 hours in a day, how wide would time zones have to be on the average?

A Universe of Orbits

Universe means all the objects in space which astronomers can see or even think might be out there. These objects include planets and moons like our earth and moon, stars like our sun, and large clusters of stars called galaxies.

In the next two chapters, you will learn more about the objects which make up the universe, and about the paths they follow. The path a sky object travels along is called its orbit. Astronomers study the orbits of sky objects in order to learn about and to predict important events such as new planets, eclipses, and tidal effects between objects in the universe.

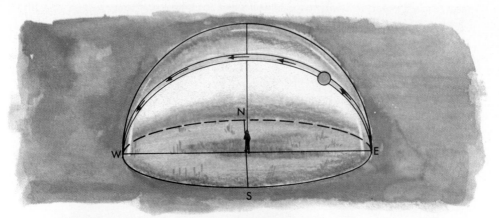

MOTION AND ORBITS

The diagram above shows the path the sun appears to follow in its daily trip across the sky. The important word in the first sentence is "appears." Most people today do not believe the sun actually goes around the earth once each day. These people say that because the earth spins, the sun appears to move through the sky. Thus, the sun's daily motion across the sky is called apparent motion.

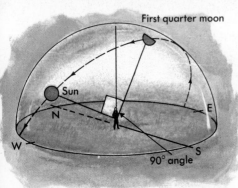

First quarter moon

Sun

N

E

W

S

90° angle

Apparent Motion. You have seen examples of apparent motion when your car or bus was stopped in traffic. If a car next to you moved ahead slowly, you may have thought for a moment that you were moving backward. Your motion, with respect to the other car, *was* backward. But with respect to the ground, your car had no motion.

The model of the sky at the left helps to describe the apparent motions in the sky. How does this model show the sun's path in summer? In winter?

The Moon's Apparent Motion. The moon's daily path across the sky is nearly the same as the sun's. The new moon crosses the sky, staying very close to the sun. The first quarter moon crosses the sky about 90° behind the sun, as shown at the left.

The full moon also follows the sun, but it is directly opposite the sun in the sky. Thus, as the sun sets, the full moon rises. The third quarter moon crosses the sky about 270° behind (or 90° ahead) of the sun.

The sky model on page 289 shows the sun and five different phases of the moon. Set up such a model. Turn the model, and use the sun's position to estimate the rising and setting times for each phase of the moon.

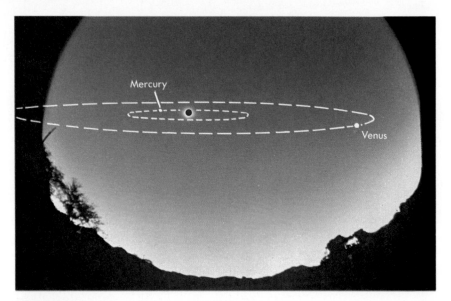

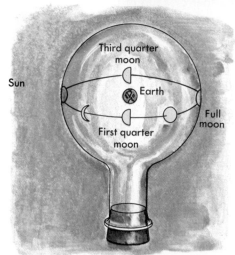

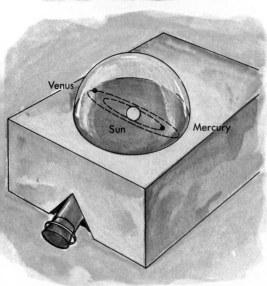

The Motions of Venus and Mercury. The photograph above shows the sun during an eclipse. At this instant, the light of the sun is blocked by the new moon passing directly in front of the sun. Note the two bright objects on opposite sides of the sun. The bright object to the right is the planet Venus, and the object to the left is the planet Mercury. These two planets are always found close to the sun. Both reflect sunlight as do the earth and moon. Why can't you see these objects in the daytime except during an eclipse?

The dotted lines on the photograph have been drawn to show the positions of these planets at other times. Note the drawings of these orbits on the sky model below. Use a wax pencil to mark these planet orbits around the sun on your sky model. Slowly turn the sky model until the sun has set. Which planet is visible in the western sky in the evening (the "evening star")? Continue turning the sky model. Which planet rises in the east in the morning before sunrise (the "morning star")?

As Venus circles the sun, it appears to move from one side of the sun to the other side in about nine and a half months. Mercury moves from one side to the other every two months. When will Mercury be an "evening star"? When will Venus be a "morning star"?

Turn the model sky to the midnight position. Look at the orbits of Mercury and Venus. Can these planets ever be seen at midnight? Explain.

Sun
Eclipsed

Mars

Jupiter

Saturn

The Motions of Other Planets. The drawing above shows the sun during an eclipse with the orbits of three other planets: Mars, Jupiter, and Saturn. These three planets, unlike Mercury and Venus, may be found either close to or far from the sun. However, these planets are always located along a line around the sky called the *ecliptic.*

Add an ecliptic to your model sky. Put on a rubber band as shown below. This rubber band should cross the sky equator at two places at an angle of 23½°. All the planets and the sun and moon are located along the ecliptic. Relocate your model sun, moon, and the orbits of Mercury and Venus so that they all fall along the ecliptic. At what point on the ecliptic would the sun be on June 21st? On December 21st?

Note the line labeled ecliptic on the sky maps on pages 274 and 275. These maps are useful in locating stars and constellations, but planets are not marked permanently on such maps. Why?

Study the star map on page 234. Why is no ecliptic shown on this map?

An Early Model of the Universe. Since early times we have observed the stars, moons, and planets which make up the universe. We have wondered exactly how they move, where they will be next month, and why they behave as they do.

The diagram on the next page shows a model proposed about 150 A.D. by Ptolemy, a Greek astronomer. Note the circular orbit that Ptolemy believed the sun followed around the earth. Use this diagram to explain where Ptolemy believed the sun would be at midnight. At noon.

According to Ptolemy, the sun went around the earth once every 24 hours. The moon went around

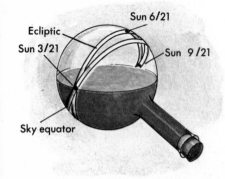

Ecliptic

Sun 6/21

Sun 3/21

Sun 9/21

Sky equator

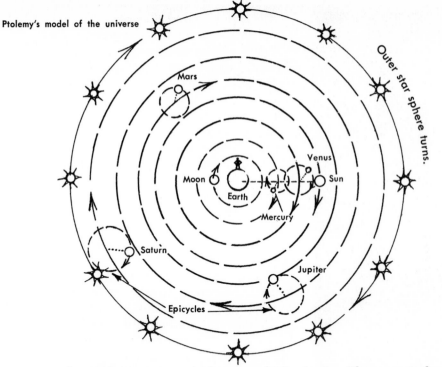

Mars

Venus

Moon ☽ ⊕ ——— ☉ Sun

Earth

Mercury

Outer star sphere turns.

Saturn

Jupiter

Epicycles

the earth once every 24 hours and 50 minutes. Thus, every day the moon became 50 minutes further behind the sun. How many days would pass before the moon was 6 hours (90 degrees) behind the sun? How many days would pass before the moon was 12 hours behind? 24 hours behind?

Can this model explain the sun's daily changes? The moon's monthly phases?

In early times, all natural happenings were thought to be controlled by the gods. Since gods were perfect, the shape and motions of the planets, stars, and sun must be perfect. The perfect shape was believed to be the sphere (as in crystal balls). Thus, all sky objects must be perfect spheres. The perfect shape for an orbit was a circle. Therefore, all objects must travel in circular orbits around the earth. Do the sun and moon in this model obey these ideas?

One problem facing Ptolemy was to devise a way to explain the way Venus and Mercury behaved. These planets stayed close to the sun. All three were godlike, and therefore must be traveling at constant speeds. But these two planets were sometimes ahead of the sun and sometimes behind it! To solve this problem, Ptolemy invented *epicycles*. Note the epicycles for Venus and Mercury. According to Ptolemy, each planet traveled around its circular epicycle at constant speed. At the same time, the center of the epicycle circled around the earth at constant speed. Which of these two planets above would be following the sun and therefore visible in the evening sky?

Early astronomers tried to predict sky happenings based on Ptolemy's theory. They soon found that Mars, Jupiter, and Saturn also did not seem to travel at constant speeds. Epicycles were added to their orbits, too.

Three phases of Venus

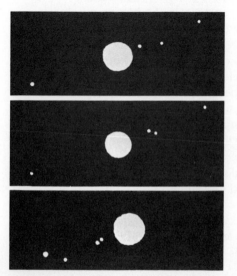

Observations and Theories. Ideas which help to explain observations are called *theories*. Ptolemy's ideas about how the universe moved are an example of a theory. Useful theories help us to predict observations of future events. For example, where will Mars be three months from now, or when will the next eclipse occur? If such predictions turn out to be accurate, the theory is a good one. However, several accurate predictions do not prove that a theory is correct. To prove that a theory is correct, all possible predictions must be accurate. How long would it take to prove that all possible predictions are accurate? Why is it difficult or impossible to prove that a theory is correct?

If one observation is made which cannot be explained or predicted by a theory, then the theory is not a good one. Either the theory must be changed or thrown out, and a new theory invented to explain both the old and new observations.

Galileo's New Observations. Scientists are always trying to make new instruments. Lasers, radio telescopes, Geiger counters, and rockets for space flights are some examples of recent inventions. Whenever someone invents a new scientific instrument, scientists eagerly try to make or buy copies of it. They use the new instrument to find out if present theories can explain all the new observations with these instruments.

In the early 1600's some Dutch lensmakers invented the telescope. This instrument made distant objects appear closer and larger. An Italian astronomer named Galileo soon heard of the instrument, and built his own telescope. Galileo observed (1) that the moon was not a perfectly smooth sphere, but had rough craters; (2) that Venus had phases much like our moon, and (3) that Jupiter had four moons in constant motion.

Study the orbit of Venus in Ptolemy's model. If Venus shines by light reflected from the sun, describe its phases. Compare your description with the observations of Venus's phases shown above.

Study the three observations of Jupiter's moons at the left. Do these moons behave as though they move in circular orbits around the earth? Propose another theory of how these moons move.

What effect did Galileo's observations with the telescope have on Ptolemy's theory?

The Sun-Centered Universe Model. Some years before Galileo pointed his early telescope at the sky, a different theory of the universe, shown below, was proposed by a Polish astronomer named Copernicus. Like Ptolemy, Copernicus believed the orbits of planets were circles.

Compare these theories in the following ways:
1. What object is at the center of each model?
2. How are day and night produced in each model?
3. Which model is simpler? In what ways?
4. How do the outer star spheres differ?
5. What phases of Venus would result from each model?

When Copernicus first proposed his model, he described it as another interesting way of explaining motions in the sky. This theory was used to predict the positions of the planets, and eclipses of the sun and moon. Copernicus's model was found to be less accurate than Ptolemy's model. However, many epicycles had been added to Ptolemy's original model to make its predictions more accurate. Thus, Ptolemy's model was more accurate in predicting, but Copernicus's slightly less accurate model was much simpler. How would Galileo's telescopic observations affect each model?

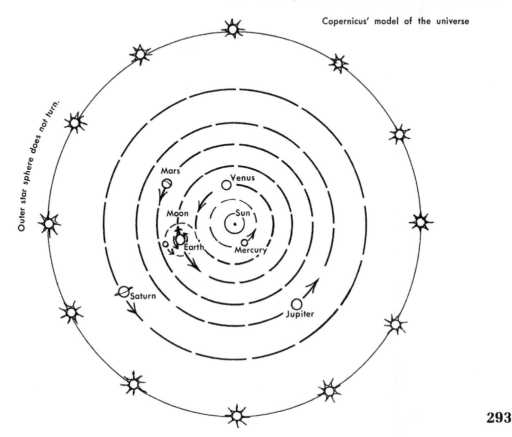

Copernicus' model of the universe

The Solar System. Today, scientists accept many of the ideas in Copernicus's theory. Most astronomers believe that the planets orbit around the sun. We know of nine such planets.

One important idea is not accepted today. We do not believe that the stars make up a giant sphere surrounding our solar system. Also, the orbits of the planets are no longer thought to be circular.

A Scale Model Solar System. Make a scale model of the sun and each planet using the table here. Use a scale of one centimeter equals 10 000 kilometers. Mercury's 4800-km diameter would be represented by a ball with a diameter of 0.5 cm. Scale models of the smaller planets can be made from modeling clay. Find or construct balls which are the correct size to represent the larger planets. How big is the sun?

Use the scale model planets to set up a scale model of part of the solar system on a large field or parking area. Use the same scale as above for the distances between planets. How far away from the model sun is the 0.5-cm ball which represents Mercury? What is the radius of the scale model orbit of Pluto?

The star which is nearest to the sun is about forty trillion kilometers away. How far away would this star be on the scale model?

Planet	Diameter (kilometers)	Distance from Sun (kilometers)
Sun	1 380 000	
Mercury	4 800	57 500 000
Venus	12 200	107 000 000
Earth	12 700	149 000 000
Mars	6 700	228 000 000
Jupiter	142 000	775 000 000
Saturn	120 000	1 430 000 000
Uranus	46 500	2 860 000 000
Neptune	44 500	4 480 000 000
Pluto	6 400	5 900 000 000

VALUE OF SCALE MODELS. Sizes and distances in the solar system are so enormous that few people can picture them in their minds. Diagrams in books give the wrong idea because they cannot be drawn to scale. Models which have been constructed to scale give the most accurate picture of large sizes and distances.

Elliptical Orbits. About 1600, a German astronomer named Kepler realized that recent observations showed that planets did not move in circular orbits. Instead, these orbits were really *ellipses.*

An ellipse is shown at the left. To draw an ellipse, stick two thumbtacks 8 cm apart into a sheet of cardboard. Place a 20-cm loop of string around the thumbtacks, and stretch the string tight with the point of a pencil as shown here. Move the pencil around the loop, always keeping the string tight. The curved line you draw is called an ellipse. The point where each thumbtack is placed is called a *focus* of the ellipse.

Change the distance between the thumbtacks. How does the change affect the shape of an ellipse? How does changing the length of the string affect an ellipse?

A long slender ellipse is said to be more *eccentric* than a nearly round ellipse. How can you draw very eccentric ellipses?

Kepler revised the idea of a sun-centered universe. He said each planet followed an elliptical orbit, with the sun at one focus. Accurate predictions were now possible, not only for planets, but also for moons, comets, and even artificial satellites.

Kepler made two other suggestions for improving the sun-centered theory. His three hypotheses have been shown to fit the evidence so well that many astronomers considered the hypotheses proven. These hypotheses became known as Kepler's Laws of Planetary Motion.

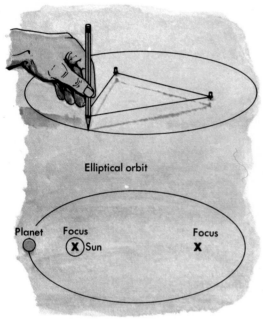

Elliptical orbit

Planet   Focus   Focus
          (X) Sun   X

THE LAWS OF SCIENCE. *Before the present century, most scientists believed that everything in the universe was governed by a few scientific laws. Most research was directed toward finding these laws. Today, scientists are no longer convinced that the universe is governed by laws. Nearly every law considered proved before 1900 has been disproved since then. Some general statements that hold for a wide range of conditions, such as Kepler's statements of planetary motion, are still called laws, but we should not think of them as perfect and unchangeable.*

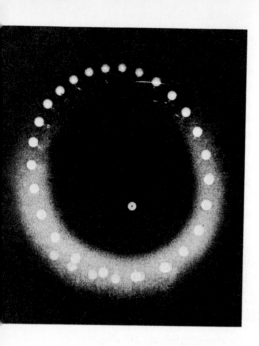

The Earth's Orbit. Stick two pins 3 mm apart into cardboard, and make an ellipse using a 19-cm loop of string. This ellipse represents a scale drawing of the earth's orbit. The scale is: 1 cm = 15 000 000 km. Mark this scale on the drawing.

The sun's diameter is 0.8 mm using this scale. At one focus of the ellipse, label a 0.8-mm circle as the sun.

The point in the orbit where the earth is closest to the sun is called *perihelion*. Locate and label perihelion. The earth is at perihelion in early January. At perihelion, the earth is about 146 000 000 km from the sun. Label this distance on the drawing.

The point where the earth is farthest from the sun is called *aphelion*. The earth reaches aphelion in early July. At this time, the earth and sun are about 151 000 000 km apart. Label aphelion and this distance on the drawing.

Motions of Planets. The photograph at the left shows a billiard ball traveling in an elliptical orbit. A rapidly flashing light recorded the ball. The light flashed at equal intervals. Where does the ball move a lot between flashes? Where does it move only a little? Where is the ball speeding up (accelerating)? Where is the ball slowing down (decelerating)?

Imagine that a planet travels once around the sun in exactly eight months. The diagram here shows the planet's position at the end of each month starting from perihelion.

Note the lines connecting the sun with the planet's position at the end of each month. These lines divide the area within the ellipse into eight parts. Kepler said that these parts all have the same area. Stated more precisely, the speeds of planets change so that a line connecting the sun and a planet sweeps over equal areas during equal periods of time. This idea is called *Kepler's Second Law of Planetary Motion*. This law holds for all sky objects which orbit other larger objects.

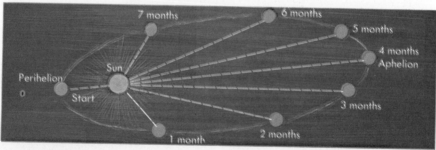

296

Orbits of Planets. Besides the nine planets, astronomers have observed other objects traveling in elliptical orbits around the sun. These include comets and minor planets.

Not all orbits in the solar system involve objects traveling around the sun. At least six planets have moons which travel in orbits around them. For example, Jupiter has twelve moons, and Saturn has nine. In addition, the planet Saturn is famous for its rings. These rings are believed to be made of millions of tiny particles traveling around Saturn.

Make scale drawings of each planet's orbit on a 30 cm × 45 cm sheet of paper. Draw the orbits of the inner planets on one side of the paper, using the data below. Draw the orbits of the five outer planets on the other side of the paper. Label each planet's orbit and the scale of the drawing.

Which planet has the least circular orbit?

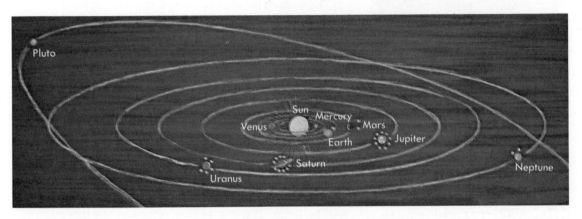

SCALE MODEL OF PLANET ORBITS
The Inner Planets (Scale: 1 cm = 15 000 000 km)

Planet	Loop Size	Distance Between Pins
Mercury	9 cm	1.5 cm
Venus	14 cm	.08 cm
Earth	19 cm	.32 cm
Mars	31.5 cm	2.7 cm

The Outer Planets (Scale: 1 cm = 300 000 000 km)

Planet	Loop Size	Distance Between Pins
Jupiter	5.2 cm	.24 cm
Saturn	9.5 cm	.48 cm
Uranus	19.0 cm	.88 cm
Neptune	28.8 cm	.32 cm
Pluto	46.7 cm	9.5 cm

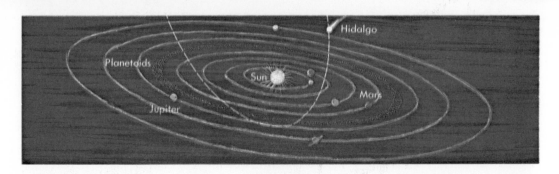

Planetoids. About 2000 minor planets, called *planetoids* (or asteroids) have been discovered. The largest one, Ceres, has a diameter of about 770 km. Most planetoids travel in orbits between Mars and Jupiter, but some travel in very eccentric orbits.

Note the orbit of the planetoid Hidalgo, shown above. In what two ways does this orbit differ from the orbits of the major planets?

Comets. The photograph at the left shows a bright *comet*. Comets are occasionally seen in the night sky close to the sun. That is, they may be low in the western sky after sunset, or low in the eastern sky before sunrise. The tail of a comet points away from the sun. Comets which are visible to the unaided eye occur less than once a year. Astronomers believe that comets are only one or two kilometers in diameter, and are composed of rock and ice particles.

Gravitational attraction between the sun and comet causes the comet to travel in a curved path. If the attraction is great, the curvature of the path will be large. The comet may return in the direction from which it came. Such a comet has been caught by the sun's pull. The comet will orbit the sun.

Some comets don't pass near enough to the sun to be pulled so strongly. The paths of these comets may be changed only slightly. These comets will not be seen again.

Comet paths

Ellipse

Halley's Comet. The orbit of the famous Halley's comet is shown on the next page. This comet was last seen in 1910. During the past 2000 years, this comet has been observed passing near the sun 28 different times. At these times it is an impressive sight in the night or morning sky. However, as the comet moves away from the sun, its brightness fades quickly and it soon disappears.

The position of Halley's comet at 9½-year intervals is shown at the top of the next page. Use

Orbit of Halley's Comet

SOME METEOR SHOWER DATES
(best seen after midnight)

Date	Name
January 3	Quadrantids
April 21	Lyrids
May 4	Eta Aquarids
July 29	Delta Aquarids
August 12	Perseids
October 22	Orionids
November 17	Leonids
December 12	Geminids

The Williamette meteorite
weighs 14 180 kg.

Kepler's Second Law of Planetary Motion to predict when the comet will be visible (near the sun) again.

Meteors and Meteor Showers. The picture at the left shows how the sky looked to millions of North Americans on the night of November 13, 1833. According to diaries and other records of that night, "shooting stars" were as thick as snowflakes. These "shooting stars" began before midnight, and increased in number until more than 20 per second could be seen.

"Shooting stars" are called meteors. If a meteor hits the earth, it is called a meteorite. The photograph below shows a large meteorite found in Oregon. Most meteors are the size of sand grains. As these particles enter the earth's atmosphere, friction with the atmosphere causes the meteor to heat up until it glows. Most meteors burn up before reaching the earth's surface.

Meteors can be seen on any clear night, but some nights are better than others. On some nights so many meteors are seen that they are called meteor showers. Astronomers believe that on these nights the earth passes through the orbit of an old comet. Debris from the comet still travels in this orbit, and the particles heat up by friction in the earth's atmosphere. Why do meteor showers occur on the same dates each year? Which shower was seen by the people in November, 1833?

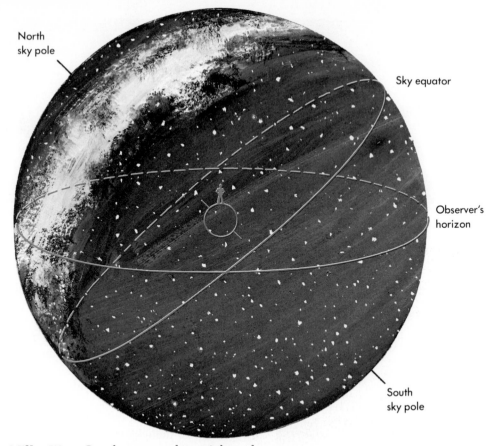

North sky pole

Sky equator

Observer's horizon

South sky pole

The Milky Way. On clear, moonless nights, sky watchers can see a whitish region across the sky called the *Milky Way*. Early sky watchers did not understand what caused the Milky Way. But today with low power binoculars you can see that the Milky Way consists of thousands of stars.

Astronomers believe that all the visible stars in the sky, including those in the Milky Way, are arranged in a lens or disk shape, like that shown below. This lens-shaped group of stars is called a

Andromeda Galaxy

Milky Way Galaxy

Direction of travel

Gravity

Sun's path

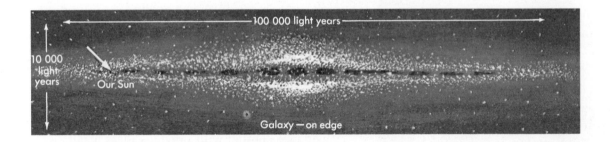

100 000 light years

10 000 light years

Our Sun

Galaxy — on edge

galaxy. **Our galaxy is called the** *Milky Way Galaxy*. When sky watchers look at the Milky Way in the sky, they are looking along the long axis of the lens-shaped Milky Way Galaxy. They can see many more stars in this direction.

The lower diagram above shows the estimated size of our galaxy in light years. A *light year* is the distance light travels in one year, about 9½ trillion kilometers. Note the position of our sun near one edge of the galaxy.

The Sun's Path. All the stars within our galaxy, including our sun, are in motion. All the stars of the galaxy are believed to be traveling in orbits within the galaxy. Our sun is believed to travel around the galaxy once every 200 million years.

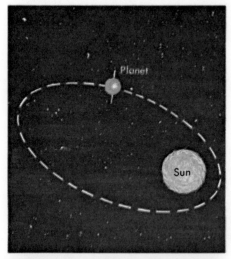

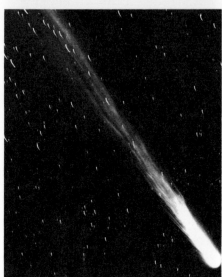

REVIEW QUESTIONS

1. How did Ptolemy's model of the solar system differ from Copernicus's model?
2. What is Kepler's Second Law?
3. Where is perihelion for the orbit shown here? Where is aphelion?
4. What force acts on the planet causing it to travel in an orbit?
5. Would the planet above be accelerating or decelerating? How do you know?
6. At what time of year is the earth at perihelion?
7. What is the object in the photograph? Of what material is the object composed? What is the shape of this object's orbit?
8. What are minor planets or planetoids? How do their orbits differ from the orbits of major planets?
9. When will Halley's comet next be visible on earth?
10. How do meteors, meteorites, and meteor showers differ?
11. Describe the sun's orbit.

THOUGHT QUESTIONS

1. Why didn't Ptolemy's theory include Uranus, Neptune, and Pluto?
2. In 1610, Galileo published his observations of the moon, using one of the first telescopes. He observed craters and mountains. How did this evidence conflict with previous ideas about objects in space?
3. When does an ellipse become a circle? When does it become a straight line?
4. How are meteors affecting the mass of the earth?
5. How would differences in the mass of the earth affect the earth's orbit?
6. In what ways is Ptolemy's model of the solar system similar to the sky model you investigated three chapters earlier?
7. Why is it winter when the earth is at perihelion?
8. Some people call planetoids asteroids. Why is asteroid a poor term for these objects?

Observing the Sky with Telescope, Camera, and Spectroscope

In 1609, the Italian scientist, Galileo Galilei, became the first person to observe sky objects with a telescope. He quickly learned that the moon has craters, Jupiter has moons, and Venus shines by reflected sunlight. As a result of his early success, other astronomers also made use of telescopes.

In 1882, a South African astronomer, Sir David Gill, became the first person to add a camera to a telescope. Not only did he get excellent photographs, but he also observed many more stars in the picture than could be seen through the telescope.

The invention of the spectroscope can be traced back to Sir Isaac Newton. In 1672, Newton first wrote that white light is really a mixture of light of different colors. By using a prism, he observed the different colors of the spectrum which make up a beam of white light. Two centuries later, the spectroscope, *an instrument for studying spectra, was invented.*

Today most new discoveries in astronomy result from observations made with spectroscopes, cameras, and telescopes.

Photographs of the central
region of Orion.
TOP, 2-minute exposure
MIDDLE, 30-minute exposure
BOTTOM, 60-minute exposure

Great Nebula

Color photograph of the constellation Orion

OBSERVATIONS WITH A CAMERA

When you look at the sky on a clear night, the faintest stars you can see are of 6th magnitude brightness. Light from 7th magnitude (fainter) stars which passes through the lens of your eye also focuses on the retina of the eye. But the image which forms is too weak to cause the retina to send a signal to the brain. As a result the brain receives no signal.

Light from a very dim source must affect the eye for about a tenth of a second if the light is to be seen. The human eye is more sensitive than photographic film. For example, a light which the eye can just barely see would not be recorded by a camera whose shutter is opened for a tenth of a second. The camera, however, has one advantage. The camera can be left open for hours. When the film is developed, the resulting picture of the light source will appear much brighter than when the eye looked at it.

Effect of Time on Star Photographs. The three photographs on page 304 show the same part of the sky. The top photograph shows the stars in the constellation Orion as you would see them on a clear night. What additional sky objects can you see in the center photograph? What objects are visible only in the bottom photograph?

OBSERVATIONS WITH SPECTROSCOPES

Suppose someone hands you an object and asks you to describe it. Your brain gets information as you taste it, shake it, weigh it, measure it, and so on. However, when astronomers try to describe stars, they can only look at them. The only information astronomers have about stars is the energy, such as light and radio waves, coming from the stars.

In spite of this limited information, astronomers have learned much about objects in the sky. They use spectroscopes to break up the incoming starlight and analyze the different shades and amounts of color coming from a star. The pattern of colors which the spectroscope produces when light is separated into its colors is called a spectrum. A spectrum produced from white light is shown above.

Temperature and Spectra. Scientists who study light tell us that all objects give off energy. Hot objects radiate a different level of energy than cold objects. The three spectra below are made by metals at three different temperatures. Each piece of

White hot spectrum

Red hot spectrum

Infrared spectrum

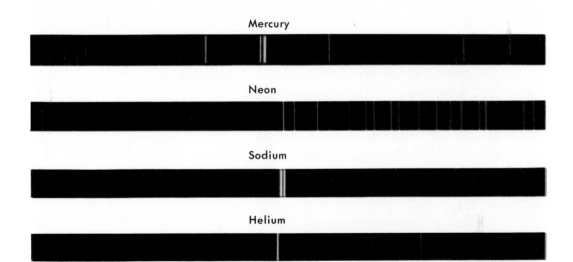

Mercury

Neon

Sodium

Helium

Spectra of some gases

metal is radiating a slightly different level of energy. The light bulb opposite has a tiny wire heated to a very high temperature. The metal becomes white hot. It gives off white light. The top spectrum on page 306 shows colors seen when white light passes through a spectroscope.

The girl in the photograph has heated a nail red hot. It is at a lower temperature than the white hot wire in the light bulb. How does its spectrum differ from that of the white hot wire?

The boy holds a bar which is not as hot as the one in the flame. He can feel the heat energy given off, but he cannot see it with his eyes. Experts on light explain that this rod is giving off energy which is beyond the red that our eyes can see. Such energy is called *infrared* or heat energy. Note its spectrum. How does the temperature of a solid metal affect the spectrum it produces? Predict the spectrum of a metal hotter than white hot.

Bright Line Spectra. Not all light is produced by heating solids such as iron, or the tungsten used in light bulbs. Light can also be produced by glowing gases. You have seen electric signs filled with neon gas, giving off a reddish light.

Some bright street lights appear bluish white. These lamps contain glowing mercury vapor. Some streets are lit with yellow lighting. These lamps contain glowing sodium vapor. How

Continuous spectrum of sunlight

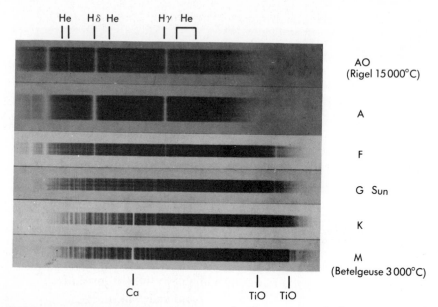

He   Hδ He   Hγ He

AO
(Rigel 15000°C)

A

F

G Sun

K

M
(Betelgeuse 3000°C)

Ca   TiO TiO

do the spectra of glowing gases differ from the spectra of glowing solids? Which spectra would you call bright line? Continuous?

Solar Spectra. The sun's continuous spectrum is shown on page 307. Does it suggest that the sun is a solid or a gas? Explain your answer.

Hot and Cool Stars. The star trail photograph at the left shows the winter constellation Orion. Are all the stars the same color? What does this suggest about the temperature of these stars? Which star do you think is cooler than most? Hotter?

The spectra above were made from stars of different colors. Blue-white stars are at the top. Red stars are at the bottom. For example, the star spectrum labeled *AO* shows the spectrum of Rigel, a bright blue-white star in Orion. Its surface temperature is about 15 000°C. The spectrum labeled *M* is from the red star in Orion called Betelgeuse. Its surface temperature is about 3000°C. How do these star spectra differ? Which have more lines, those from hot stars or cool stars? Which photograph shows our sun's spectrum?

Giant and Dwarf Stars. Scientists know that if a solid such as a piece of wire is heated to a high temperature, it will give off light. If a larger piece is heated to the same temperature, the spectrum will be the same. More light will be given off because of the increased area giving off light.

This same idea can be applied to stars. Astronomers use the idea of parallax to estimate how far away the closer stars are. When they take into account star distances, they find that most stars of the same temperature are about the same size. Thus, they give off the same amount of light. There are some important exceptions. A good many red stars give off hundreds and thousands of times more light than other red stars. What must be true of the size of these stars?

Some blue-white stars give off only ¹/₁₀₀ as much light as other blue-white stars. What must be true of the size of these stars?

The Doppler Effect. The drawing at the right shows an imaginary bird adding energy to a pond. The bird is hovering in one spot and dropping rocks into the pond at regular intervals. Describe the movement of the energy waves.

Suppose the bird always drops stones at the rate of one every two seconds. At what rate will the waves reach the person on shore?

In the next drawing the bird is flying slowly toward the person on shore. It is dropping stones at the same rate as before. How does the pattern of waves differ? Why? Will the waves still reach the observer at the same rate? Explain. How would the waves arrive for an observer on the opposite shore? This change in rate is called the *Doppler effect*.

Now imagine you are the observer. You know this bird always drops stones at the rate of one every two seconds. How can you tell if the bird is flying toward or away from you by observing the rate at which the waves reach the shore?

Doppler Effect and Spectra. Scientists believe that light has wavelike properties. Each region of the light spectrum consists of waves of a slightly different length. Red light has longer wavelengths than blue light. When spectral lines of sky objects are shifted toward the red end of the spectrum, the arriving waves are longer than expected. Astronomers conclude that the light source is moving away from us. If the spectral lines are shifted toward the violet, the waves are shorter than expected, and the light source is moving toward us.

Study the Saturn spectrum. What do the shifts in the spectrum suggest? How is Saturn moving?

Astronomers find that nearly all star spectra are shifted toward the red. This is called the *red shift*. Thus, light waves seem to be arriving at a slower rate than expected. What does this suggest about the motion of other stars relative to us?

Spectra of Saturn (B) and its rings (A and C) with a reference spectrum.

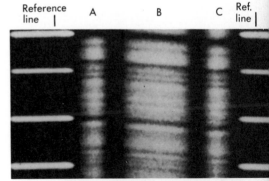

June 1959   May 1972

HAPPENINGS IN OUR GALAXY

When you look at the clear night sky, you see hundreds of stars, some bright and some dim. Thus, if most people are asked to describe our galaxy, they picture millions of stars of different brightness, but all of them unchanging. This is not accurate. Astronomers know that the sky has many different objects which are constantly changing.

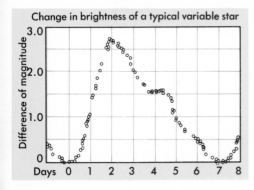

Change in brightness of a typical variable star

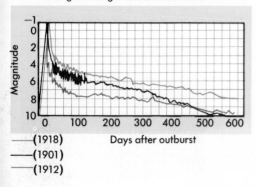

Changes in brightness of three novas

——— (1918)
——— (1901)
——— (1912)

Days after outburst

Variable Stars. The two photographs above show the same part of the sky on different nights. What is different about the star at which the arrows point? The graph at the left shows how one star changes in brightness. Astronomers have classified such *variable stars* into several different groups. There are over 500 variable stars in our galaxy which belong to the same group as the star shown above. These variable stars increase and decrease in brightness by 2½ times about once a week.

Several thousand variable stars are known in our galaxy. Most are classified according to how rapidly they change in brightness, and how great the change is. Within each group, the stars usually have similar spectra, and give off the same amount of light. How can this be used to determine the distance to such a star? How can this be used to determine the distance to other galaxies?

Novas and Supernovas: The Spectacular Variables. In 1572, a Danish astronomer observed a new star never seen before which suddenly increased so much in brightness that it was as bright as Venus. This star could even be seen in daytime. Such stars which suddenly increase in brilliance by over a hundred thousand times are called *supernovas*. They are rare events occurring only about once a century.

310

1908

1915

1920

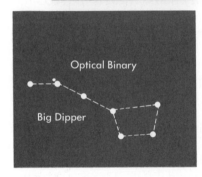

Optical Binary

Big Dipper

About once each decade astronomers observe a star increasing in brightness by ten thousand times. These stars are called *novas*. The changes in brightness of these novas are shown on the graph on the opposite page.

Astronomers believe that star brightness depends on the star's temperature, size, and distance from us. Discuss in class how each of these factors acting separately might result in a nova or supernova. Which possibility seems most likely?

Binary Stars. The Big Dipper, shown below, can be recognized by most students. Have you ever noticed that the middle star in the handle has a companion star next to it? These two stars are really not very close to each other. One star is much closer to us than the other. Such double stars are called *optical double stars*, because our eye sees them as close together, even though they are not.

The illustration at the left also shows a view of the brighter star of the optical double star in the Big Dipper. Such double stars are called *telescopic binaries*. These stars have been observed since 1650. They slowly travel around each other. To be more exact, they both revolve around a common center of mass, much like the earth and moon do. Note the photographs above taken over a period of 12 years. What is happening?

Note the two spectra below. These spectra were made from the light of the brighter telescopic binary in the drawing, called Mizar. The upper spectrum shows Mizar's typical starlight. The lower spectrum shows pairs of lines, each consisting of a line shifted to the red and a line shifted to the violet. Astronomers say Mizar is really two stars moving around a common center. At times, one star is moving toward us and one star is moving away from us. Such stars are called *spectroscopic binaries*. Thousands of telescopic and spectroscopic binaries have been observed and catalogued. Why do these stars produce spectra with single lines at certain times?

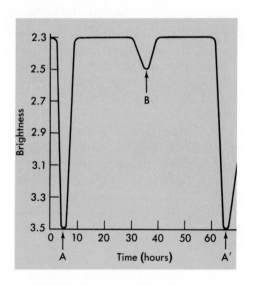

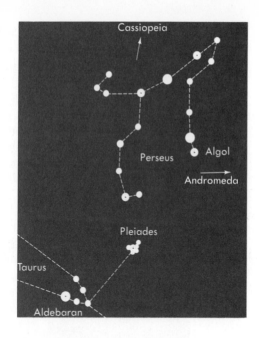

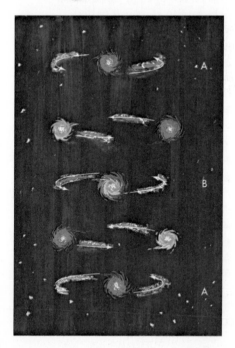

Eclipsing Binary Stars. Algol is the second brightest star in the constellation Perseus. The graph above at the left shows the changes in Algol's brightness during a 60-hour period. Because Algol changes in brightness, it is classified as a variable star. Because its spectrum sometimes contains pairs of lines, it is believed to be a binary star. The pairs of lines, however, are not identical. One set of lines looks like those found in the spectrum of a red star. The other set of lines looks like those found in the spectrum of a much hotter star. Propose an explanation for the two different types of spectra.

The spectral lines of Algol are not always equally bright. For example, at the time labeled *A* on the graph, only one spectrum can be seen. This is the spectrum of the red (cooler) star. At the time labeled *B*, the spectrum of the red star is much fainter. The explanations for these observations are shown in the diagrams at the left. At time *A* in the graph above, the fainter red star is passing in front of the hotter, brighter star. As a result, the amount of light coming from this binary star is much less. At time *B* on the graph, the hotter, brighter star is passing in front of the cooler, red star. Suggest a reason for showing the hotter star as smaller than the red star.

Star Clusters. The photograph on the next page was taken of a faintly visible star in the

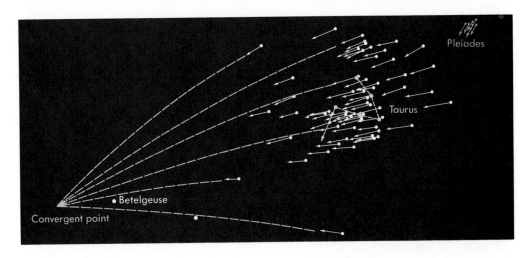

Pleiades

Taurus

• Betelgeuse

Convergent point

constellation Hercules. The photograph reveals that this "star" is really thousands of stars. The star cluster easiest to see in the winter sky is the group called the Pleiades (or Seven Sisters). A telescope or binoculars will reveal that there are hundreds of fainter stars grouped around the brighter stars which you can see.

The diagram above shows the direction in which the stars in the Pleiades are moving. The length of the arrows shows how far the stars will move in 60 000 years. The diagram also shows the direction of movement of another star cluster in the V-shaped constellation Taurus. Estimate when this cluster of stars will arrive at the point in the sky just to the left of the star Betelgeuse.

Nebulae: Gas and Dust Clouds. Certain regions of our galaxy contain vast amounts of dust or clouds of gas. These regions are called *nebulae*. Many nebulae are as large as constellations, but they are too faint to be seen. Cameras using long time exposures make these nebulae visible to us. The four photographs of nebulae on page 315 were made with the aid of telescopes and long time exposures.

One way to recognize gas in the galaxy is to photograph a region first with blue sensitive film and then with red sensitive film. If gas is present in the galaxy, it will scatter the blue light so that less light reaches the blue sensitive film than the red sensitive one. It is this idea which explains why the clear sky appears blue to us. The blue part of the sun's light is scattered from all regions of the sky to our eyes.

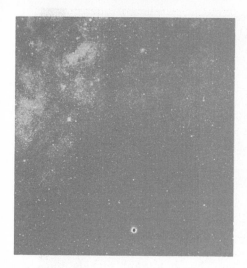

Blue filter

Red filter

The two photographs at the left show the Milky Way taken along the plane of our galaxy. The upper film was sensitive only to blue light. Why are there fewer stars in this picture? This region of the sky not only has the greatest number of stars, but also the greatest amount of gas and dust.

The photographs on the next page show several types of nebulae. At the upper left is the Great Nebula in Orion. It can be seen in the sky as the middle "star" in the sword hanging from Orion's belt. As you can see from the photograph, it does not look like a star when seen through a telescope. Longer time exposures show that this dust cloud extends over a region larger than the whole constellation. This nebula shines by reflecting the light from nearby stars.

If no stars are nearby, then such dust clouds tend to absorb the light from stars behind them. Under these conditions, the nebula appears dark, as in the Horsehead Nebula shown at the lower right on the next page.

The nebula at the lower left is called a *planetary nebula.* The light from such nebulae is not just reflected light. These nebulae have small, but very hot stars located at their centers. These stars are so hot that most of the light they give off is ultraviolet light. Our eyes cannot see ultraviolet light. This light is absorbed by the atoms in the nebula which then give off the energy as visible light. This process is called *fluorescence.*

The upper right photograph shows another type of nebula. Suggest how it probably gives off light.

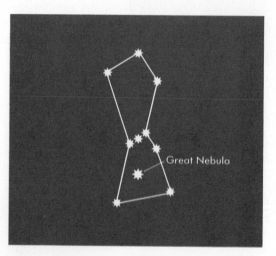

Great Nebula

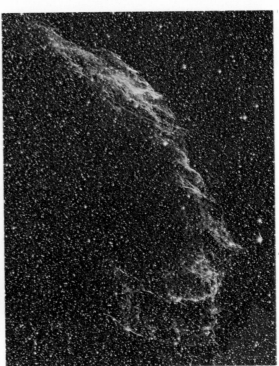

Great Nebula of Orion

Veil Nebula

The Ring Nebula, a planetary nebula

Horsehead Nebula

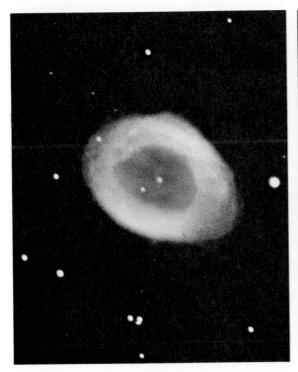

Clouds of
Magellan

BEYOND OUR GALAXY

Astronomers are constantly making better telescopes and using more sensitive film. They use aids not only to see objects further away, but also to see nearby objects more clearly. With each improvement, astronomers find more and more stars and nebulae like the ones closer to us. They also discover new types of sky objects further away. As they learn more about these new objects, astronomers will better understand how stars such as our sun begin, grow old, and die.

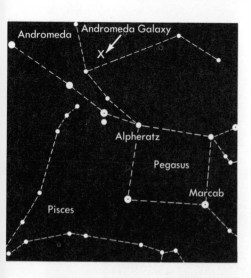

The Clouds of Magellan. When the great explorer, Magellan, sailed to the Southern Hemisphere, he observed two faint clouds of light in the southern night sky. These Clouds of Magellan are shown in the photograph above.

In the 20th century, large telescopes have shown that these clouds are more than nebulae. Today's telescopes show that the clouds consist of all the different kinds of stars found in our galaxy, plus nebulae and great amounts of gas. These clouds are our closest neighbors, beyond our own galaxy. The diagram on page 317 shows the distance to scale of the Clouds of Magellan in space. Estimate the distance from the center of our galaxy to the Clouds of Magellan.

Galaxies Beyond Our Galaxy. Nearly everything we see in the night sky is a part of our galaxy. There are two objects in the night sky which can be

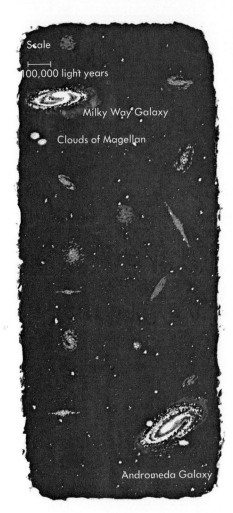

Great Spiral Galaxy

seen without aid and which are not part of our galaxy. The closest of these, the Clouds of Magellan, can only be seen in the Southern Hemisphere. In the Northern Hemisphere, a very faint haze is visible in the constellation Andromeda. This haze was thought to be a nebula when observed through earlier telescopes. The photograph above, taken with a very large telescope, shows that this object is a galaxy, much like our own Milky Way Galaxy. It is called the Andromeda Galaxy (or the Great Spiral Galaxy).

This galaxy is in an ideal position for viewing from the United States. The galaxy passes nearly directly overhead close to the Great Square of Pegasus, as shown on page 316. On very clear nights, it appears as a faint haze to the eye, or when viewed through binoculars or a low power telescope. This faint haze is the furthest point in the universe you can see without a telescope or other aid. Estimate its distance using the scale drawing above.

Sa

Sb

Sc

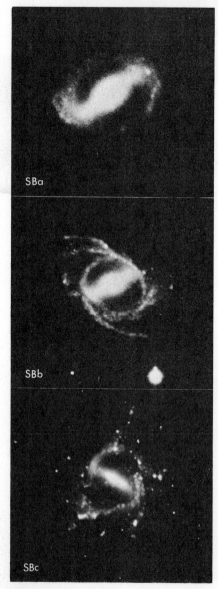

SBa

SBb

SBc

Classifying Galaxies. Today, astronomers have identified over 10 000 different galaxies on their photographs. These galaxies have been classified into four major groups. About 60% of the known galaxies are classified as *spiral galaxies*. These spirals are further classified as Sa, Sb, or Sc, as shown in the photographs above. What does the a, b, or c indicate about the spiral?

About 18% of the galaxies are classified as *barred spirals*. These barred spirals are further classified as either SBa, SBb, or SBc, as shown in the photographs at the left. What do these letters indicate?

Another 18% of the galaxies are classified as *elliptical galaxies*. These galaxies are further divided into eight subgroups labeled E0 through E7. E0, E5, and E7 are shown in the photographs below. How do galaxies in each subgroup differ?

The remaining 4% of the galaxies are *irregular galaxies* of the type pictured at the top of the next page. These galaxies are sometimes classified by the symbol (Ir). You have previously seen photographs of the Clouds of Magellan and the Andromeda Galaxy. Classify each of these galaxies using the system above.

The Local Group. As astronomers study

E 0

E 5

LOCAL GROUP MEMBERS

Name	Type	Distance (light years)
Milky Way	Sb	—
2 Clouds of Magellan	Ir	180 000
5 elliptical	E's	200–800 000
NGC 6822	Ir	950 000
3 ellipticals	E's	1 100 000
IC 1613	Ir	1 500 000
Andromeda Galaxy	Sb	1 500 000
M32 and NGC 205	E's	1 500 000
M33, IC10, 342	Sc's	1 800 000

An irregular galaxy

galaxies, they learn that galaxies are not uniformly distributed throughout space. Instead, galaxies seem to cluster together in groups. There are about twenty galaxies in the cluster which includes our galaxy. This grouping is called the *Local Group*. The chart above lists 18 other known galaxies in the Local Group. Note the galaxies labeled M32 and NGC 205. These galaxies are companions of the Andromeda Galaxy and can be seen in photographs of the Andromeda Galaxy earlier in this chapter. What two companions does our galaxy have?

The Local Group has a shape similar to the shape of our galaxy. Our galaxy and the Andromeda Galaxy are on opposite sides of the Local Group, as shown at the right.

Beyond the Local Group. Our largest telescopes and best film cannot show us the most distant stars. Instead, we have photographs which show galaxies each of which undoubtedly contains millions of stars. The photograph below at the right shows a region of the sky in the constellation Corona Borealis. How many galaxies can you count? This cluster of galaxies is believed to be about 700 million light years from us. If someone could see our planet from that distant galaxy, what would be their view of life on our earth?

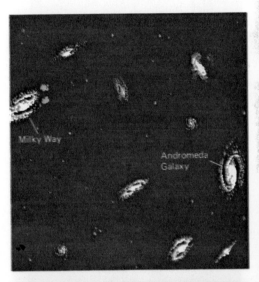

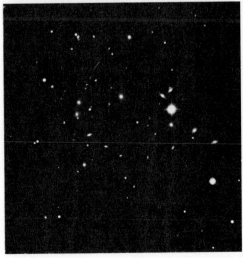

319

Quasars: A New Focus for Astronomy. In 1932, an American radio engineer noticed that there was always radio "noise" on a particular wavelength of the shortwave radio band. Then the observer noticed that at the same time each day the noise got worse. By changing the direction of the antenna, the observer concluded that the "noise" was radio waves coming from space. With this discovery, the field of radio astronomy was born. The top photograph shows modern radio telescopes for focusing radio waves from space.

Radio telescopes now scan the skies, mapping regions which give off intense radio signals. In the 1960's, radio astronomers located points in the sky emitting powerful radio signals. When astronomers looked at these regions, they could detect only very faint spots of light. These objects were named quasi-steller radio sources, or *quasars*.

Spectroscopes indicate that quasars give off light shifted so far to the red that the quasars must be moving away from us at 90% of the speed of light! This huge red shift also means these objects must be further away than anything else we have seen. The light from these objects must have left the quasar when the universe was just beginning. A great deal more remains to be learned about these baffling objects.

REVIEW QUESTIONS

1. How can a camera see objects in the night sky which your eyes cannot see?
2. What is the Doppler effect?
3. How do the spectra from hot solids and glowing gases differ?
4. What is a supernova? How often is one seen?
5. What is a spectroscopic binary?
6. In what ways are the stars in star clusters similar?
7. How do astronomers see dust in our galaxy?
8. By what three different ways are the nebulae at the left made visible to us?

THOUGHT QUESTIONS

1. Why were many of today's galaxies once thought to be nebulae?
2. Why might the study of a star 5 billion light years away tell us about how our sun began?

THE ATMOSPHERE

The Ocean of Air

The air that surrounds us stretches upward many kilometers, becoming thinner and thinner until it blends with the space between planets. The height of the atmosphere is sometimes given as 50 to 60 kilometers; often it is given as 500 to 600 kilometers. However, there is no definite upper boundary to the atmosphere, and there can be no exact figures to accurately describe its thickness.

The atmosphere is sometimes referred to as an ocean of air, because it is similar to the salt water ocean. There are currents in the atmosphere as there are currents in the ocean. While the atmosphere is a mixture of water vapor, other gases, and solid particles, the ocean is also a mixture of water, gases, salts, and solid particles. In addition, you and I, like the animals in the ocean, must live in and breathe these mixtures.

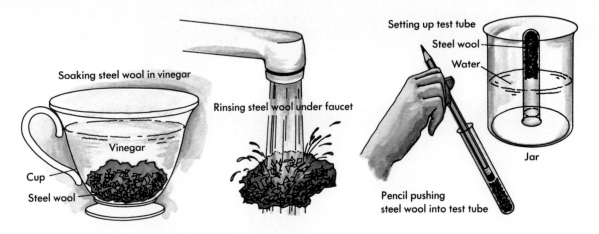

Soaking steel wool in vinegar

Cup

Vinegar

Steel wool

Rinsing steel wool under faucet

Setting up test tube

Steel wool

Water

Jar

Pencil pushing
steel wool into test tube

GASES IN THE AIR

You know the air must be composed of several gases, because you know that your body changes the air you breathe in. Your breath and the breath of all breathing animals adds moisture and carbon dioxide to the air. However, carbon dioxide and water vapor are only a part of the mixture of gases we call air. Oxygen and nitrogen make up 99 percent of the volume of dry air. Carbon dioxide and several other gases are present as about one percent of the total volume. The amount of water vapor can vary from 0 to 4 percent.

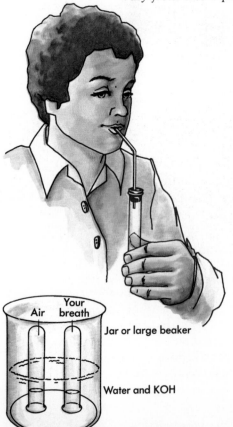

Air

Your
breath

Jar or large beaker

Water and KOH

Separating Gases from the Air. One way to find out what gases are in the air is to find a way to remove one of the gases. Soak a piece of steel wool in vinegar for 10 minutes. Rinse the vinegar-soaked steel wool in clear water, and push it all the way to the end of a test tube. Place the test tube with the open end down into a jar of water. Mark on the jar the level of the water inside the test tube. What happens to the water level inside the test tube after several hours? Overnight? What happened to the steel wool? Propose an explanation for what happened.

Measure the change in the water level, and calculate how much of the air reacted with the steel wool. Chemists tell us that the only gas in air which reacts with steel wool and forms rust is oxygen. From your experiment, how much of the air is oxygen?

Oxygen of the atmosphere reacts with many elements in such common processes as burning, rusting, and respiration. Make a list of other processes in which oxygen plays a part. Name the elements or compounds involved and the products formed.

Keep potassium hydroxide away from your face and eyes. The diluted solution will not hurt your skin, but wash your hands well after the experiment is set up.

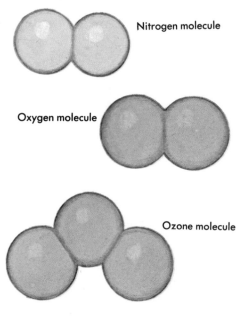

Nitrogen molecule

Oxygen molecule

Ozone molecule

Composition of air

78% Nitrogen

21% Oxygen · 1% Other gases

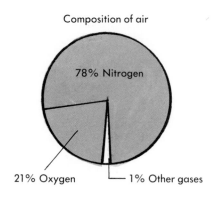

To test for another gas in the air, add two tablespoonsful of concentrated potassium hydroxide solution to 250 ml of water in a jar. Put a clean, dry test tube with the open end down into the jar. For comparison, fill another test tube with your breath. Do this by inserting into a clean test tube a two-hole stopper with a glass tube in one hole. Blow through the glass tube to replace the air inside with your breath. Quickly remove the stopper, and put the test tube with the open end down into the jar, as shown on page 324. Mark the level of the solution inside each test tube on the outside of the jar. What has happened to the level of the solution in the test tubes after several hours? How can you explain the change in the levels? Which test tube had the greater change in the level of the solution? Why? Chemists tell us that the only gas in the air which reacts with potassium hydroxide is carbon dioxide. Which test tube had more carbon dioxide?

Ozone. A peculiar odor often noticed near electrical equipment is caused by the gas ozone. Ozone is a special form of oxygen. Each molecule has three atoms of oxygen instead of the usual two.

Ozone is even more active chemically than oxygen. It is used to bleach cloth, to kill bacteria, and to remove odors. It is also unstable, breaking up readily to produce oxygen gas.

Electric sparks can change oxygen gas to ozone. This is the method used in ozone manufacturing plants.

Ozone is rare in the low levels of the atmosphere. At high levels, it is an important part of the atmosphere because it absorbs those rays from the sun (ultraviolet) which are harmful to many living things.

Nitrogen. Nitrogen makes up about four-fifths of the earth's atmosphere. The gas remaining in a test tube of air after oxygen has been removed by rusting is nearly pure nitrogen.

Nitrogen gas is not very active chemically, and plays little part in most chemical processes. Nevertheless, nitrogen is very important and is found in many compounds, including all proteins.

Nitrogen gas combines with oxygen in lightning bolts, producing nitric oxide which unites with rain and falls to earth. The nitrogen compounds produced by this process dissolve in soil water and are used by plants.

325

Small amounts of nitrogen are removed from the atmosphere by some of the microorganisms which live in the soil. These microorganisms produce nitrogen compounds such as proteins. When plant and animal proteins decay, other nitrogen compounds form. Ammonia, a compound of nitrogen and hydrogen, is one of the odors easily recognized around manure piles. Ammonia gas enters the atmosphere from decaying matter, but much of the gas dissolves in rainwater and washes into the soil.

Other Gases. The atmosphere contains only a small amount of *carbon dioxide*, much less than 1%. Nevertheless, this tiny amount is very important to the living things of the world. It is used by green plants to manufacture their food supply.

Almost 1% of the atmosphere is made up of *argon*. The atmosphere also contains tiny amounts of *neon*, *helium*, and other rare gases. Argon, neon, and helium are not chemically active. For this reason, argon is used in incandescent electric lamps so that the filaments cannot oxidize at high temperatures.

Electrical Excitation. Gases exposed to high voltages change electrical energy to light and produce a glow. Each type of gas glows with its own special color. The red-orange glow of neon is probably the most familiar, because this gas is used in so many advertising signs.

Argon glows with a deep violet light which contains much ultraviolet radiation. Special argon glow lamps are used as a source of ultraviolet light for studying fluorescent minerals.

Nitrogen glows with a light blue color. Electric sparks in air show this color. The color of lightning is due chiefly to nitrogen.

Sunset in dusty atmosphere   Sky color in clean atmosphere

DUST IN THE ATMOSPHERE

The atmosphere contains an astonishing amount of dust. For example, a cubic centimeter of classroom air may contain 3 million bits of solid material. The total weight of all this dust is small, but because of their great numbers, the particles play an important part in many earth processes. Our world would be a much different place without dust.

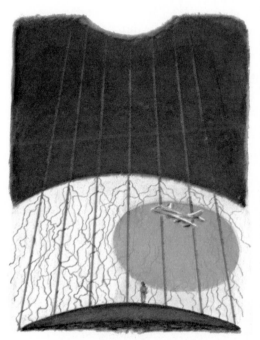

Dust and Sky Color. The picture at the beginning of this chapter shows an almost black sky overhead. The familiar blue appears only below the level of the airplane. This is because the airplane is above the part of the atmosphere that contains dust.

Very fine dust scatters light. For example, sunbeams can be seen when sunlight is reflected from dust particles in the air. Blue light is scattered more than red light. Thus, the atmosphere separates colors of light passing through it.

The daytime sky appears blue because the blue part of sunlight reaches our eyes after having been scattered by fine particles in the atmosphere. Red and orange light, on the other hand, pass directly through the atmosphere with little scattering. The sun seems to have a red-orange color especially at sunrise and sunset. This happens because more red-orange light is scattered when light travels a longer path through the atmosphere.

Particles in Air. To find out how many particles are floating in the air, make an air sampler by folding a piece of cardboard 10 cm square in half. Make two surfaces for catching particles by cutting out 2-cm square pieces of wire screening. Wind double-sided masking tape around each

327

Sampler for collecting particles

Wire screening with double-sided tape wrapped around it

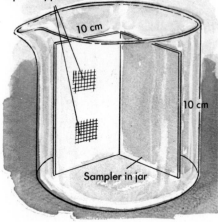

10 cm

10 cm

Sampler in jar

TABLE PARTICLE SIZES
(in microns)

Virus	0.01–0.1
Bacteria	1–25
Fog	5–60
Dust	1–1000
Smoke	0.1–1
Rain	500–1000

1 micron = 1/1000 mm

square. (Be careful not to get dirt and fingerprints on the tape.) Stick the two squares to the cardboard. Fold the cardboard shut and staple. Place the sampler in an envelope until you are ready to use it.

Choose a location where you want to sample the air. Open the sampler, and place it in a large jar or can to protect the sampler from dust blowing up from the ground. After 24 hours, collect the sampler and examine the squares with a microscope to see what has fallen on it.

Count the number of particles in at least ten squares of the screening. Compute the average number of particles per square, and estimate the number of particles that fell on your 2-cm square sampler. Compare your results with those of your classmates. On a map of your town mark the places where samples were taken, and note how many particles were collected by each sampler in a 24-hour period. Explain why some areas have more particles in the air than others.

Other Pollutants. In addition to dust particles, the air also contains unwanted gases that cause damage to buildings, metals, and fabrics. Make up samplers as before, but this time cut two windows 2 cm on a side in the cardboard. Stretch a piece of nylon material across one, and aluminum foil across the other. Use tape to hold them in place. Put these samplers at various locations around town. After two weeks collect the samplers, and examine them under the microscope. What changes do you observe?

The Automobile. One major source of air pollution is the internal combustion engine which moves automobiles, trucks, buses, and airplanes. Collect some exhaust from an automobile by holding a large clear plastic bag near the exhaust pipe of a car while the motor is running. (Be careful not to touch the exhaust pipe as it is likely to be very hot.) When you have a bagful, tie the bag closed. Select two healthy plants.

Put the bag with the exhaust in it around one of the plants. Put another bag with clean air around the second plant. Observe the plants, recording whatever changes you see. Why is it necessary to put a bag with clean air around the second plant?

The internal combustion engine is considered by most to be the most serious polluter of our atmosphere. About 80% of the families in the

MOTOR VEHICLE REGISTRATIONS IN UNITED STATES
(in millions)

Type of vehicle	1950	1955	1960	1965	1970	1975	1980 (est)
Cars, taxis	40.3	52.1	61.7	75.2	84.4	107.1	120.0
Trucks, buses	8.8	10.5	12.7	15.1	18.1	26.2	31.8
Motorcycles	0.4	0.4	0.6	1.4	2.5	4.9	8.2

United States own cars. Many families have more than one. Study the chart of motor vehicle registrations in the years since 1950. Make a graph of the total number of registrations since then. Predict the number of registrations in 1985 and 2000 if the present growth rate continues. Which prediction is likely to be more accurate? What circumstances might upset your predictions?

Below is a table of the output of air pollutants from motor vehicles operating at 40 km/h. Determine how much your car pollutes the atmosphere in one week? How many kilometers per liter of gas does the car average? How many liters does the car use each week? How many grams of each pollutant does the car emit each week?

How many autos are there in your town? How much pollution do all of them push into the atmosphere? What effects do these pollutants have on humans, plants and other living things? How long do the pollutants stay in the atmosphere? What eventually happens to them?

AVERAGE OUTPUT OF POLLUTANTS FROM MOTOR VEHICLES
OPERATING AT ABOUT 40 KM/H

Source	Pollutant	Quantity of Output (grams/liter of fuel)
Gasoline engines	Carbon monoxide	284.9
	Nitrogen oxides	22.2
	Hydrocarbons	15.8
	Solid particles	1.3
	Sulfur oxides	1.0
Diesel engines	Hydrocarbons	21.6
	Solid particles	13.3
	Nitrogen oxides	12.0
	Carbon monoxide	7.1
	Sulfur oxides	4.8

Sources of Dust. The pictures on this page suggest some of the many sources of dust found in the atmosphere. Many metric tons of soil and sand may be blown from plowed fields and from deserts during a wind storm. A single volcanic explosion may blast thousands of metric tons of powdered rock into the air. Dust from a volcanic explosion may remain in the atmosphere for over a year.

The ocean also adds large quantities of salt particles to the atmosphere. Spray is whipped upward, water evaporates from the solution, and bits of salt remain to be carried away by the wind.

Fires of all kinds add particles in the form of soot. These particles may be pure carbon, but usually contain certain other chemicals as well.

Plants add countless billions of spores and pollen grains to the air's load of dust. Some of these cause great discomfort to people who suffer from hay fever.

Dust from Outer Space. We think the earth collects many metric tons of dust from outer space each day. Particles that pass near the earth are pulled to it and fall into the atmosphere. Where these particles come from is not known, but it is suspected that they may have been parts of comets.

Some particles fall fast enough to be heated white-hot by the friction between them and the air. These glowing particles are called *meteors*.

WATER IN THE ATMOSPHERE

The atmosphere holds an enormous amount of water in the form of vapor. The condensation of some of this vapor during a single summer storm may produce a rainfall of three centimeters over several thousand square kilometers. To gain an idea of this much water, measure off a square 30 meters on a side. Imagine that this is the base of a tank 30 meters high. Three centimeters of rainfall on a single square kilometer would be sufficient to fill this tank to overflowing.

Rate of Evaporation. Evaporation is a heat-absorbing process which tends to cool the surroundings. The more rapid the evaporation, the greater the cooling effect.

The rate of evaporation can be studied by noting the amount of cooling produced during the process. Two thermometers are usually used. One measures the temperature of the air; the other is covered with a wet cloth, and shows the drop in temperature caused by evaporation.

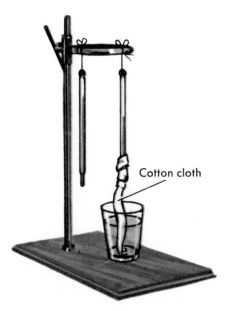

Cotton cloth

Make a wet-bulb thermometer by wrapping the bulb with one end of a strip of cotton cloth. Dip the other end of the cloth in water. Fan the thermometer before taking a reading.

Take wet-bulb and dry-bulb readings in several places, such as a warm room, a damp basement, a sunny field, and a woodland. What do the results tell you about the rate of evaporation in each place?

Evaporation and Climate. Evaporation affects climate strongly by cooling moist surfaces. Damp soil is usually cooler than dry soil. A grass-covered field is usually cooler than a bare field. The effect is

331

especially noticeable during hot weather when air near a lake is cooler than air a few km away.

Saturation. A mass of air may contain so much water vapor that evaporation seems to stop. For each molecule of water that enters the air, another is forced out and the total remains the same. Such air is said to be *saturated*.

Air within a closed jar which contains a little water is soon saturated. Air in holes in the ground, in forests, and near bodies of water is saturated or nearly saturated much of the time.

How do the readings of wet- and dry-bulb thermometers compare in saturated air? Why do we feel hotter in saturated air than in dry air?

Temperature and Saturation. Test the effect of temperature on saturation, using strips of cobalt chloride paper, as shown at the left. This paper turns pink as it absorbs water from moist air. When it gives off moisture to dry air, it turns blue again.

Breathe into a bottle a few times to moisten the air inside. Hang a strip of cobalt chloride paper in the bottle, closing it with a rubber stopper. Set the bottle in hot water for a few minutes, and note the color of the paper. Then set the bottle in cold water, and note the color of the paper. How does temperature affect the saturation of the air?

The Saturation Curve. The weight of water vapor in saturated air has been carefully measured for different temperatures. The results are given in the graph on the next page in grams of water per cubic meter of air.

How much water can each cubic meter of air in your classroom contain at its present temperature? Weigh out the amount of water.

Relative Humidity. Air is not always saturated. Often it contains less water vapor than it can hold.

Suppose the air in your classroom is at 25°C. The saturation curve shows that air at this temperature can hold 22 grams of water per cubic meter. If the air actually contains only 11 grams of water per cubic meter, it is 50% saturated.

Such air is often described as having a *relative humidity* of 50%. Completely saturated air has a relative humidity of 100%. What is the relative humidity of air that contains 20 grams of water vapor when it can hold 50 grams per cubic meter?

Hygrometers. Relative humidity is determined by instruments called *hygrometers*. The most

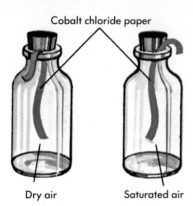

Cobalt chloride paper

Dry air   Saturated air

COBALT CHLORIDE PAPER. Add 70 grams of cobalt chloride to 100 ml of water. Dip thin strips of paper towels into the red solution. Place the wet strips on other pieces of paper toweling until they are dry. Then cut the small strips into small test pieces. Stopper the solution when it is not in use. To test for water, add drops of liquid to the dry blue test pieces. If the cobalt chloride strips turn red, the liquid contains water.

common type of hygrometer makes use of wet- and dry-bulb thermometers.

The two thermometers read the same when the air is completely saturated, because water evaporates and condenses at the same rate. Thus there is no water loss and no cooling. The difference increases as the air becomes drier and evaporation speeds up.

Values of relative humidity are calculated from the table below. Suppose that the dry-bulb thermometer reads 21°C, and the other reads 13°C. The difference between the two readings is 8°C. The table shows that the relative humidity is 39%. The air contains about two-fifths of the moisture it can hold at 21°C.

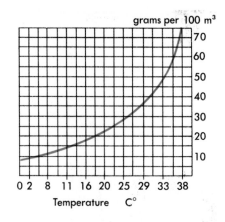

TABLE OF RELATIVE HUMIDITY IN PERCENT
Difference between Dry-Bulb and Wet-Bulb Thermometers, °C

	0.5	1.0	1.5	2.0	2.5	3.0	3.5	4.0	4.5	5.0	5.5	6.0	6.5	7.0	7.5	8.0	8.5	9.0	9.5	10.0
10	94	88	82	77	71	66	60	55	50	44	39	34	29	24	20	15	10	6	—	—
11	94	89	83	78	72	67	61	56	51	46	41	36	32	27	22	18	13	9	5	—
12	95	89	84	78	73	68	63	58	53	48	43	39	34	29	25	21	16	12	8	—
13	95	89	84	79	74	69	64	59	54	50	45	41	36	32	28	23	19	15	11	7
14	95	90	85	79	75	70	65	60	56	51	47	42	38	34	30	26	22	18	14	10
15	95	90	85	80	75	71	66	61	57	53	48	44	40	36	32	27	24	20	16	13
16	95	90	85	81	76	71	67	63	58	54	50	46	42	38	34	30	26	23	19	15
17	95	90	86	81	76	72	68	64	60	55	51	47	43	40	36	32	28	25	21	18
18	95	91	86	82	77	73	69	65	61	57	53	49	45	41	38	34	30	27	23	20
19	95	91	87	82	78	74	70	65	62	58	54	50	46	43	39	36	32	29	26	22
20	96	91	87	83	78	74	70	66	63	59	55	51	48	44	41	37	34	31	28	24
21	96	91	87	83	79	75	71	67	64	60	56	53	49	46	42	39	36	32	29	26
22	96	92	87	83	80	76	72	68	64	61	57	54	50	47	44	40	37	34	31	28
23	96	92	88	84	80	76	72	69	65	62	58	55	52	48	45	42	39	36	33	30
24	96	92	88	84	80	77	73	69	66	62	59	56	53	49	46	43	40	37	34	31
25	96	92	88	84	81	77	74	70	67	63	60	57	54	50	47	44	41	39	36	33
26	96	92	88	85	81	78	74	71	67	64	61	58	54	51	49	46	43	40	37	34
27	96	92	89	85	82	78	75	71	68	65	62	58	56	52	50	47	44	41	38	36
28	96	93	89	85	82	78	75	72	69	65	62	59	56	53	51	48	45	42	40	37
29	96	93	89	86	82	79	76	72	69	66	63	60	57	54	52	49	46	43	41	38
30	96	93	89	86	83	79	76	73	70	67	64	61	58	55	52	50	47	44	42	39
31	96	93	90	86	83	80	77	73	70	67	64	61	59	56	53	51	48	45	43	40
32	96	93	90	86	83	80	77	74	71	68	65	62	60	57	54	51	49	46	44	41
33	97	93	90	87	83	80	77	74	71	68	66	63	60	57	55	52	50	47	45	42
34	97	93	90	87	84	81	78	75	72	69	66	63	61	58	56	53	51	48	46	43
35	97	94	90	87	84	81	78	75	72	69	67	64	61	59	56	54	51	49	47	44

Dry-Bulb Temperature (°C)

Find the relative humidity of your classroom. Measure out the water which the air can hold and the water which it actually holds.

The Dew Point. The saturation curve shows that cool air can hold less water vapor than warm air. Therefore, if a mass of unsaturated air is cooled enough, it becomes saturated. If cooled still further, moisture may condense out as dew or fog. The temperature at which condensation begins is called the *dew point*.

Suppose that a mass of air at 30°C contains 15 grams of water vapor per cubic meter. Refer back to the saturation curve on page 333 to find out whether the air is saturated or unsaturated. To what temperature can the air be cooled before moisture begins to condense? What is the dew point of the air?

Measuring Dew Point. Set a thermometer into a shiny can half full of water. Add small amounts of ice a few pieces at a time, stirring the mixture to cool the water evenly.

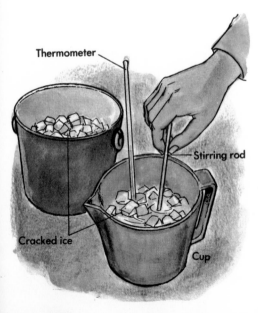

Thermometer

Stirring rod

Cracked ice

Cup

Watch the outside of the can, but do not breathe on it. When the metal appears misty, note the temperature of the water. This temperature is the dew point of the air around the can.

The dew point of very dry air may be below the freezing point. If dew does not form on the can when the water inside is 0°C, add salt to the ice-water mixture. The temperature of the mixture can be lowered several degrees in this way.

Find the dew point of the air in several places, such as the classroom, a damp basement, near a sunny wall, and in a hole in the ground.

Dew. A pitcher of ice water cools the surrounding air by conduction. If the air is chilled below its dew point, water vapor condenses on the outside of the pitcher as *dew*. Why do ice water pitchers collect more dew in summer than in winter?

Dew collects on other cold objects in the same manner—on bottles taken from a refrigerator, on windows during cold weather, and on cold water pipes. Discuss other situations in which you have seen dew form and the conditions that made condensation possible.

Dew forms outdoors on objects that are cool. Cooling is most rapid on clear nights. Roofs, automobiles, rocks, and leaves become cold, thus chilling the surrounding air. If the temperature of the air drops below the dew point, water condenses on the cool surfaces.

Frost. The dew point of the air may be below freezing. Then, if the air is chilled enough, water condenses out in the solid state rather than in the liquid state. The result is *frost*.

Frost is not frozen dew as many people believe. Frozen dew collects first as drops of water which later cool below the freezing point. Frost forms only below the freezing temperature.

Fill a bottle with hot water and let it stand for two minutes

Pour out most of the water

Water molecules leaving the air in the solid state collect in patterns, because forces between the molecules are stronger in some directions then in others. Thus, frost grows crystals, molecule by molecule, producing delicate designs of beauty.

A Fog in a Bottle. A *fog* is a cloud of tiny drops of water which has condensed on dust particles in the atmosphere. Produce a fog in a bottle by the method shown here. What is the source of the moisture in the air? Is the air in the bottle very dry or nearly saturated? What cools the air below its dew point?

Find out whether the presence of dust affects the condensation of moisture. Repeat the above experiment after holding a lighted splint in the bottle. What kind of dust is added by the flame? What effect does the dust have?

Radiation Fogs. The earth's surface radiates heat into space on clear nights. The overlying air is then cooled to the temperature of the earth by conduction and settles into hollows. A fog may form in this cool air. Such a fog is called a *radiation fog*.

Put an ice cube on top of the bottle

Add smoke and replace the ice

335

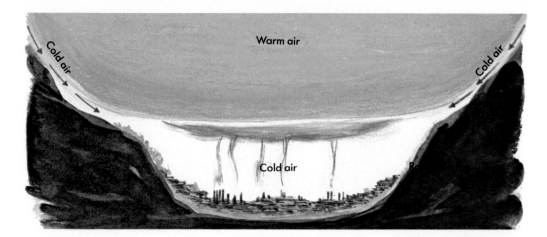

Study the diagram above. Explain why the cold air settles. Why does smoke tend to rise to a certain level and then spread out? Why are radiation fogs more common in early autumn than during other seasons? Why are there no radiation fogs on windy nights?

Advection Fogs. A mass of moist air may be chilled below its dew point when blown across a cold surface. The fog formed under these conditions is called an *advection fog*. Explain the cause of the fog on the top of the mountain in the picture to the left.

During winter, advection fogs are common when masses of moist air from warm bodies of water are blown across cold land surfaces. In summer, advection fogs are common when masses of moist air are blown across cold ocean currents, such as the Labrador Current.

Making a Cloud. A sudden increase in air pressure raises the temperature of the air. A sudden decrease in pressure lowers the temperature.

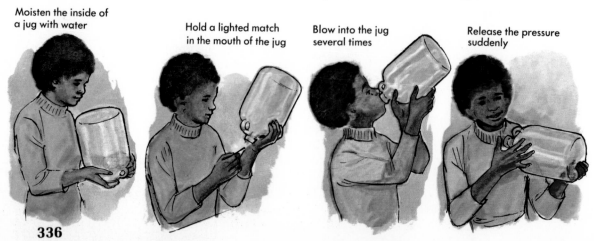

Moisten the inside of a jug with water

Hold a lighted match in the mouth of the jug

Blow into the jug several times

Release the pressure suddenly

Moisten the inside of a glass jug with water. Hold a lighted match in the jug to add dust particles. Then place the mouth of the jug against your lips and force air into the jug. Release the pressure suddenly, as shown on page 336.

Repeat the process of increasing and decreasing the pressure several times. Describe what happens as the pressure is increased and what happens as the pressure is decreased. Explain the effects in terms of changing temperature.

Table salt in metal can lid

Hygroscopic Dust. Some substances attract moisture more strongly than others. Such substances are called *hygroscopic* or "water-seeking." Microscopic crystals of salt from the surface of the oceans are an example of hygroscopic dust.

Clouds and fog may form on hygroscopic dust particles even when the humidity is less than 100%. Therefore, such particles are important in causing weather changes.

Test different kinds of dust in the jug used for the above experiment. Try several kinds of smoke, chalk dust, fine soil, and the like. Add salt particles by heating salt in a metal can lid as shown above. Be sure to change the air completely in the jug between each trial. Which kinds of dust produce the thickest clouds?

How a Cloud Is Formed. A cloud is formed in a rising mass of air. The air mass expands as it moves upward into regions of lower pressure, its pressure decreases, and its temperature drops.

If the temperature drops to the dew point of the air, moisture begins to condense on dust particles. The bottom of a cloud marks the level at which the dew point of the air is reached.

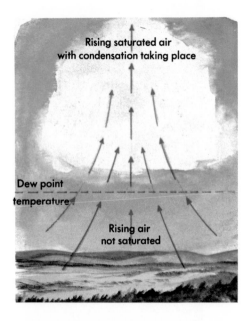

Rising saturated air with condensation taking place

Dew point temperature

Rising air not saturated

The air continues to rise and cool. Additional moisture condenses, and produces the upper portion of the cloud. Very moist air may produce a cloud thousands of feet thick.

Shapes of Clouds. The shape of a cloud depends chiefly upon the direction in which an air mass is moving. The picture at the right shows a cloud formed in a mass that is moving almost straight up. Such a cloud is called a *cumulus* cloud.

A warm air mass may slide upward over a colder mass much as it may slide up a slanting roof. Clouds formed in such a sliding mass are usually spread out over large areas. The higher clouds tend to be thinner than the lower clouds.

Cumulonimbus clouds

Rain. Rain is commonly explained as being formed when drops of water in a cloud grow large enough to fall. Such an explanation is greatly oversimplified; the process of rainmaking is but poorly understood.

Few scientists today believe that drops of water in a cloud grow larger by taking additional water vapor from the air. Nevertheless, none of the other theories has as yet been proven.

One popular theory suggests that raindrops begin as ice crystals which form in the tops of very tall clouds, like the one at the left. These crystals can then cool air if blown downward into moist regions, adding water by condensation until large enough to fall as snow or rain. There is much evidence that rain is sometimes produced in this way.

However, recent studies have shown that rain may fall from clouds that are warmer than the freezing point of water. New theories suggest that large drops may collect small drops by bumping into them. This process has not yet been observed in clouds.

Ice crystals   Water droplets

OVERSIMPLIFICATIONS. *Explanations are commonly oversimplified, either because of ignorance or because of a desire to prevent confusion. However, an oversimplified explanation is not scientifically honest. It may suggest that a certain problem was easily solved, or it may conceal a problem that is as yet unsolved. It is well not to accept completely any explanations for which supporting evidence is not presented.*

SUMMARY QUESTIONS

1. Why is the atmosphere sometimes referred to as the ocean of air?
2. How do you know there is oxygen in the air? About what percent of the air is oxygen?
3. Name other gases found in air.
4. What kinds of pollutants are you likely to find in the air in and around cities?
5. How has the internal combustion engine become our most serious air polluter? What is being done about it?

Mt. Everest 8 700 m
Pikes Peak 4 230 m
Sea level

DENSITY AND PRESSURE

Air has weight; in other words, it is pulled toward the center of the earth by gravity. Therefore, the air near the earth is under pressure because of the weight of the air above it.

Air is also elastic. The portion of the atmosphere near the earth's surface is squeezed together because it is under pressure; this air is more dense than air at high altitudes. If we could see particles of air, we would find them closer together at sea level than we would on a mountain top.

Measuring Density. Fill a football or basketball with air until the ball is hard. Weigh the ball as accurately as possible. Collect the air from the ball, as shown in the illustration at the left, and measure the volume of the air to the nearest liter. Then weigh the empty ball.

Calculate the weight of the air in the ball. Then calculate the weight of one liter of this air. From this figure, determine the weight of one milliliter. The density of air is expressed as the weight of one milliliter of air. What is the density of the air in the ball?

Compare the result with the density of air as given in an encyclopedia. Why are there differences between the two figures?

339

Egg

Burning match

Egg acts as valve

Effects of Air Pressure. Try the following activity to demonstrate the effects of differences in air pressure from one place to another. Use a milk bottle or a jar with a neck that is just too small to allow a hard-boiled egg to pass through. Drop a lighted match into the bottle, and place a peeled, hard-boiled egg in the mouth. How does the lighted match affect the air in the bottle? What happens when the air in the bottle cools? Where will the air pressure be greater, inside or outside of the bottle? Why does the egg slip into the bottle?

To remove the egg from the bottle, tip the bottle up, as in the illustration, and blow into the bottle past the egg. The egg acts as a valve allowing the air you blow to enter, but preventing air from escaping. Where will the air pressure increase? Why does the egg slip out?

Measuring Atmospheric Pressure. Atmospheric pressure is measured with a device called a *barometer*. One type of barometer consists of a glass tube, one meter long, filled with mercury. The tube is placed upside down with the open end in a dish of mercury, as shown here. The atmosphere pushes on the mercury. The greater the atmospheric pressure, the higher the level of the mercury in the tube will be. Why?

Atmospheric pressure cannot hold up a mercury column one meter high. Therefore, some of the mercury runs from the tube leaving a partial vacuum in the top of the tube. How is the height affected if a little air or water vapor is present in the top of the tube?

A pressure of 13.6 grams per square centimeter can hold up one centimeter of mercury. What is the pressure in your classroom?

340

Aneroid Barometers. Mercury barometers are used for accurate measurements, but they are not easily carried about. Therefore, barometers which contain no liquids have been invented. Such barometers are called *aneroid* barometers. A common form is pictured at the right together with a view of the inside.

An aneroid barometer contains a small can A from which as much air as possible has been removed. A strong spring E keeps the can from being crushed. An increase in atmospheric pressure squeezes the sides of the can together, moving the lever D and pulling on the chain C. The chain turns the axle B to which the hand of the barometer is attached.

Examine an aneroid barometer and find the parts described. Keep a record of daily changes in the barometer readings.

Elevation and Pressure. The device shown below is a very sensitive barometer that indicates tiny changes in atmospheric pressure. Set up the device as pictured, or use a thermos bottle in place of the bottle to reduce the effect of temperature changes. Explain why a change in atmospheric pressure makes the drop of liquid move.

What is the effect of lifting the barometer from the floor to the top of a step ladder? From the bottom of a stairway to the top?

Explain why the water drop moves, in terms of the air pressure inside and outside the barometer.

Pressures at High Altitudes. The decrease in atmospheric pressure shown by a sensitive barometer being carried up a flight of stairs continues at higher ascents. At 5400 meters, a mercury column in a barometer stands at about 38 centimeters, half the normal reading at sea level. At 32 kilometers, the atmosphere holds up about one cm of mercury. At 80 km, the weight of the overlying air is too small to affect an ordinary barometer.

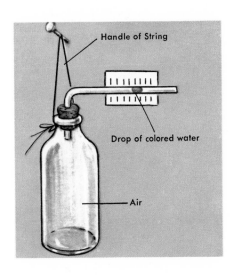

Handle of String

Drop of colored water

Air

The balloon shown on page 342 is only partly filled with gas, but it is ready to ascend. The gas in the balloon will expand as it rises to regions of lower pressure. Why? At 5400 meters, the volume of the gas will double; at 12 000 meters, the volume will be six times as great. What would happen if the balloon were full at the start?

Man at High Altitudes. Many people suffer from altitude sickness if they ascend several

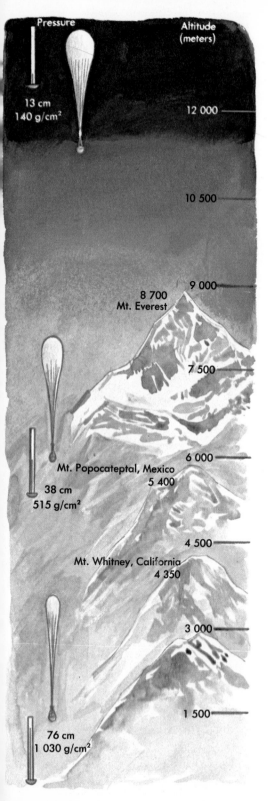

Pressure

13 cm
140 g/cm²

Altitude
(meters)

12 000

10 500

9 000

8 700
Mt. Everest

7 500

6 000

Mt. Popocateptal, Mexico
5 400

38 cm
515 g/cm²

4 500

Mt. Whitney, California
4 350

3 000

1 500

76 cm
1 030 g/cm²

thousand meters within a short time. The exact cause of the sickness is not known, but it is probably due to decreased atmospheric pressure. Recovery is usually rapid after a few days of adjustment to low pressure.

Above 6000 meters, people are affected by another condition — lack of oxygen. This problem will be discussed later in the chapter.

At the low pressures found above 9000 meters, gases dissolved in the blood come out of solution and form bubbles. These bubbles may cut off the flow of blood in the blood vessels, causing great pain, unconsciousness, and even death. Pressurized cabins or pressurized suits are needed to maintain suitable conditions around the body at such altitudes.

Altitude and Density. A person who stands on top of a 5400-meter peak is above one-half of the earth's atmosphere. Each cubic meter of the air at this height weighs half as much as at sea level. The person takes in only half as much air with each breath. An "empty" bottle contains about half as many molecules of air.

Airplanes are affected in two ways by the decreased density at high altitudes. They meet less air resistance and, therefore, forward motion is easier. On the other hand, there is less air to provide an upward force on the wings. They must go very fast to maintain altitude. Why do most commercial planes fly below 12 000 meters?

Rockets, however, do not depend upon the atmosphere for lift. High altitude travel is an advantage because air resistance is low.

The Oxygen Supply. At high altitudes, each cubic meter of space contains less air and,

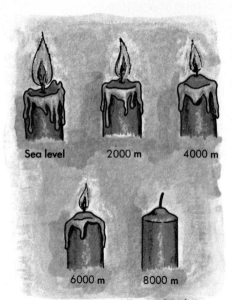

Burning candles at different altitudes

Sea level   2000 m   4000 m

6000 m   8000 m

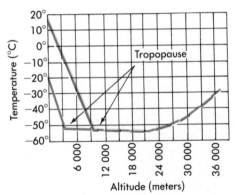

therefore, less oxygen. However, the percentage composition remains about the same.

The picture here shows candle flames at different altitudes. The rate at which oxygen is supplied to a flame decreases as the density of the air decreases. The chemical reaction slows down and less heat is produced. At 7500 meters, the reaction is too slow to maintain a flame. Why is life above 6000 meters difficult? Why were oxygen masks used during the conquest of Mt. Everest?

Temperatures in the Atmosphere. The graph at the left gives temperatures of the atmosphere at different altitudes. The red line shows atmospheric temperatures above a position where the ground level temperature is 15°C. The blue line shows temperatures above a position where the ground level temperature is −20°.

According to the graph, what is the lowest temperature a pilot can expect to find in the upper atmosphere? At what levels may he find this temperature? What is the average loss in temperature per 3000 meters of altitude along the red line? Along the blue line?

The Tropopause. Scientists divide the lower atmosphere into two layers, the troposphere and the stratosphere. The altitude at which the air temperature stops changing is the boundary between the two layers, called the tropopause.

The graph shows that the tropopause occurs at about 10 000 meters above the ground if the ground temperature is 15°C. What is the altitude of the tropopause if the ground temperature is colder?

Is the tropopause lower near the equator or near the poles? Is the tropopause at a lower altitude in the summer or in the winter at your location? Use the diagrams below to check your answers.

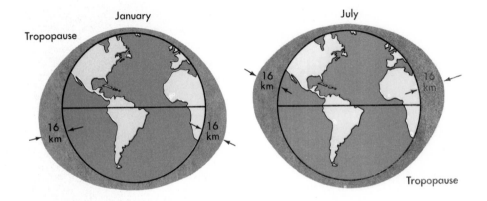

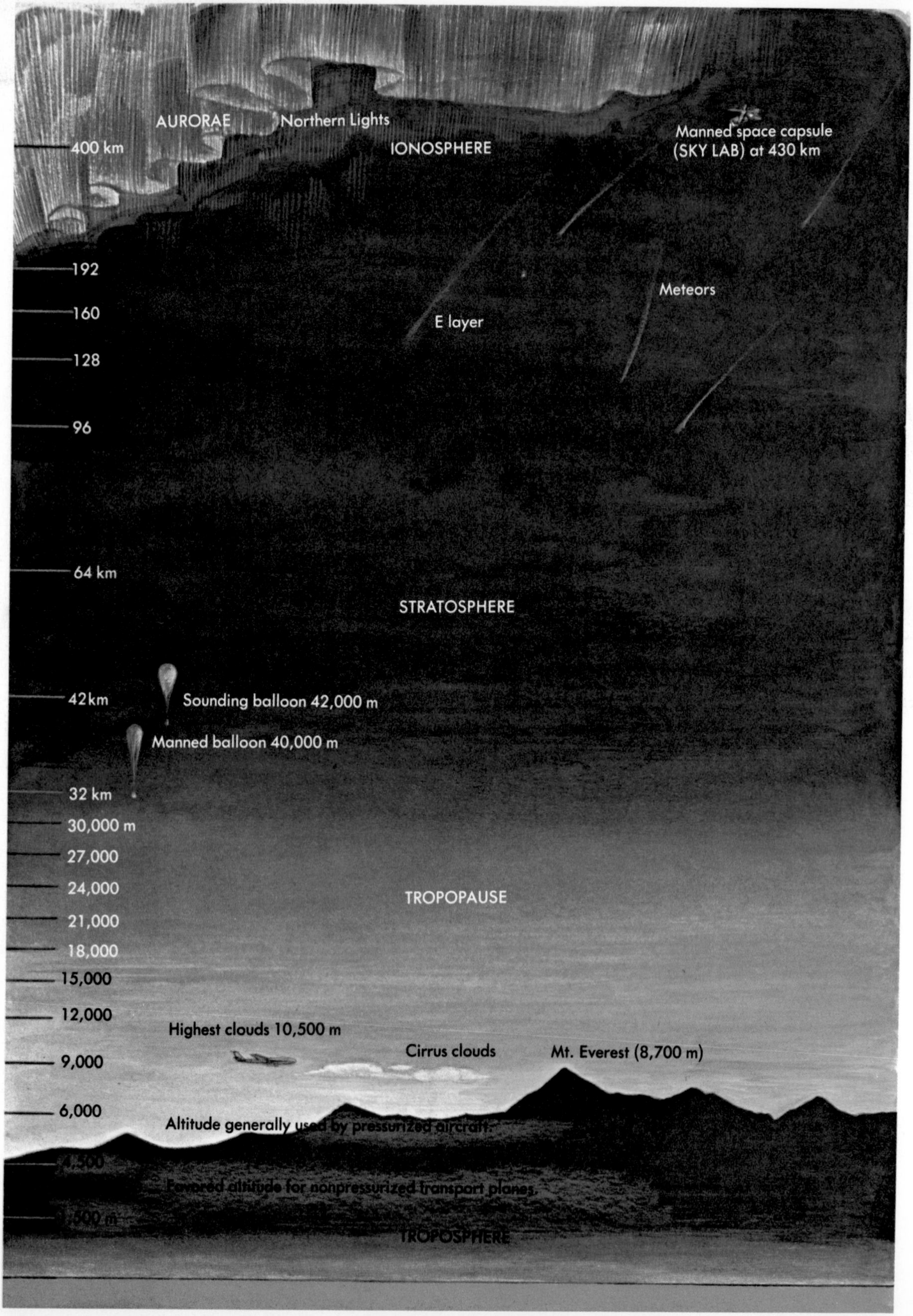

AURORAE
Northern Lights
400 km
IONOSPHERE
Manned space capsule
(SKY LAB) at 430 km
192
160
128
96
Meteors
E layer
64 km
STRATOSPHERE
42km
Sounding balloon 42,000 m
Manned balloon 40,000 m
32 km
30,000 m
27,000
24,000
21,000
18,000
15,000
12,000
TROPOPAUSE
Highest clouds 10,500 m
Cirrus clouds
Mt. Everest (8,700 m)
9,000
6,000
Altitude generally used by pressurized aircraft.
4,500
Favored altitude for nonpressurized transport planes.
1,500 m
TROPOSPHERE

The Upper Atmosphere. The picture on page 344 shows some conditions in the upper atmosphere. What is the highest level at which clouds are found? In which region do most meteors burn? Where is ozone plentiful? At what level does the stratosphere blend into the ionosphere? In which region are the northern lights produced?

The Troposphere. The lower region of the atmosphere receives most of its heat from the sun-warmed earth. Air touching the surface is warmed by conduction. It expands and rises by convection. As this air rises, it loses heat by radiation, cools, and finally settles back to earth.

The level at which upward circulation ends is the imaginary boundary between the troposphere and the stratosphere. Thus, the troposphere is a region of weather, with rising air currents, condensation, clouds, rain, and snow.

There are strong winds called *jet streams* near the tropopause. These blow over large areas and are unlike the familiar winds of the lower troposphere.

Explain in terms of convection currents why the tropopause may be at different altitudes, depending upon season and latitude.

The Stratosphere. The stratosphere lies above the region of convection. There are no clouds, precipitation, or storms. Temperatures stay low.

Discuss the advantages and disadvantages of air travel within the stratosphere as compared with the troposphere. Study the picture on page 323. Discuss the conditions in the photograph.

The Ozone Layer. The upper part of the stratosphere contains a great deal of ozone. It is thought that ultraviolet radiation from the sun produces ozone by breaking up oxygen molecules which reunite as ozone. According to this theory, most of the ultraviolet radiation is absorbed and does not reach the earth's surface.

Recent explorations with rockets have shown that sunlight in space does not contain as much ultraviolet as expected. Theories about the ozone layer have since been changed.

The Ionosphere. The ionosphere is so named because many of the particles in this region have become ions by gaining or losing electrons. Most of the ions are produced by ultraviolet and X-ray type radiations in sunlight. Other ions may be produced by charged particles emitted by the sun during solar flares.

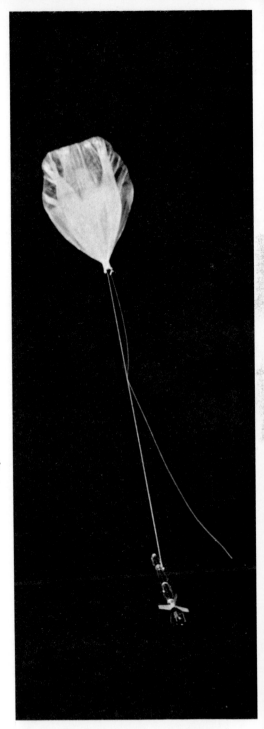

Conditions in the upper stratosphere, where balloons cannot go, are explored by rockets carrying weather instruments. Parachutes lower the instruments back to earth.

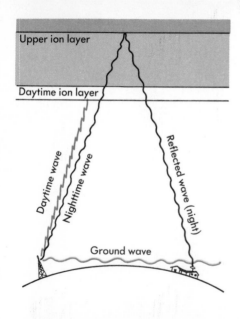

Upper ion layer

Daytime ion layer

Daytime wave

Nighttime wave

Reflected wave (night)

Ground wave

The ionosphere plays a major role in short-wave radio transmission. At night, radio waves are reflected from a very high layer, as shown at the left. These can be picked up by distant receiving stations. During the day, however, sunlight produces an additional lower layer of ions. This layer absorbs some of the energy of the waves and reduces the short-wave reflection.

REVIEW QUESTIONS

1. What conditions cause a low fog?
2. What conditions might cause a fog on a mountain top?
3. Why is the sky usually blue?
4. Why is the sky often red at sunrise and sunset?
5. What is the purpose of the instrument on the left?
6. How does this instrument change when carried up a mountain?
7. What are the four most plentiful gases in the atmosphere?
8. What is meant when we say that the relative humidity is 50%?
9. What is ozone?
10. What are five sources of dust in the atmosphere?
11. What is normal atmospheric pressure at sea level in (a) grams per square centimeter, and (b) centimeters of mercury?
12. What is the boundary between the troposphere and the stratosphere called?
13. Why is half of the earth's air concentrated in the lower 5.5 km of the atmosphere?

THOUGHT QUESTIONS

1. Why does the dew point change from day to day?
2. What is the air pressure in grams per square centimeter on a mountain where a barometer reads 50 cm of mercury?
3. Why are most artificial satellites orbited at altitudes greater than 500 kilometers?
4. Why do eyeglasses often fog over in winter weather?
5. Why is the sky black in this photograph of the moon's landscape?

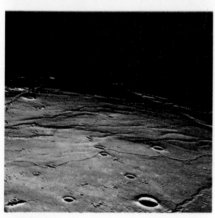

346

2

Heating the Atmosphere

The sun supplies the earth with nearly all of its heat energy, and yet the sun does not heat the atmosphere directly. Most of the sun's radiant energy passes through the atmosphere without warming it up. However, when the radiant energy strikes the earth's surface, the surface becomes warmer. The warm surface radiates energy which does not pass through the atmosphere. The atmosphere traps this energy and is heated by it. The atmosphere is heated from the warmth of the earth.

Our planet is remarkable because surface temperatures all over the earth do not vary by many degrees. This fact is largely due to the circulation of the atmosphere which carries heat with it as it circulates. To better understand the atmosphere, it is important to know how energy affects it.

ENERGY FROM SUNLIGHT

Our atmosphere circulates constantly, transporting heat and water vapor from one place to another to produce the conditions we call weather. All this atmospheric movement requires energy which is provided by the sun in the form of radiation, both visible and invisible. However, radiant energy cannot directly set air in motion; it must first be changed to heat energy in order to produce convection. Thus, the study of weather begins with the change of radiant energy to heat energy.

Absorption of Radiant Energy. Pour water into a shallow tray, and measure the temperature of the water in Celsius degrees. Set the tray in the sunlight for ten minutes. Measure the temperature change. Then measure the volume of the water in milliliters. Calculate the calories absorbed by the water. One calorie is the amount of heat needed to warm one milliliter of water one Celsius degree. Use this equation:

$$\frac{\text{Calories}}{\text{absorbed}} = \frac{\text{Volume in}}{\text{milliliters}} \times \frac{\text{Temperature}}{\text{change in C}°}$$

Energy Value of Sunlight. Meteorologists estimate that sunlight entering the atmosphere can provide about two calories per minute for each square centimeter of surface. This value is called the *solar constant*, though it is not exactly constant since radiation from the sun varies somewhat.

Measure the area of the tray used in the last experiment. How much heat was absorbed by each square centimeter of surface? How much heat did each square centimeter absorb per minute?

Radiometers. The *radiometer* shown here is helpful in studying radiation. The device has four vanes which rotate in a partial vacuum. Each vane

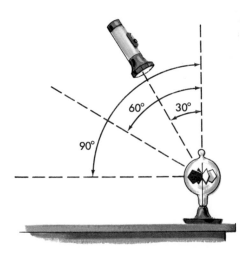

Angle	Revolutions Per Minute	Percent
90°		100%
60°		
30°		
0°		

has a shiny side which reflects radiant energy, and a black side which absorbs it.

According to accepted theory, when air molecules touch the black surfaces, they absorb heat, speed up, and fly away, giving a backward push to the vanes. Set a radiometer in sunlight. Which way does it revolve?

Effect of Inclination. Shine a lighted flashlight directly at and level with a radiometer. In this position the light strikes the vanes at 90°. Set the light far enough away so that the revolutions of the vanes per minute (r/min) can be counted.

Hold the flashlight beam so that it makes a 30° angle with the vanes. Count the r/min. Repeat at other angles. Compare the r/min at other angles with the r/min when the light strikes the vanes at 90°. Why should the distance of the flashlight from the radiometer be kept the same at each angle?

Assume that 100% of the energy is absorbed by the vanes when they are at 90° to the light beam. Calculate the percentage received in the other positions. Graph the results.

Study the diagrams below. Discuss the changes in energy received by the surfaces during the day and during the year.

What percentage of solar energy arrives on March 22nd at the top of the atmosphere at 30° North compared to the amount arriving at the equator? Compare the solar energy arriving on this date at 60° North and at the equator.

Wavelength is the distance between the beginning of one vibration and the beginning of the next.

Nature of Solar Radiation. Energy from the sun includes both visible radiation and radiation to which our eyes are not sensitive. Visible radiation is made up of the colors seen in the rainbow.

Paint the bulbs of two thermometers black. Produce a spectrum with a glass prism. Place the bulb of one thermometer just beyond the red end of the spectrum, and the other bulb just beyond the violet end. Compare the readings after three minutes. Repeat, reversing the thermometers. Why? Test other regions of the spectrum.

One theory of radiation suggests that energy is carried by vibrations. According to this theory, sunlight is made up of vibrations having many wavelengths. Red light has a longer wavelength than orange light, and orange light has a longer wavelength than yellow light. The blending of colors in the spectrum shows that the change in wavelength is gradual.

The invisible part of the spectrum just beyond the violet is called the *ultraviolet*. The invisible part just beyond the red is called the *infrared*. Which has the greater heating effect? Longer wavelengths?

Energy of Infrared Radiation. Obtain an infrared lamp and a standard incandescent lamp of the same rating, about 300 watts. Produce a spectrum from the light of each lamp. What colors are produced?

Screw the incandescent lamp into a reflector, and place it far enough from a radiometer so that the vanes turn slowly. Count the revolutions per minute. Replace the incandescent lamp with the infrared lamp. Count the revolutions again.

What conclusions can you draw about the heating effect of infrared radiation? Which conditions in this experiment are not well controlled? How may these affect your results?

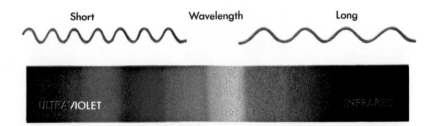

Short   Wavelength   Long

ULTRAVIOLET   INFRARED

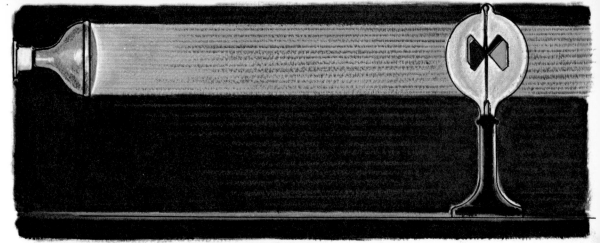

Energy Distribution in the Spectrum. Our eyes are most sensitive to yellow and green light. We may think that most of the sun's energy is transmitted at these wavelengths. Actually, most of the energy is transmitted at longer wavelengths.

The graph shows the distribution of the energy in sunlight. At which wavelengths is the radiation strongest? Compare the energy of the different colors of visible light. Compare the energy transmitted by radiation in the infrared and the ultraviolet wavelengths.

The total amount of solar energy transmitted to the earth per minute is two calories per square centimeter, as measured perpendicular to the sun's rays. How much of this energy is transmitted by all the infrared radiation? By all the visible light? What wavelengths transmit the rest of the solar energy?

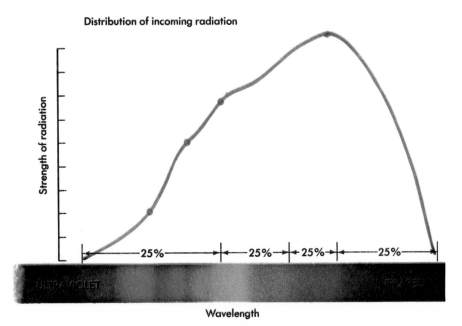

Distribution of incoming radiation

Strength of radiation

25%   25%   25%   25%

ULTRAVIOLET   INFRARED

Wavelength

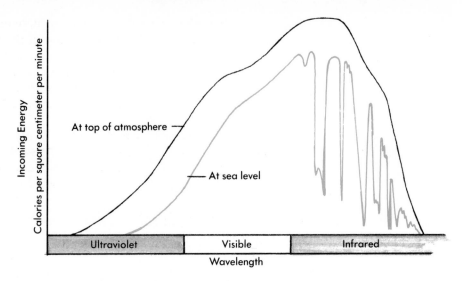

At top of atmosphere

At sea level

Incoming Energy

Calories per square centimeter per minute

| Ultraviolet | Visible | Infrared |

Wavelength

HOW THE ATMOSPHERE IS HEATED

Most solar radiation passes through the atmosphere. Therefore the atmosphere is heated very little by the direct absorption of sunlight. Most atmospheric heat is gained from the sun-warmed surfaces of the earth. Heat is transferred from the soil to the air by conduction, convection, and radiation. The atmosphere also gains heat energy when water evaporates, although this energy does not appear as heat until the vapor condenses as rain or snow.

Atmospheric Absorption. The graph above compares the amount of solar energy which reaches the top of the atmosphere with that which arrives at sea level. The atmosphere absorbs a little energy and reflects some. Estimate the percentage of the solar constant which reaches sea level.

Notice that atmospheric absorption is not the same at all wavelengths. Infrared radiation at some wavelengths is almost completely absorbed. Water vapor, carbon dioxide, and ozone are responsible for absorption at special wavelengths.

Discuss the nature of the radiation to be expected on a high mountain peak where most of the atmosphere lies below it. Discuss the differences in incoming radiation to be expected during dry days and humid days at sea level.

Absorption by Solids. Obtain two shiny metal cans of the same shape and size. Cover half of one can with candle soot or dull black paint.

Fill both cans with water at 20°C. Set them in sunlight with the black side of one can toward the sun. Measure the temperature change in each can after 15 minutes. What has happened?

Shiny surfaces reflect most of the radiant energy which falls on them. Dark, dull surfaces absorb most of the energy and change it to heat. No known substance is a perfect absorber, but soot is among the best. It can absorb about 97% of the radiation falling on it.

Conduction to Air. Air is a very poor heat conductor. Nevertheless, the atmosphere gains some of its heat by conduction from warm surfaces. The apparatus shown below is helpful in studying this transfer. Notice the shimmering seen on the screen. It occurs as the air is heated by the hot object. You have probably seen this effect before. What were the conditions when you saw it?

According to one theory, air molecules touching a hot surface gain energy and move faster. They may give some of this energy to molecules they strike. Thus a thin layer of air is heated.

What happens to this heated air? How far from the hot object is the heating effect visible? In which direction is the effect greatest?

Energy Transfer by Evaporation. An enormous amount of heat can be removed from a surface by the evaporation of water. About 600 calories are needed to evaporate one milliliter of water.

This energy passes into the air with the water vapor. However, air temperatures do not rise immediately, because the energy of the vapor is not in the form of heat. Only when the vapor condenses is the energy released as heat, to be added to the heat of the atmosphere.

Hot object

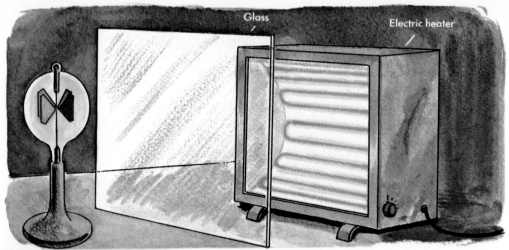

Glass

Electric heater

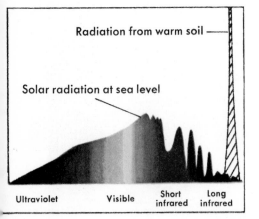

Hot Body Radiation. Fill a metal can with water at 50°C. Hold your hand near the can. At what distance can you feel warmth? Can a radiometer detect any radiation? Fill the same can with water at 100°C. At what distance can you feel warmth? Can the radiometer detect any radiation?

Test an electric iron in the same way. Test an electric heater that has a visible heating element. What conclusions can you draw about radiation from hot objects?

The photograph at the left was taken at night with film sensitive to infrared radiation. What does the photograph show you about the nature of radiation from hot objects?

Short- and Long-Wave Infrared. The graph shows the nature of solar radiation reaching the earth and the nature of radiation from warm soil. What are the chief differences in the solar radiation and the radiation from the earth? What effect does the temperature of an object have on the wavelength of the energy it is radiating?

Discuss the nature of the radiation from an electric iron, an electric toaster, an infrared lamp, and an incandescent lamp. Test these sources of radiation using a radiometer. Why should you hold the radiometer the same distance from each source? Which radiate mostly infrared of long wavelengths? Which radiate infrared of both short and long wavelengths? Is the radiometer more sensitive to short or long infrared waves?

Hold a pane of window glass between a radiometer and each of the sources in the preceding paragraph. Which can pass more readily through window glass, long- or short-wave infrared radiation?

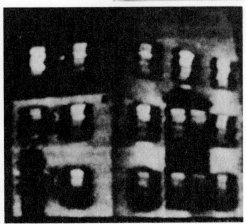

Radiation from warm soil

Solar radiation at sea level

Ultraviolet   Visible   Short infrared   Long infrared

The Greenhouse Effect. Mount two thermometers on folded strips of black cardboard, as shown here. Set both in bright sunlight with two thermometers on the shady sides of the strips. Cover one with a glass jar. Compare the temperature changes in the two thermometers.

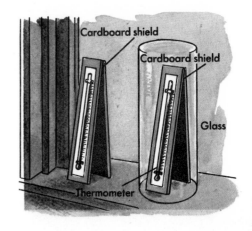

What happens to the radiant energy falling on the two black cardboard shields? In what two ways does energy leave the cardboard shields? In what two ways does the jar reduce the loss of energy?

Study the photograph of the greenhouse. What type of infrared radiation enters the greenhouse? What happens to this energy? What type of infrared radiation is given off by the soil and other objects? Why does this type of radiation escape from the greenhouse more slowly than solar radiation enters?

The Atmospheric Heat Trap. Our atmosphere serves the earth much as glass serves a greenhouse. Air does not absorb most of the short-wave radiation in sunlight. However, air absorbs nearly all the long-wave radiation given off by warm rocks and soil. What happens to the long-wave radiation absorbed by air?

Why is air at sea level generally warmer than air at higher levels? Why does the air on a high mountain cool off faster at night than air in a valley? Discuss the possible effects of changes in the amount of water vapor or carbon dioxide in the air. Discuss the effect of a loss of air into space.

Effects of Air Pollution. Some scientists believe that air pollution may be affecting the weather. As our population grows, and as our demand for more energy increases, we burn more oil, coal, gas, and other fuels. More burning means more carbon dioxide released into the atmosphere. This results in more of the earth's radiation being trapped, and thus increases the greenhouse effect. Even a slight rise in the earth's temperature would produce great changes in the climate of the earth.

Other scientists argue that the dust particles, soot, and gases sent into the atmosphere act as a screen which reflects sunlight. If less sunlight is reaching the earth's surface the temperature may go down. A drop of only a few degrees could start another ice age.

There are definite indications that air pollution affects local weather. Studies in several cities seem to show that the rainfall these cities receive is related to the amount of pollution around them.

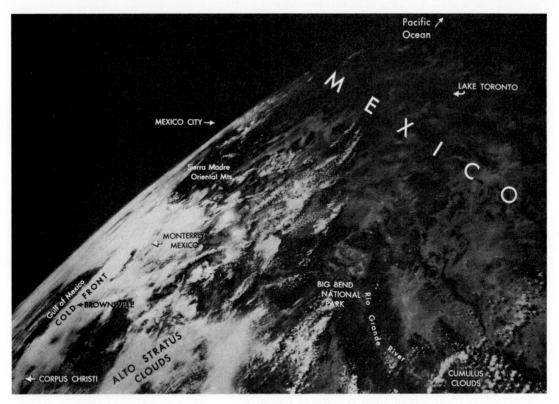

Reflections from Clouds. Tests have shown that a dense cloud reflects about 76% of the radiant energy falling on it. Discuss the climatic effect of the clouds shown in the photograph. Estimate the number of calories being reflected from each square centimeter of the thickest clouds.

Meteorologists estimate that, for the earth as a whole, about 38% of the incoming solar energy is reflected. Astronomers estimate that the moon reflects only about 7% of the solar energy received, Venus reflects about 76% of the solar energy. Discuss the reasons for these different values.

Use a radiometer to compare incoming energy on a clear day and when there is a heavy cloud cover. Also test the incoming energy when the sky is covered with cloud layers of different thicknesses.

Atmospheric Scattering. Dust and water droplets scatter a small part of the solar energy passing through the atmosphere. Part of this scattered energy passes back into space and is lost. The remainder is scattered toward the earth.

Set a radiometer where it is shaded from direct sunlight but receives light from other parts of the sky. Explain the results.

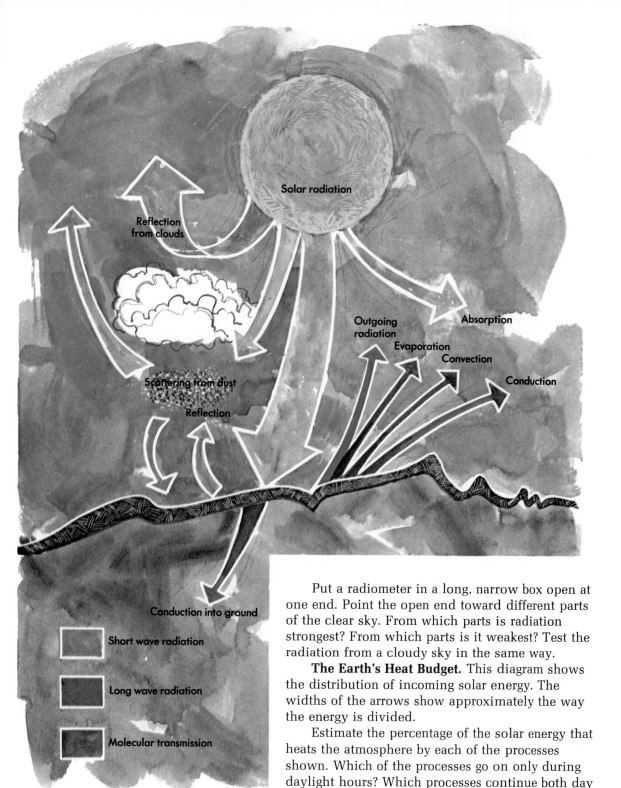

Solar radiation

Reflection
from clouds

Outgoing
radiation

Absorption

Evaporation

Convection

Conduction

Scattering from dust

Reflection

Conduction into ground

Short wave radiation

Long wave radiation

Molecular transmission

Put a radiometer in a long, narrow box open at one end. Point the open end toward different parts of the clear sky. From which parts is radiation strongest? From which parts is it weakest? Test the radiation from a cloudy sky in the same way.

The Earth's Heat Budget. This diagram shows the distribution of incoming solar energy. The widths of the arrows show approximately the way the energy is divided.

Estimate the percentage of the solar energy that heats the atmosphere by each of the processes shown. Which of the processes go on only during daylight hours? Which processes continue both day and night?

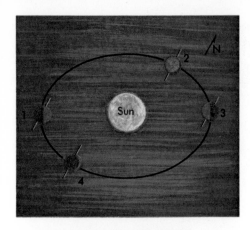

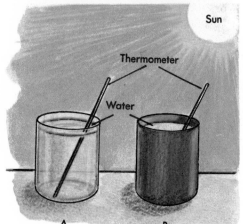

REVIEW QUESTIONS

1. In what form does solar energy reach the earth?
2. What is the effect of the clouds on incoming solar energy?
3. What is the season in the Northern Hemisphere of the earth at each position in the diagram?
4. At which wavelengths does the sun transmit most of its energy?
5. How is water vapor related to the transfer of heat energy into the atmosphere?
6. How does the earth's atmosphere act like the glass on a greenhouse?
7. How does air pollution affect the weather?
8. Explain how the temperature will change in can A and in can B in the diagram.

THOUGHT QUESTIONS

1. Why does the sun's radiant energy have to be changed to heat energy to set the atmosphere in motion?
2. How are water vapor, carbon dioxide, and ozone involved in absorbing the sun's energy?
3. Explain how the energy from sunlight heats the water in each of the two cans in the diagram.
4. Use the graph below to answer these questions:
   a. At which latitude is the most solar energy absorbed?
   b. At which latitude is there the greatest difference between the amount of energy radiated from the earth and the amount of solar energy received?
   c. At which latitude is there more energy radiated from the earth than solar energy received?

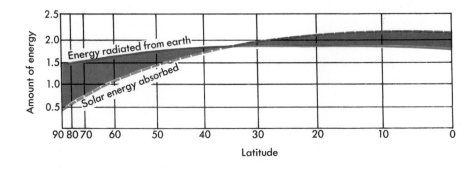

3

Motions of the Atmosphere

Anyone approaching our planet for the first time would be impressed by our atmosphere. They would have to wonder how life could exist on a planet whose atmosphere is constantly moving. You have felt it moving as a gentle breeze and sometimes as a strong wind. The air moves in response to numerous forces.

From your studies in the previous chapter you know that some parts of our planet absorb more energy than others. Winds, through their motions, tend to spread the sun's heat energy over the entire surface of the planet.

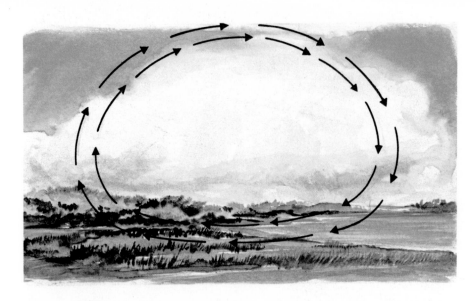

CAUSES OF WINDS

Winds are caused by unequal heating of the atmosphere. Small, local winds may be set in motion by the heating of a bare field during a sunny afternoon, or by the cooling of a mountain peak at night. Continental winds are produced as land masses warm up or cool off during the changing seasons. The major winds are mostly the result of differences in radiation received and absorbed in equatorial and polar regions.

Convection Cells. A convection current in the atmosphere, made up of a complete circuit of moving air, is often called a *convection cell*. Produce a model convection cell in a box, as shown below. Use a 100-watt lamp in a metal can as a heat source. Make the current visible with smoke.

Heated air expands and becomes less dense. Where in the box is the air being heated? Why does the air rise above the light bulb and sink when it reaches the end of the box away from the bulb?

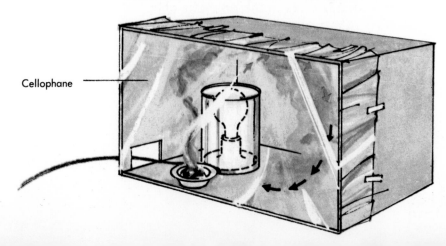

Cellophane

A convection cell should be explained in terms of differences in density of warm and cool air. Explain the convection cell at the top of page 360. At what time of day should such a cell be most active? When should it disappear?

Mountain Winds. On calm, clear nights, mountain tops usually cool faster than the valleys below. Why? During the night, cool air drifts down the mountain sides. Describe the convection cell that is set up by this circulation.

At what time of day would you expect such a mountain breeze to begin? When would you expect the wind to be strongest? When would you expect it to stop?

Mountain winds are usually strongest in gullies which lead down the mountain sides. Explain why.

Winds may blow all day down valleys that contain glaciers. Why? At what season would you expect glacial winds to be most common?

Land and Sea Breezes. The diagrams to the left describe conditions along a coast during periods of good weather. Which type of surface along a coast heats up more during the day? Why? Which type of surface cools faster at night? Why? What happens to air temperatures above these surfaces by day and by night?

Explain the convection cells described by the arrows. What is the direction of the breezes across the shore by day, and by night?

When would you expect the daytime circulation to be strongest? When does it end?

Land breeze

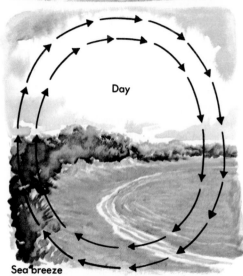

Day

Sea breeze

361

Left: Fairweather cumulus
Right: Thunderstorm cumulus

Freezing temperature

Dew point

When is the nighttime circulation strongest? When does it end?

Clouds in Convection Cells. Air rising through the atmosphere expands because the atmospheric pressure decreases as it rises. As it expands, it becomes cooler. What happens if temperatures drop below the dew point?

Most clouds in convection cells are cumulus clouds, as shown above. Where is air rising? How does the relative humidity change from ground level up into a cumulus cloud? Explain the flat bottom of a cumulus cloud in terms of temperature and dew point. Explain the puffy tops.

Convection of dry air may be gentle because the rising air loses energy and slows down. Convection is usually faster in moist air because condensation releases energy to the air. Which of the cumulus clouds shown above indicates faster circulation? Which indicates higher relative humidity at ground level? Which cloud may produce rain?

Thunderstorms. Rapid convection may cause gusty storms with lightning and hail. Ice crystals form when moist air is blown up into the freezing zone. The crystals which fall below the freezing zone then serve as centers for condensation. These particles may fall as rain, if the ice melts on the way down. Or they may be blown back up into the freezing zone several times and become hail.

Water drops become electrically charged when torn apart by the wind. The smaller parts may be blown into another part of a cloud, carrying their charges with them. Thus, different parts of a cloud may have different charges. What may happen then?

Hurricanes. Hurricanes are very destructive storms formed only in tropical regions. The photograph shows a hurricane over the West Indies. What evidence is there that the hurricane is a whirling storm? Does the storm whirl in a clockwise or a counterclockwise direction?

According to one theory, a hurricane develops when warm, moist air begins to rise. The air expands, cools, and condensation takes place. Condensation releases heat which is added to the rising air, making it rise faster. The earth's rotation causes the moving air to whirl. The center around which the air is blowing is the eye of the storm.

According to this theory, why does a hurricane form only over water? Why does it die out soon after moving inland?

Study the diagram. Describe a hurricane in terms of diameter, wind direction, and clouds.

A recording barometer made the graph here as a hurricane passed by. Describe the changes in atmospheric pressure. Why is there lower pressure in the eye of the storm?

West Indian hurricanes usually travel toward the northwest while in the tropic zone, and curve northeast after entering the temperate zone. Make a study of recent hurricanes, mapping their courses.

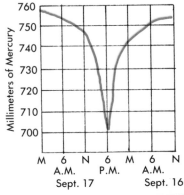

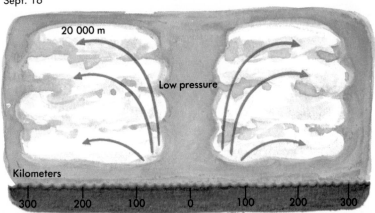

Dust devils

Vortex Formation. A column of gas or liquid which flows with a whirling motion is called a *vortex*. A vortex is often seen as water flows down a drain. As heated air rises, it may form a vortex.

Cut the ends from a metal can. Then cut the can from top to bottom, producing equal halves. Arrange the pieces around lighted birthday candles, as shown at the left. Note the slits between the pieces.

Watch the flames while you adjust the pieces of metal. Adjust by changing the sizes of the slits and the diameter of the cylinder which is formed.

Why does air move upward in the cylinder? Why does it whirl? How do you know when a vortex develops? Try to change the direction in which the vortex whirls.

Whirlwinds. Dust devils are often seen over dry, level plains or deserts during calm, summer afternoons. They usually arise suddenly, travel a short distance, and die out in a few minutes. These small whirlwinds are not tornadoes.

Whirlwinds form when air next to the ground becomes very hot. This layer does not rise as a whole, because cool air cannot get underneath it. However, some hot air may leak upward over mounds or rocks.

Compare a whirlwind with the vortex you produced over candles. What is the source of heat? Why does a column of air rise? What starts the column whirling? Why may the column whirl in either direction?

Why are whirlwinds most common in level country over dry soil or sand in the afternoon? Why are whirlwinds uncommon in cloudy weather?

Developing tornado

Tornadoes. A tornado resembles a large whirlwind, but there are many differences. A whirlwind starts at ground level during calm, dry weather; it whirls in either direction, and it rarely lasts very long. A tornado starts at cloud level during hot, moist weather; it almost always whirls counterclockwise (in the Northern Hemisphere), and it may last an hour or more while traveling many kilometers.

Tornadoes usually travel along thunderstorms. The funnel sometimes touches the ground, and other times retreats into the clouds. In general, the tornado moves from the southwest to the northeast. Although the tornado may move at a rate of 30 to 60 km/h, the winds may exceed 120 km/h.

Tornadoes often form when a mass of cold air blows over warm, moist air. It is believed that the warm air breaks up through the cold air, and is given its whirling motion by winds at high levels blowing in opposite directions.

Tornadoes gain their high speed because water vapor condenses in the rising air. This releases heat that makes the air rise still faster. The vortex whirls faster, grows downward, and draws in more warm air and water vapor. Thus, the vortex continues to gain energy until it becomes very dangerous.

Waterspouts. A waterspout is a vortex of air rising over water. The visible part is a hollow column of the cloud. It is not a column of water, as many people believe.

Some waterspouts seem to be tornadoes; they start at high levels. Others are like whirlwinds; they start at water level. Much remains to be learned about the behavior of waterspouts.

Waterspout

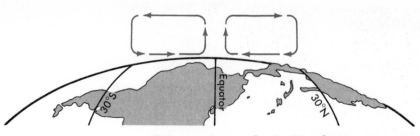

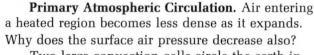

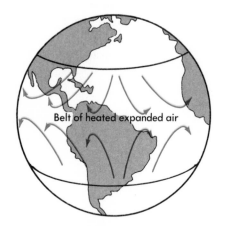

Belt of heated expanded air

Earth

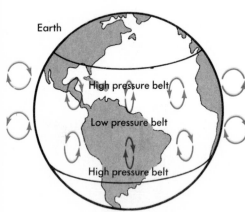

High pressure belt

Low pressure belt

High pressure belt

Primary Atmospheric Circulation. Air entering a heated region becomes less dense as it expands. Why does the surface air pressure decrease also?

Two large convection cells circle the earth in the equatorial region. What is the cause of the low pressure belt where the two cells join? Explain the circulation shown above. How does the position of the cells change between January and July? How does the position of the cells change between July and January? What causes this change in position?

Secondary Circulation. A high pressure belt is produced on each side of the convection cells. This high pressure results in part from descending air. The high pressure is also due to the increased density of the cool, dry air.

Most of the air entering the high pressure belts continues to circulate in the primary convection pattern. However, some spreads out poleward, thus setting up additional circulation patterns.

Wind Belts. The earth has high pressure regions at the poles, where the air is cold and dense, in addition to the high pressure belts. Air flowing outward from all of these regions of high pressure makes up the major winds of the earth.

Effect of the Earth's Rotation. Use a world globe to study the effect of the rotation of the earth on the path of an object moving along a north to south line. Find the 90° W longitude line on the globe. With your finger trace the line from the North Pole to the equator. Through which provinces and states does the line pass?

Point your finger along the 90° W longitude line in Minnesota without touching the globe. Move your finger directly south while someone else slowly turns the globe from west to east. Is your finger pointing at the same longitude as before? Where is your finger pointing?

When winds blow from north to south they move in the same direction as your finger. What path would winds blowing from south to north

366

take? Repeat this activity using the 60° W longitude line through South America.

In 1835, Gaspard Coriolis stated that the earth's rotation had an effect on the motion of all things. He explained that winds tend to turn toward the right from a straight path in the Northern Hemisphere. In the Southern Hemisphere winds turn to the left.

Note the convection cells in each hemisphere as shown along the edge of the globe below. Use the Coriolis effect to explain why the air in these cells does not blow directly north and south.

This globe shows the location of the wind belts in April and October. About how many degrees do they shift northward between April and July? Where are the belts in January? In which belt or belts is your home during the different seasons?

SUMMARY QUESTIONS

1. What is the cause of winds?
2. How does the breeze at the seashore change from daytime to nighttime?
3. What is the Coriolis effect?

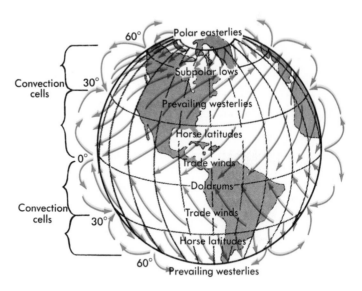

Mountain   Marsh bayou

WEATHER SYSTEMS

Most lands in the latitude of the United States have very changeable weather. Air from high pressure regions on either side enters these lands, blowing in opposite directions, and setting up numerous storms. Weather may be cool, warm, clear, or cloudy depending upon the type of air that is present. Weather patterns are very complicated, and accurate predictions are difficult to make.

Air Masses. Bodies of air that remain over warm seas for some time become warm and humid. Bodies of air that remain over frozen land become cold and contain little moisture. After a body of air has become uniform throughout, it is called an *air mass*.

The climate of North America is affected by the masses shown on the map. What qualities are suggested by the names of these masses? How do the masses gain these qualities?

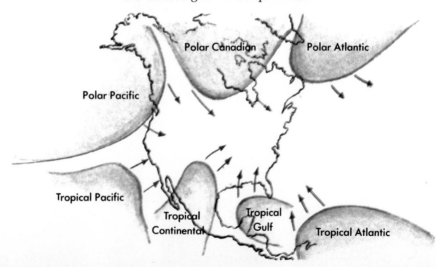

Polar Canadian   Polar Atlantic

Polar Pacific

Polar Pacific

Tropical Pacific

Tropical Continental   Tropical Gulf   Tropical Atlantic

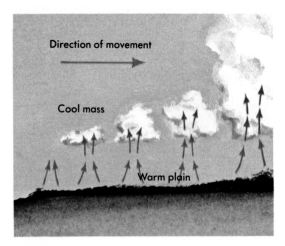

Direction of movement

Cool mass

Warm plain

At times, a part of a mass may squeeze outward in the direction of the arrows on the map. Discuss the possible effect of this air on the region it crosses.

Describe the probable qualities of the air masses above the regions shown in the photographs on the opposite page. How can you recognize these masses as they pass by?

Changes in Air Masses. Air masses may change as (1) the bottom layer is warmed or cooled by the earth's surface, and (2) the entire mass moves up or down slopes.

Study the diagram above. What is happening to the moving mass? Why do cumulus clouds often form? What is happening to the mass shown in the photograph above? How do you know?

What happens as a mass crosses a surface cooler than itself? Which picture on these two pages represents such a condition? Explain what happens.

The illustrations below show a common condition in mountainous country. Explain the diagram and photograph. Why is this condition common in the southern Appalachians in winter?

Warm moist mass

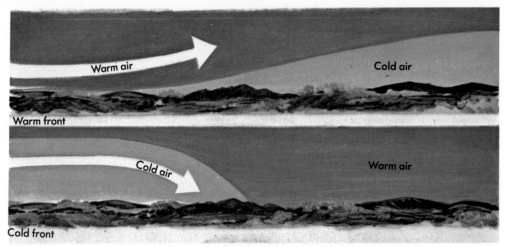

Warm air · Cold air

Warm front

Cold air · Warm air

Cold front

Fronts. Warm and cold air do not mix easily because of their different densities. Thus two unlike masses are separated by an invisible boundary that can be located by thermometer readings.

A moving warm mass tends to ride up over cooler air ahead of it, as shown in the diagram. Why? The invisible boundary, which is the front of the warm mass, is called a *warm front*.

Explain why a moving cold mass tends to push under warmer air. The boundary between the cold air and the warm air is called a *cold front*. Why?

Discuss the advantage to weather forecasters in knowing the location of a mass and its front, together with the speed and direction of the mass.

Cyclonic Storms. Temperate zone weather is marked by a succession of low pressure systems moving from west to east a few days apart. Air spirals counterclockwise toward the center of these systems, giving them the name *cyclones*. A cyclone is a system of revolving air masses.

Temperate zone cyclonic storms begin along the

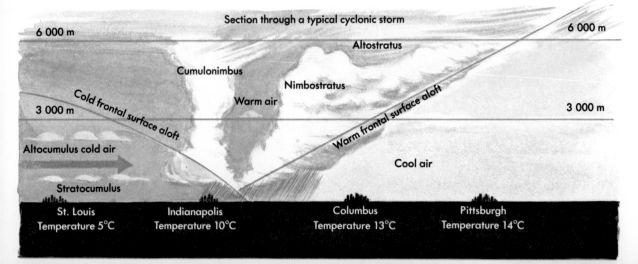

Section through a typical cyclonic storm

6 000 m   6 000 m

Altostratus

Cumulonimbus

Nimbostratus

Cold frontal surface aloft · Warm air

3 000 m   3 000 m

Warm frontal surface aloft

Altocumulus cold air

Cool air

Stratocumulus

St. Louis	Indianapolis	Columbus	Pittsburgh
Temperature 5°C	Temperature 10°C	Temperature 13°C	Temperature 14°C

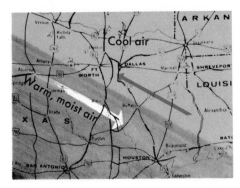

1. The two currents of warm and cold air.

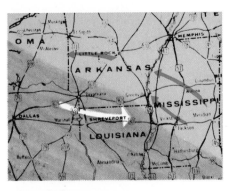

2. A ripple begins.

boundary between warm westerly winds and cold easterly winds. The two sets of winds slide past each other without mixing. However, according to modern theory, the winds set up waves along their boundary much as waves are produced on water.

These diagrams explain how a cyclonic storm develops. Note that the storm is moving eastward.

1. Cold easterlies and warm westerlies slide past each other without mixing. Why?

2. A ripple begins as upper level winds cause an area of low pressure to form.

3. The ripple becomes a wave. Air masses move around the ripple because of the Coriolis effect. This motion produces a vortex of low pressure.

4. The storm system contains both warm and cold masses. What happens to the air at each front?

5. The cold front is overtaking the warm front. What is the direction of the air in each quarter of the storm?

6. The warm air has been lifted above ground level by the cold air. The storm will soon die out. How far has it moved since it began?

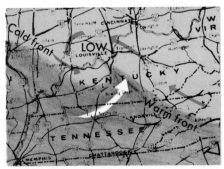

3. A warm and cold front develop.

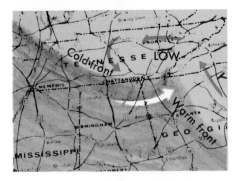

4. The cold front is overtaking the warm front.

5. The cold front overtakes.

6. The storm is dying out.

Warm Front Weather. Air, warmed by the ground, expands, rises, and slides over a cold air mass. This expanding warm air cools, and clouds form. Why does rain fall as a warm front approaches?

The diagram shows the types of clouds that form along a warm front. Describe these clouds.

Rainfall along a warm front is usually heaviest when the warm front is at low altitudes. As the warm air slides upward along the front and cools, it has less moisture and the rain decreases.

High wispy cirrus clouds are usually the first sign of an approaching warm front. Why are cirrus clouds made up of ice crystals? Describe the change in clouds as the warm mass continues to approach.

A warm front is usually part of a cyclonic storm. Study the diagrams on the preceding page. What is the usual wind direction before and after a warm front passes? What changes may be expected in pressure, temperature, and humidity? List signs used to identify an approaching warm front. What predictions can be made after it is identified?

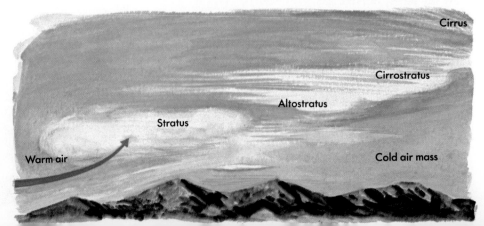

Cold Front Weather. The long red arrow below shows what happens to warm air ahead of an advancing cold front. Suggest two ways the warm air mass would be cooled.

The lifting warm air cools below its dew point, producing clouds. This condensation process releases heat. Why might this process cause the cloud to rise? Such clouds often contain high winds, and sometimes hail and lightning.

What is the usual wind direction before and after a cold front passes? What changes may be expected in pressure, temperature, and humidity?

The air behind a cold front is often dry. If it is, the skies usually clear and good weather follows. If the air is moist and is warmed at the bottom by passing over warm surfaces, the sky may remain covered with heavy clouds. However, these clouds may not produce rain or snow.

Suppose that you have seen a weather map showing that a cold front is approaching. What predictions can you make?

Cold air mass

Warm air mass

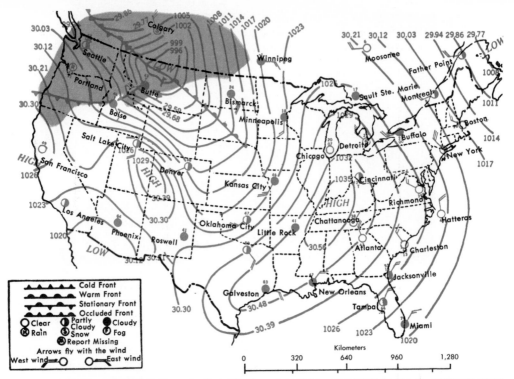

WEATHER MAPS AND FORECASTING

Some weather changes can be predicted from clouds and wind directions. Dependable forecasting is impossible without a knowledge of approaching weather systems. The Weather Service collects information about conditions over the continent and uses it to prepare maps like this one. The probable movement of weather systems can be determined from a series of maps.

Isobars. The heavy lines on weather maps are called *isobars*, meaning lines of equal pressure (*iso*—equal, and *bar*—pressure). A barometer reads the same when carried along an isobar, if kept at the same level above the sea.

The number on an isobar gives the barometric pressure in inches of mercury and in *millibars*. One thousand millibars are nearly equal to the average pressure at sea level. What isobars on the map indicate places with average atmospheric pressure? Where are pressures above average? Below average?

When pressure differences are great and close together, winds will blow hard from the area of high pressure to the area of low pressure. How do the isobars help a meteorologist to determine the direction of winds, and where winds are blowing the hardest?

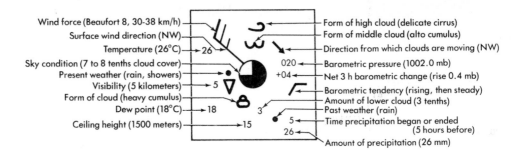

Wind force (Beaufort 8, 30-38 km/h)
Surface wind direction (NW)
Temperature (26°C) — 26
Sky condition (7 to 8 tenths cloud cover)
Present weather (rain, showers)
Visibility (5 kilometers) — 5
Form of cloud (heavy cumulus)
Dew point (18°C) — 18
Ceiling height (1500 meters) — 15

Form of high cloud (delicate cirrus)
Form of middle cloud (alto cumulus)
Direction from which clouds are moving (NW)
020 — Barometric pressure (1002.0 mb)
+04 — Net 3 h barometric change (rise 0.4 mb)
Barometric tendency (rising, then steady)
Amount of lower cloud (3 tenths)
3 — Past weather (rain)
5 — Time precipitation began or ended
(5 hours before)
26 — Amount of precipitation (26 mm)

Station Data. Circles on a weather map show weather stations. Information obtained at each station is printed around the circle, shown above.

The map on the opposite page is simplified by leaving out some of the data. Study the map key. Describe weather conditions in San Francisco, Chicago, New Orleans, Boston, and your area.

Locating Fronts. A meteorologist first writes the station data on a map and draws the isobars. Next the fronts are located, using wind direction, temperature, humidity, and cloud forms. How does knowledge of wind direction help to locate a front? How are warm and cold fronts shown on a map?

Weather Forecasting. The map below shows the usual paths of weather systems. The heavy lines show paths of high pressure. Light lines show paths of cyclonic storms. The map also tells the average distance traveled daily by weather systems.

Assume that the weather systems mapped on the opposite page behave in average fashion. Predict how these systems will move in the next 48 hours. Forecast the weather for cities along their paths.

Study daily weather maps and predict the movements of weather systems. Forecast the weather for your area and check your predictions.

REVIEW QUESTIONS

1. Why does condensation of water vapor increase the circulation in a convection cell?
2. In what direction do weather systems move across North America?
3. What conditions occur before an approaching warm front? Cold front?
4. Study the weather map below. What was the wind direction in New York City?
5. What was the sky condition in Galveston, Texas?
6. What type of air mass was over the Great Plains?
7. What type of front is shown between Chattanooga and Buffalo?

THOUGHT QUESTIONS

1. Use the map below to predict the weather in Atlanta, Georgia, during the next 24 hours.
2. What predictions can be made for Detroit?
3. What happens to a maritime air mass from the Pacific Ocean as it crosses the Rockies?

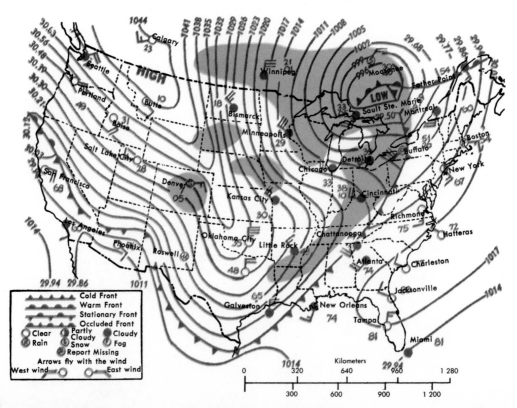

4

Microclimates

Weather changes from day to day, even from hour to hour; but in any one place, the changes tend to follow a pattern. One region may have a long hot summer and a cool winter; another region may have a long cold winter and a cool summer. The average weather pattern of a region is called its climate.

Climate is usually described for a large area, such as a state or region. However, the climate in a very small place, such as this stream, is important for the organisms living there. Weather conditions in these small areas are called microclimates.

Describe two places in the above photograph where the microclimates may be different. How can the plants help you to determine the microclimate of a place?

White soil (white sand)   Light gray soil   Dark gray soil   Black soil (powdered charcoal)

THE NATURE OF CLIMATE

Climate is usually described in terms of atmospheric conditions, such as air temperature, humidity, rainfall, wind, sunshine, and clouds. These weather factors are important to all organisms living in the open air. However, soil temperature, moisture, and gases are equally important to organisms living in the soil. Water temperature and dissolved gases are important to organisms living in ponds, streams, and oceans. Therefore, biologists often speak of soil climates and water climates as well as atmospheric climates. Climate may be thought of as the average conditions found in any place where organisms live.

Absorption of Solar Energy. Nearly all of the earth's energy comes from the sun in the form of radiation. Some of this energy is absorbed by land and water, raising their temperatures.

The amount of radiant energy absorbed by a surface depends upon the nature and color of the surface. Candle soot can absorb 97% of the energy in sunlight. Polished silver may absorb less than 3%.

Dark soils and light soils may differ greatly in temperature after a few hours of sunlight. Fill three or more pans with soil of different colors, made by mixing different amounts of powdered charcoal and white sand. Set the pans in sunlight. Which soil becomes warmest?

Temperatures of Sunlit Surfaces. An ordinary thermometer is not satisfactory for obtaining surface temperatures, because the bulb may be partly in the air above the surface or partly in the material below the surface. The device at the top of page 379 gives more satisfactory results.

A cardboard tube shields the thermometer so that the bulb is not heated by sunlight. How is the bulb heated? What is the purpose of the insulation?

ABSORPTION OF SOLAR ENERGY
BY VARIOUS SURFACES
(Sun directly overhead)

Deep water	93%
Asphalt	91%
Dark soil	85%
Forests	82%
White sand	80%
White concrete	75%
Grassy meadows	70%
Old Snow	45%
New Snow	20%

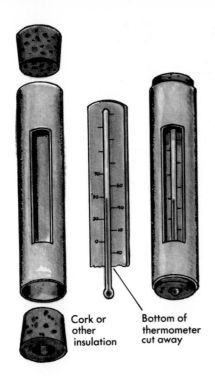

Cork or other insulation

Bottom of thermometer cut away

What may prevent this device from giving exact surface temperatures?

Take the temperature of surfaces that have been in sunlight for several hours. Take the temperature of similar surfaces that have not been in sunlight. List the surfaces in order of their ability to change solar energy to heat.

Reflected Radiation. The table on the opposite page lists approximate percentages of solar radiation absorbed from common surfaces. The energy not absorbed is reflected. What percentage of the solar energy is reflected in each case?

Why may a person become sunburned quickly while skiing? Why is the climate at the base of a south-facing wall usually warmer than the climate a meter away? Explain why a leaf growing over white sand may receive more than its usual share of solar energy.

Land and Water Temperatures. How do land and water compare in their ability to absorb and hold heat energy? Using a lamp, two similar bowls, four thermometers, soil, and water, set up an experiment like the one illustrated below. (As you prepare this investigation, think of the variables which you will want to control.)

Position the thermometers to measure the temperature just above and just below the surface of the soil and water. Turn the light on for ten minutes, and record the temperature of each thermometer every minute. Turn off the light, and continue to record the temperature of each thermometer for the next ten minutes. On a single **graph**, plot the temperature of each thermometer **against** the time. Does the air become warmer over **the water** or over the soil? Over which does the air **heat up** faster? Which received more energy? Which **lost heat** energy faster when the light was turned off?

What are your conclusions to this experiment? In what way do your conclusions help you to explain why the climate of coastal areas tends to be less extreme from summer to winter than inland areas?

Absorption by Chlorophyll. Sunlight is made up of radiations of many wavelengths, including infrared and ultraviolet. Over half of the energy in sunlight is carried by the infrared wavelengths. The remainder is carried by visible and ultraviolet radiations.

Bulb under soil

Water

Soil

Some surfaces, such as asphalt, are able to absorb almost all wavelengths equally. Other surfaces are good absorbers of certain wavelengths but are poor absorbers of others.

Your eyes tell you that green leaves do not absorb all wavelengths. What portion of visible radiation is reflected to your eyes? What portion of visible radiation is absorbed?

The photograph above was made with film sensitive only to infrared radiation. The photograph is light where infrared radiation fell on the film. What do the dark areas represent?

Study the photograph. What surfaces reflect a good deal of the infrared? What surfaces absorb much of the infrared? Why do buildings appear light on some sides and dark on others? Why is the sky dark?

Physicists estimate that chlorophyll reflects 44% of the solar energy falling on it. How does this reflection affect the temperature of a leaf? How does it affect the climate beneath the leaf?

Discuss the effect of green grass on soil temperatures. Compare the surface temperatures of bare soil and grass-covered soil in sunlight.

Atmospheric Heating by Conduction. The photographs below show what happens around heated objects. A special lens system makes dense air appear red or yellow, and thin air appear green.

The left photograph shows the air around a soldering iron. What is happening to the air? Which way is it moving? Why?

The photograph at the right describes conditions above a hot, flat surface. Note that a layer of hot air clings to the surface without rising immediately. Cool air must get beneath this layer to push it upward. The layer may cling to the hot surface for seconds and even minutes.

The right photograph also shows a second layer of hot air which separated from the surface and is rising. Explain the shimmering often seen above hot surfaces, such as pavements in sunlight.

Explain why the climate one centimeter above a sunlit surface may be much different from the climate a meter higher. Why may small animals be killed while crossing a sunlit surface while larger animals are not?

Some trees, such as sugar maples, grow well in open fields, but their seedlings do not live through their first summer in the same places. Explain this in terms of climate. Where might the seedlings find a suitable climate for survival?

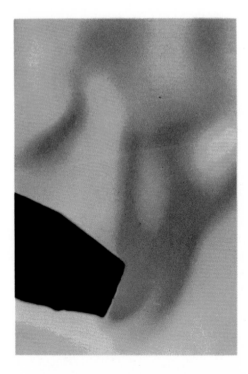

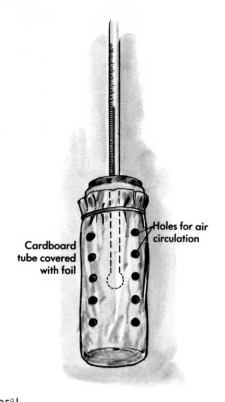

Cardboard tube covered with foil

Holes for air circulation

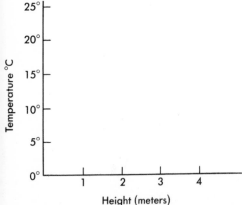

Temperature °C

25°
20°
15°
10°
5°
0°

1   2   3   4

Height (meters)

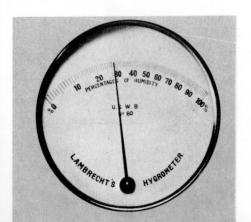

The heated layer over a desert or dry plain occasionally becomes a few meters thick during sunny afternoons. Why does this happen only in calm weather? Explain the importance of this layer in determining the types of plants living there.

Measuring Air Temperatures. A thermometer measures only its own temperature. To determine air temperatures, we must make certain that the thermometer is at the same temperature as the air.

The thermometer shown here has been given a shield to make it more suitable for determining air temperatures in sunlight. What is the purpose of the shield? Why must the thermometer be shielded below as well as above? How does the aluminum foil help? What is the purpose of the holes in the shield? Why may the thermometer still fail to provide exact air temperatures?

Make a shielded thermometer and determine air temperatures at different levels over sunlit surfaces on a calm day. Compare temperatures over asphalt, concrete, bare soil, grass-covered soil, and water. Plot the measurements on a graph. Try to explain differences in the curves. How may your attempts to measure air temperatures change the conditions you are studying?

Changes in Relative Humidity. Air can hold more water vapor as its temperature rises. Why does the relative humidity tend to increase as the temperature drops? Why does the relative humidity tend to decrease as the temperature rises unless there are puddles, ponds, or other water present?

What may happen to the relative humidity over rocks and bare soil during calm, clear days? Measure the relative humidity over bare surfaces in sunlight and shadow, using a hygrometer like the one shown. Explain the differences. Compare the readings with those over grass and water.

SUMMARY QUESTIONS

1. How may the climate of a place be described?
2. Why is it necessary to shield a thermometer to measure the air temperature precisely?
3. What effect does a tree have on the climate at its base?
4. What is the effect of temperature on the amount of water vapor a mass of air can hold?

MICROCLIMATES IN SOIL

Soil is made up of solid particles separated by spaces filled with gases and water. In the spaces are found countless millions of tiny organisms, many of which play important parts in making soil suitable for plant growth. Therefore, the climate within the soil is of great importance to many living things, including ourselves.

Cooling of the Soil. Soil loses heat by radiation, by conduction to the air above, and by evaporation of water. Explain why a cold wind cools the surface of the soil rapidly.

Evaporation removes enormous quantities of heat, about 600 calories per milliliter of water (8500 calories per tablespoonful). Measure the surface temperature of bare dry soil in sunlight. Sprinkle the soil with water having the same temperature as the soil. Note the temperature changes at the soil surface. How does the relative humidity above the soil affect the cooling rate?

Radiation Cooling. The surface of soil radiates heat both day and night. During what part of the day is the outgoing radiation likely to be greater than the incoming radiation? What happens to the soil during these hours? Why does the soil never stay hot for very long?

The radiation from soil is in the long wavelength region of the infrared spectrum, and is called *long infrared*. This type of radiation is readily absorbed by water vapor; the energy is added to the atmosphere as heat. Some of this heat may be transferred back to the soil by conduction.

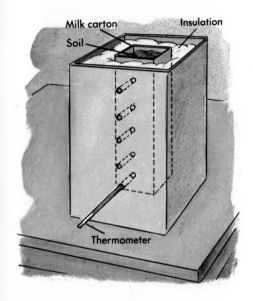

Milk carton

Insulation

Soil

Thermometer

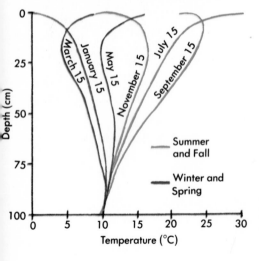

Radiant energy that is not absorbed by the air may be reflected back to the soil by clouds or by the leaves of trees. The remaining energy may be lost to space.

Why does soil cool faster when the air is dry than when it is moist? Explain why soil cools faster during clear nights than during cloudy nights.

Discuss the surface temperatures of desert soils during the day and at night. Discuss the temperature changes in woodland soils.

Heat Conduction in Soil. Part of the heat absorbed at the soil's surface is conducted downward. The rate of conduction depends upon the nature of the soil.

Test the conductivity of soil with the apparatus shown at the left. The milk carton of soil is insulated at the sides and bottom with glass wool or vermiculite. Temperatures are taken at different levels through holes in the side.

Fill the carton with soil that has been kept in a cold place. Set the apparatus in sunlight, and take the temperatures every half hour. Plot each set of readings on a graph.

Test two or more soils at the same time; for example, wet sand, dry sand, loam, and humus. Also test the rate at which soils lose heat by setting cartons of warm soil in a cold place.

Explain the graph of soil temperatures in terms of the results of your experiment. These measurements were made where winters are not very cold.

The Soil Atmosphere. Open spaces in soil contain a mixture of gases that may be different from air. What is the usual percent composition of air?

Decay processes and organisms in the soil usually require oxygen and produce carbon dioxide. These processes and organisms also give off small amounts of other gases, such as ammonia and hydrogen sulfide. What changes might you expect to find in air that is trapped in the soil?

Soil particles are usually covered with a film of liquid water. Therefore, the relative humidity of gases in soil is nearly 100% all of the time.

Why do soil organisms need a change in soil atmosphere? In what types of soil is a change of atmosphere easiest? What agents may bring about the change?

Climate in a Burrow. Why does the temperature in an animal's burrow change little from day to day? Why do summer temperatures remain low? How can a burrowing animal escape freezing temperatures?

Why is the relative humidity in a burrow generally high? Why does the relative humidity change little from hour to hour?

What are some other places that have a climate much like that in an animal's burrow?

Measure the temperature and relative humidity in burrows. Attach a combination thermometer and hygrometer to a long stiff wire and push it into a burrow. Wait for the pointers to stop moving. Take out the instrument and read it quickly.

Sunshine

Muddy water

Clear water

Water

Coil of wire

Ice

Alcohol lamp

MICROCLIMATES IN WATER

Climates in ponds and streams may seem simpler than those in air, chiefly because there are fewer changes, and these changes take place more slowly. Nevertheless, water is an unusual substance compared with other liquids. Temperature changes in water cause some surprising effects that have great influence on the climate both within the water and on the land nearby.

Heat Absorption. Visible light can penetrate many meters of clear water, but most infrared is absorbed within two meters. In shallow water, some solar energy may reach the bottom. What happens then? Why is shallow water often warm in the summer?

Test how mud or other suspended particles affect heat absorption. Fill a jar with a suspension of dark mud. Fill another jar with clear water. Set the two jars in bright sunlight, and measure the temperature changes during a period of 15 minutes.

What is the effect of the suspended particles on the amount of energy absorbed from sunlight? Explain what has occurred.

Conduction by Water. Ponds and lakes receive most of their heat at the surface. What are two ways by which surface water can be heated?

Heat may be transferred from the surface layer to lower levels by conduction, convection, and by mixing caused by wind action. Test how water conducts heat as shown above at the right. Decide from the results of this experiment whether water is

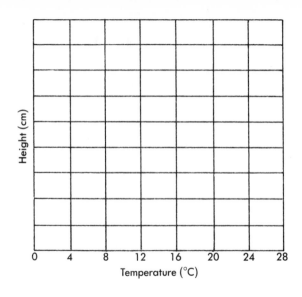

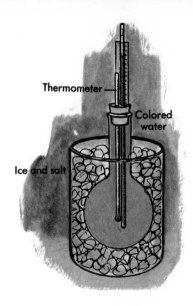

Thermometer

Colored water

Ice and salt

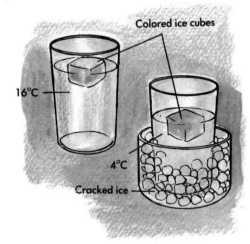

Colored ice cubes

16°C

4°C

Cracked ice

a good or poor conductor of heat, and whether much heat can be transferred to the bottom of a pond by conduction.

Volume Changes in Water. Set up the apparatus shown above. Fill the flask with colored water warmed to 30°C. Force in the stopper until the water rises high in the tube. Then set the flask in an ice-salt mixture.

Keep a record of the height of the water in the tube as the temperature drops. Watch with special care as the temperature approaches freezing. Graph the results. At what temperature does the volume of water seem to be smallest?

Density Changes in Water. What effects do volume changes have on density? What happens to the volume of water as it cools? When was water most dense in the last experiment?

Contraction of the glass flask affects the results of this experiment. Suitable correction for this contraction gives a figure of 4°C for the temperature at which water is most dense.

Make ice cubes from colored water. Lower one of these cubes into water at 16°C. Watch the movement of colored water as the ice melts. Explain its movement in terms of density.

Set a glass of water into a jar of ice cubes until its temperature is 4°C. Lower a colored ice cube into the glass. Explain the movement of the colored water.

Summer Conditions in Lakes. Describe your experiences with water temperatures while

387

swimming in lakes. Where was the water warmest? Is the change from warm to cold layers gradual or sudden?

Lakes receive solar energy only at the top. What happens to the density of the water as it is heated? Why does the warm water stay at the top? Why does the deeper water warm up very slowly?

Explain why the bottom of a deep lake may stay at 4°C all summer.

Autumn Cooling. Lakes lose heat at the surface by long infrared radiation and by conduction to the air above. What happens to the density of the surface water as it cools. What does the cooled surface water do?

The circulation of water set up in this way is known as *convection*. Convection in a lake is a much more rapid process than conduction. One or two days of cold weather can chill a lake to a temperature uncomfortable for swimming, but many days are needed to warm the lake through the same range of temperatures.

The diagrams on page 389 show how convection cools a lake from top to bottom. Explain the diagrams. Why does convection stop when the lake has cooled to 4°C?

The Autumn Overturn. Most of the oxygen dissolved in lake water is absorbed at the surface from the air. The upper layers of a lake usually contain all the oxygen they can hold.

During the summer, there is little mixing of the warm, upper layers with cold water at the bottom.

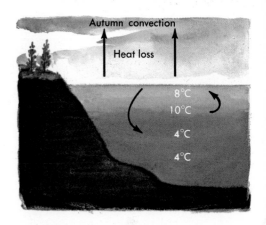

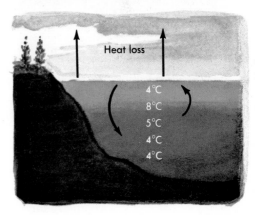

Oxygen in lower levels is gradually used up by organisms and by chemical decay processes. Sometimes fish can no longer live in the deep parts of lakes.

When autumn comes, water at the surface settles, carrying oxygen with it. When all the water is at the same density, winds can mix the lakes from top to bottom. In what ways does this period of autumn overturn have important effects on the climate of deep lakes?

Winter Conditions. Convection stops when a lake has cooled to 4°C throughout. If surface water cools below 4°C, it remains at the top. Why? If the water freezes, the ice remains at the top. Why?

Why is there little mixing of surface and bottom layers during cold winters? What effect does ice have on mixing? What is the probable temperature at the bottom of a deep lake during the winter. What may happen to the oxygen supply?

Spring Warming. Describe the process by which an ice-covered lake warms up to 4°C. When does mixing by wind action become possible? What is the importance of the spring overturn? How does a lake warm up beyond a temperature of 4°C?

SUMMARY QUESTIONS

1. What are three ways by which soil can lose heat?
2. What will happen to the temperature of a surface from which water is evaporating?
3. What type of radiation is absorbed by water vapor?

MICROCLIMATES IN AIR

Microclimates in air are apt to be much more changeable than microclimates in soil or water. Temperatures sometimes change 30°C within 24 hours. Relative humidity may drop from 100% at dawn to 20% at noon. Also, winds often play an important part in these microclimates.

Not all microclimates are equally changeable. Some are sheltered from wind and direct sunlight. Some are near bodies of water which slow down rates of heating and cooling. Nevertheless, most microclimates in air undergo a wide range of conditions from season to season, if not more often.

Climates on Slopes. The small valley shown above runs east and west. The climate on one side is often much different from that on the other side. Why may the temperatures be different? Why may the relative humidity be different? When are the climates on the two sides of the valley most nearly alike? When are they most different from each other?

Study the climates on the two sides of a valley or small hill. Measure soil temperatures, air temperatures, and relative humidity during both cloudy and sunny days. Explain the differences in your measurements.

Effect of Wind. Discuss the effect of wind on microclimates. How may a breeze affect temperatures and moisture in a microclimate? What other changes can be caused by wind?

Discuss the effect of wind upon organisms in a microclimate. How may a breeze affect the temperature of organisms and the moisture content of their bodies? In what ways may wind benefit an organism, and in what ways may it harm an organism?

Study wind velocities in open fields with the device shown here. Each thread and each weight must be as nearly alike as possible. Find out how bushes, high grass, large rocks, and ridges of soil affect wind velocity. Discuss the reasons why some microclimates are not greatly affected by wind, even during severe storms.

Climate on a Rock. Take the surface temperature of a sunlit rock on a windless day. Take the temperature of the air close to the surface. Repeat the measurements for a similiar rock in the shade.

Why do temperatures change rapidly on bare rock? How does the relative humidity change? Why is a rock often covered with dew at night? Why is there little water at other times? How may wind affect the climate on a rock? Discuss some of the problems organisms have trying to live on bare rocks.

Few other microclimates are as difficult for organisms as bare rock. Nevertheless, certain plants, called lichens, thrive on bare rock. The photograph shows lichens common in temperate regions. Other types of lichens may be the only plants in polar regions and on mountaintops.

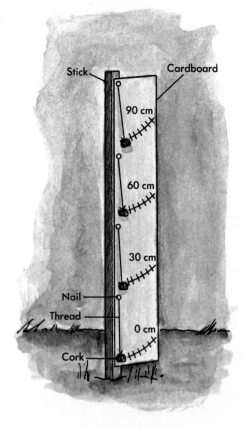

Climates near Water. Water temperature rises much less than that of air or soil while absorbing the same amount of heat. Water has the greatest heat capacity of any natural substance. How does this capacity affect its cooling rate?

How does evaporation affect the temperature of water? When is this effect greatest?

Discuss the expected climate in places near water. What is the effect of the water on air and soil temperatures as summer comes? What is the effect on relative humidity?

What is the effect of water on climate as winter comes?

If possible, measure soil, water, and air temperatures, together with relative humidity at the edge of a pond or stream. Compare the climate with that of a nearby dry field.

Snow Cover. Snow has an important influence on microclimates. Snow is an excellent insulator. Snow also reflects most of the short-wave solar radiation falling on it while radiating long-wave infrared rapidly. A great deal of heat is required to melt snow, about 80 calories per gram.

Discuss the effect of snow on soil temperatures during the winter. Discuss the effect of snow on air temperatures directly above it during the day and during clear nights. What effect may snow have on the coming of spring?

Fields and Woodlands. A single woodland or open field usually includes several types of microclimates. There may be places with bare soil and places with plant cover of various heights. The amount of shade differs from place to place. The slope differs in direction and steepness. There may also be special microclimates under rocks and logs, within hollow trees, and in cracks in the bark of trees.

The photograph below shows an area with a great range of microclimates. Identify some of them.

Investigate the microclimates in a small area near your school. First, survey the area and list all the microclimates you can identify. Then choose one or more of these microclimates for detailed study. Make measurements under different weather conditions and times of day.

Describe the microclimate you have chosen. Give the direction and steepness of the slope. Identify the type of soil. Make drawings showing the type of plant cover, if any. Estimate the amount of sunlight in terms of hours per day.

Measure soil temperatures at several levels beneath the surface. Measure wind velocity at several levels above the surface. On a calm day, measure air temperatures and relative humidities at several levels above the surface.

Repeat measurements on cloudy days and on rainy days if possible.

Compare your findings with those of classmates making studies of other microclimates within the chosen area. Which microclimates have the greatest range of conditions? Which microclimates have the smallest range of conditions?

REVIEW QUESTIONS

1. What is a microclimate? How can it be described?
2. Name three kinds of surfaces that absorb at least 50% of the sunlight falling on them.
3. What happens to the solar energy falling on the grass in photograph A?
4. What happens to the solar energy falling on the sand in photograph B?
5. Why is the water at the top of the pond at A often colder than the water at the bottom during the winter months?
6. When does the wind mix the pond water most easily?
7. Why does the air above the sand in B become very hot during the summer months?
8. Why does the air above the sand in B become very cold during the night?
9. Why do air temperatures change less from hour to hour in the climate of A than in the climate of B?
10. Name three ways by which soil loses heat.
11. What can you tell from a photograph made with film sensitive only to infrared radiation?
12. Why does the air shimmer over a hot road?
13. What are three ways by which heat may be transferred through the water in a pond?
14. At what temperature is water most dense?
15. Describe the changes in the volume of a sample of water as it is cooled to freezing.
16. How is relative humidity affected by changes in air temperature?
17. What happens to air when it is heated?

THOUGHT QUESTIONS

1. What are the characteristics of a surface that absorbs most of the energy of the sunlight falling on it?
2. Why do many desert animals live in underground burrows?
3. How do the temperatures in desert soils compare with forest soils at night?
4. What effect do the seasons have on a pond which freezes over in the winter?
5. Why do places near water tend to have more constant climates than places far from water?

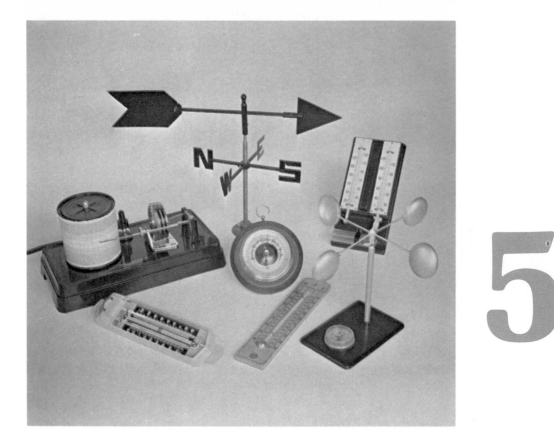

Investigating on Your Own

All of us are amateur meteorologists. We make direct observations by looking at the sky for signs of rain, snow, or clearing. We note the wind, how hard it is blowing, and from what direction. We make indirect observations when using instruments to measure weather conditions. By reading the temperature on a thermometer we can decide how to dress for going outdoors. We make weather forecasts, and plan our daily activities based on what we think the weather will be.

We are unable to observe some changes in the weather except by the use of instruments. Air pressure changes constantly, as does the amount of humidity in the air. To keep accurate records of these changes, we use a barometer to measure air pressure, and a hygrometer to measure humidity.

This chapter will show you how to set up a weather station for keeping records of the daily changes. The chapter will also give you information useful in building the instruments.

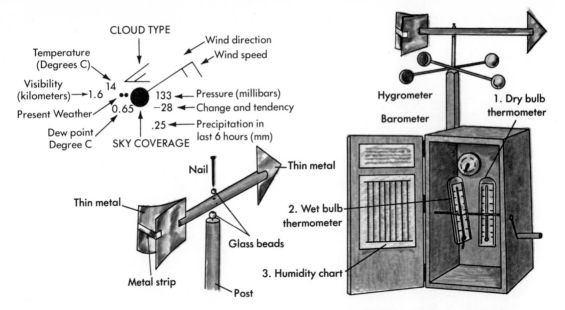

CLOUD TYPE

Temperature
(Degrees C)
Visibility
(kilometers) →1.6

14

Wind direction
Wind speed

Present Weather

Dew point
Degree C

0.65

.25

133 ← Pressure (millibars)
−28 ← Change and tendency

Precipitation in
last 6 hours (mm)

SKY COVERAGE

Hygrometer

Barometer

1. Dry bulb
thermometer

Nail

Thin metal

Thin metal

2. Wet bulb
thermometer

Glass beads

Metal strip

3. Humidity chart

Post

SETTING UP A WEATHER STATION

If weather observations are made at the same time each day, they can easily be compared. From the record of such observations you will be able to identify conditions which repeat in regular patterns. When conditions are reported accurately and regularly, the patterns can be identified and used to predict the weather.

Set up a weather station and keep a daily record of conditions, using the same form shown on weather maps. The thermometer and hygrometer should be shaded from sunlight while measuring conditions in open air. They may be kept in a ventilated aluminum or white box. The wind vane and anemometer should be in the open as high as possible. The barometer may be indoors.

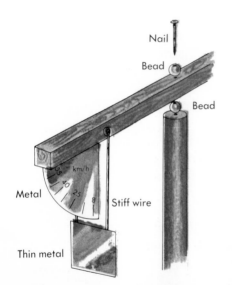

Nail

Bead

Bead

km/h

Metal

Stiff wire

Thin metal

Making a Wind Vane. Slip pieces of thin metal into slots sawed into a round stick. Fasten them with brads. Balance the vane on a knife edge, and bore a hole through the balance point. Pivot the vane on a nail with a glass bead for a bearing. Mount the vane where buildings or trees do not interfere with the movement of the wind.

Making a Deflection Anemometer. Shown here is a deflection anemometer. The deflection of the swinging plate depends upon the speed of the wind. A frozen food dish supplies the swinging plate; a coat hanger supplies the wire. Hold the anemometer out the window of a moving car and mark the position of the deflection plate at different speeds. Use the anemometer to determine wind speeds. Keep a record of your observations.

Making a Rain Gauge. Pour water to a depth of two centimeters into a large, straight-sided can. Then pour the water into a tall, straight-sided bottle. Mark the depth of the water on a strip of adhesive tape, and divide the tape below the mark into 20 equal parts. A metal funnel can be put inside the can to reduce evaporation.

Set the can outdoors away from buildings and trees. After a rain, pour the water from the can into the bottle and read the amount of rainfall to the nearest millimeter.

Making a Nephoscope. A nephoscope is used to determine the speed and direction of clouds. Find a flat piece of glass approximately 20 cm by 20 cm and paint one side black, or use a mirror that size. Draw a circle with a diameter of 15 cm on the unpainted side. Put a dot in the exact center of the circle. The circle should be marked off in degrees and labeled N, E, S, and W.

The base of the nephoscope can be made out of 2-cm thick plywood cut 3 cm larger than the glass (23 cm × 23 cm). Glue the glass or mirror painted side down to the base.

A small sighting post may be made of a piece of 2-cm thick dowel attached to a block of wood so that it can stand alone. The top of the post should be 7.5 cm above the bottom surface of the mirror.

Set the nephoscope down so that south on the circle actually points north. Use a magnetic compass to position it exactly. Make sure it is level by checking with a carpenter's level.

Determine the direction in which the clouds are moving. Place the sighting post in a position around the circle so that the clouds appear in the center of the circle. Watch the motion of a cloud as it moves across the mirror. Note where the cloud moves off the circle.

To calculate the speed of a cloud, it is necessary to know the cloud's height above the ground. The table in the margin gives some approximate heights of clouds. Your local Weather Service office will give you the exact height of the clouds at the time you call. Use a watch with a second hand to determine the time it takes for the cloud to travel from the center to the edge of the circle. Estimate the speed of the cloud by using the formula:

$$\text{Speed} = \frac{\text{cloud height}}{\text{time}}$$

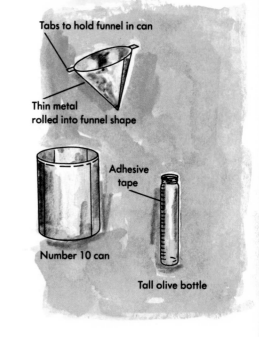

Tabs to hold funnel in can

Thin metal rolled into funnel shape

Adhesive tape

Number 10 can

Tall olive bottle

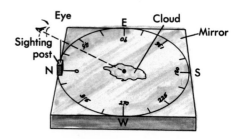

Eye — E — Cloud — Mirror

Sighting post

N

S

W

Compass

Level

CLOUD HEIGHTS

Type of Cloud	Meters
Fair weather cumulus	500–1500
Stratus	2000–4000
Altocumulus	5000
Cumulonimbus	2000–8000
Cirrus	8000–9000

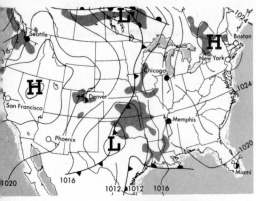

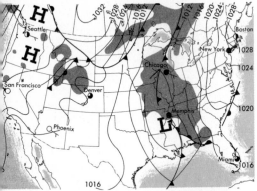

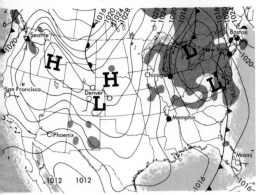

Weather Maps. There are many Weather Service Stations scattered across the North American continent as well as some on the oceans. Each day they send information on the local weather conditions to a central office of the National Weather Service. The information includes such data as temperature, dew point, wind direction and speed, barometric pressure, and cloud cover. The National Weather Service uses these data and the data received from satellites orbiting the earth to prepare a map showing the weather across the continent each day.

The maps at the left show the weather for three days in a row. Locate the centers of the cyclones (L) and anticyclones (H). Identify the air masses and the fronts. Compare the shapes of the cyclones to the diagrams of the development of cyclones found in Chapter 3. Identify the stages you see on these weather maps. Where is snow or rain falling?

Predicting from Weather Maps. On the latest map, note the location of each cyclone (Low) and anticyclone (High). How far have the centers moved in 24 hours? In what direction have they moved? Where would you expect them to be in 24 hours?

Compare the development of cyclones on these maps to the typical cyclone diagrammed in Chapter 3. What do you predict the stage of each storm will be on the fourth day? What regions are likely to experience rain or snow?

Outline a map of the North American continent similar to these weather maps. Sketch in the conditions you predict for the fourth day. Sketch the fronts as you think they might appear using the predicted positions of the cyclones and anticyclones. Shade in those areas where precipitation will probably occur. Try to predict the wind speed and direction, temperature, cloudiness, precipitation, and pressure trends for your own area.

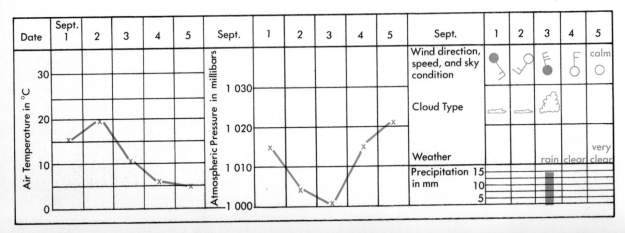

EXPERIMENTAL RESEARCH

1.   Make a series of color photographs of the sky just after sundown for two or more weeks. Study daily weather maps and try to explain the differences in the photographs.

2.   Set up a mercury barometer inside a large bottle that is closed with a two-hole stopper. Change the pressure inside the bottle, and study the changes in the column of mercury. (**CAUTION:** Mercury vapor is toxic.) How much air will a vacuum pump remove?

3.   Produce ozone with a "sterilamp," connecting the lamp to a toy transformer and enclosing the lamp in a bottle. (**CAUTION:** Do not look directly at the lamp, because it gives off ultraviolet light.) Test the effect of the ozone on a culture of bacteria.

4.   Produce ozone as described in Number 3, and test its effect on the odor of onion which has been rubbed on glass.

5.   Make a wet- and dry-bulb hygrometer like that shown at the right. Measure the relative humidity of the air in one place at the same time each day for two weeks or more. Relate your findings to weather changes.

6.   Measure the water that falls in a straight-sided can during a single rain storm. Calculate the water that fell on a square kilometer.

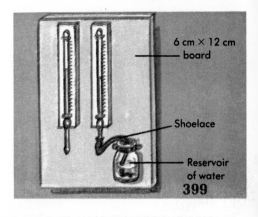

6 cm × 12 cm board

Shoelace

Reservoir of water

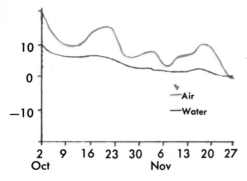

10

0

—10

— Air

— Water

2   9   16   23   30   6   13   20   27
Oct   Nov

7. Collect a large quantity of snow, melt it, and filter out the dust. Study the dust with a microscope, a strong magnet, and a Geiger counter.

8. Leave a large, open jar outdoors on a cold, damp day until a hygrometer reads the same inside as outside. Then cover the jar, bring it into a warm room, and note the change in the hygrometer inside the jar.

9. Keep a daily record of air temperatures and stream temperatures for a month or more. Graph the results, and compare the variations in the two curves. Look for relationships between them.

10. Make a study of the plant zones in mountainous country. Take color photographs of the plant types and prepare a talk based on your pictures.

11. Obtain average hourly temperatures and relative humidities from a local weather station. Plot these on graph paper and study the two curves for relationships.

12. Study the microclimates in a cave for a period of several weeks or months. Measure temperatures, relative humidities, and wind velocities, if any.

13. On a clear, windless night, take air temperatures at 10-cm intervals from soil level up to a height of two meters. Repeat the measurements on a cloudy, windless night. Prepare a diagram that shows the temperature distribution.

14. Set up a thermistor circuit as shown below to measure temperatures in microclimates.

Thermistor (50 ohms)

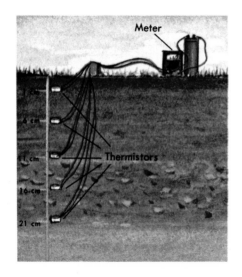

(Thermistors are sold in radio supply stores.) Electrical resistance of a thermistor decreases as its temperature increases, allowing more current to flow. Dip the thermistor in water of known temperatures to calibrate the meter readings. Then use the thermistor to obtain temperatures at different depths in a pond.

15. Bury five thermistors in soil at depths of 1 cm, 6 cm, 11 cm, 16 cm, and 21 cm. Lead the connecting wires to the surface where they can be connected to a meter and dry cell, as shown. Take readings at hourly intervals throughout the day, and throughout the night, if possible. Plot the results as shown.

16. Keep a record of temperature and humidity in an animal's burrow for several weeks. Attach the instruments to a long, stiff wire that can be pushed into the burrow.

17. Make a water sampler like that shown below, and obtain water temperatures at different depths from the top to the bottom of a lake. Repeat the measurements at weekly intervals. In winter, chop a hole in the ice if necessary. Make a series of charts showing the temperature distribution.

18. In very cold weather, weigh a large pan of snow, leave it exposed for 24 hours, and then reweigh it. Calculate the amount of water that changed directly from the solid state to a vapor (sublimed).

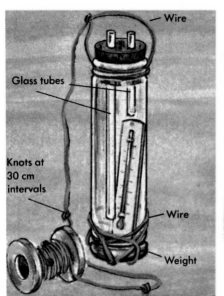

OTHER INVESTIGATIONS AND PROJECTS

Diameter of Raindrops	Approximate Height
.5 mm	300 m
1.0	1 000
2.0	2 000
3.0	4 000
4.0	5 000
5.0	7 000
6.0	10 000

1. Visit a weather station to find out how weather conditions are measured and how weather maps are prepared.

2. Study raindrops by covering the bottom of a flat pan with flour to a depth of 2 cm. Take the pan outdoors during a rainstorm. What did the raindrops do to the flour? Wait for several dozen drops to hit. Use a spoon to put the flour into a strainer. Shake the strainer and allow the loose flour to fall through. Study the small balls of flour remaining in the strainer. How were they made? What shape are they? What size are they? Use the chart at the left to estimate from what altitude the raindrops fell.

3. Use clay to make a model of a mountain range. Show with tufts of cotton the regions where clouds form as wind blows across the range. Use paper cutouts to show the distribution of the different types of vegetation on each side of the mountain.

4. Find out how temperatures are obtained in the lower levels of deep oceans and lakes.

5. Make a map showing the distribution of rainfall in North America.

6. Prepare a report on some of the theories that have been proposed to explain changes in climates, such as the coming and ending of the recent Ice Age.

7. Set up a demonstration like that at the left to show that slope affects the amount of heat absorbed from radiant energy.

8. On a very cold day, measure temperatures above, below, and within snowbanks. Repeat the measurements on a warm day. Make a diagram that shows the distribution of temperatures.

9. Make a model of a warm front or a cold front. Use a sheet of transparent plastic to represent the front, and use tufts of cotton to represent clouds. Dip the bottom of tufts in dilute black ink for thick, dark clouds. A suggested model is shown at the top of the next page.

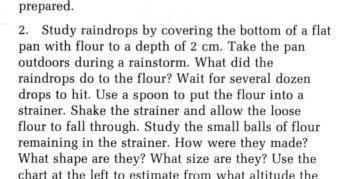

Support

Reflector

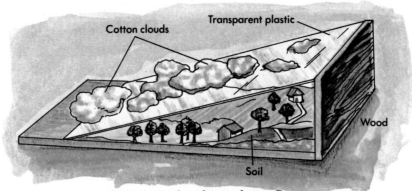

Cotton clouds

Transparent plastic

Wood

Soil

10. Make the sunshine recorder shown here. Cut a round box lengthwise in half, and tack it to a block. Make a cover for the open side and punch a smooth hole in the center. Paint the inside of the recorder a dull black.

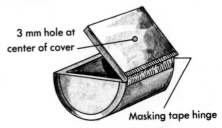

3 mm hole at center of cover

Masking tape hinge

Place a sheet of photographic paper or blueprint paper in the box, fitting it to the bottom. Set the box where sunlight falls on it all day. Remove the sheet and develop it in a cupful of water and a teaspoonful of table salt. (Use plain water for blueprint paper.) The streak caused by sunlight on the paper is broken whenever clouds cover the sun.

11. Report on the history of the National Weather Service (formerly the U.S. Weather Bureau).

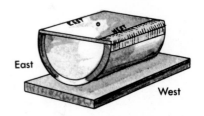

East

West

12. Make detailed notes of conditions before, during, and after a thunderstorm. Use sketches to show cloud forms. Write a report explaining the conditions you have observed.

13. Find pictures of instruments used for measuring weather conditions. Prepare a bulletin board exhibit describing how the instruments operate.

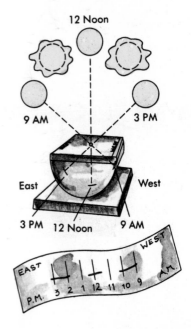

12 Noon

9 AM

3 PM

East

West

3 PM   12 Noon   9 AM

EAST   WEST   A.M.

P.M.   3 2 1 12 11 10 9

14. Make a chart that shows the paths of cyclonic storms and high pressure systems by seasons across North America.

15. Keep an hourly record of wind velocity and direction while at a beach during a period of good weather. Use graphs to show wind conditions during this period.

16. Collect weather sayings, such as "Red sky at night, sailors' delight; red sky in the morning, sailors take warning." Test some of these to find out if they are dependable.

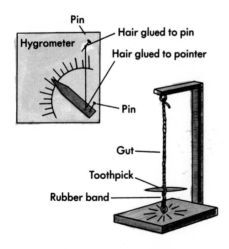

Hygrometer

Pin
Hair glued to pin
Hair glued to pointer
Pin
Gut
Toothpick
Rubber band

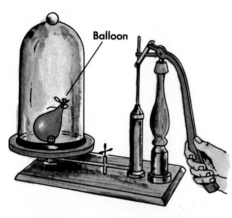

Balloon

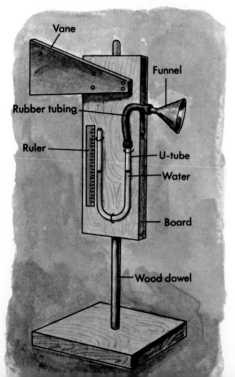

Vane
Funnel
Rubber tubing
Ruler
U-tube
Water
Board
Wood dowel

17. Keep a record of the time when cirrus, cirrostratus, altostratus, and stratus clouds appear in advance of a warm front. Find out how long each type precedes rain or snow.

18. Keep a record of weather forecasts given in newspapers or radio programs and compare them with the actual weather. Calculate the percentage of correct forecasts. (Note that some forecasts may be correct except for the time when the weather pattern arrives.)

19. Find out about careers in meteorology. Prepare a chart describing the jobs available, the background necessary, and the duties performed.

20. Make a hair hygrometer. Remove grease from the hair with strong soap or a weak lye solution. To mark the scale, compare the hair hygrometer with a wet- and dry-bulb hygrometer.

21. Make a hygrometer using a gut violin string.

22. Read about the uses of ozone, and the methods by which it is prepared commercially.

23. Make a circle graph showing the composition of dry air.

24. Prepare a bulletin board exhibit of pictures which show some of the more important sources of dust in the atmosphere.

25. Demonstrate the effect of pressure on a balloon. Put a partly inflated balloon under a bell jar and pump out as much air as possible.

26. Make the pressure type of anemometer shown here. Note that the upright board pivots on a dowel so that the vane keeps the funnel pointing into the wind. Calibrate the anemometer by holding it out the window of an automobile.

27. Prepare a report on hail, describing it and how it is formed.

28. Prepare a report on various forms of precipitation, such as sleet, snow, and ice. Decide how each is formed. Collect pictures that show these different forms.

29. Investigate the attempts to produce rain by seeding the clouds with dry ice and other chemicals.

Major Climates

The general climates of Maine, Kansas, Arizona, Hawaii, and Alaska differ greatly. Within each state a variety of climates can also be found. Geographers may classify climates by zones, such as polar or tropical. Climates may also be described by such terms as hot, dry, and mild. Biologists often describe climates in terms of plant life, such as rain forests or tundra.

Latitude and Climate. All points on an isotherm have the same air temperature. The black lines on the map are isotherms, and show the average temperatures over the world in January. Where are the hottest and coldest regions? Why are these regions over land instead of water? Describe the general temperature distribution between the poles and the equator. Compare the temperatures at 30°, 40°, and 60° north of the equator with the temperatures at 30°, 40°, and 60° south of the equator. Why are the temperatures at southern latitudes different from those at northern latitudes? Discuss the appearance of a similar map for July.

What conditions besides temperature should be considered when describing a climate? Describe the climate where you live so that a stranger would understand. In describing your climate do you include the kinds of plants which grow around you?

Often the various climates take their names from the typical plant life of the area. If you were to travel from the equator northward on a continent, you would find yourself passing through zones where the plants gradually change from one kind to another.

The general zones of plant life between the equator and the poles are shown on page 406. In some regions local conditions may prevent the development of certain zones. Refer to an atlas for the distribution of the zones. In which zone do you live?

Equator   Pole

Tropical forest / Desert or grassland / Deciduous forest / Coniferous trees / Tundra / Snow and ice

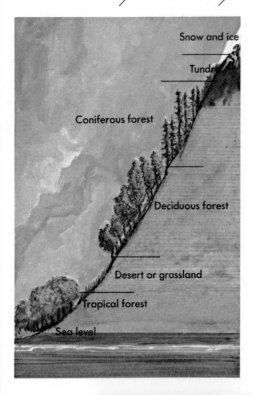

Snow and ice

Tundra

Coniferous forest

Deciduous forest

Desert or grassland

Tropical forest

Sea level

Altitude and Climate. The photograph below shows a shield volcano in Hawaii. The top is covered with snow most of the year; few plants grow there. Some of the high valleys are much like the tundras of northern Alaska; they contain low plants but no trees. Farther down the slope is a belt of coniferous trees, and below that is a belt of deciduous trees. The foothills are dry and covered with grass. Broadleaf evergreen trees in the foreground are not native, but survive as long as they are given water.

The diagram at the left shows the belts of plant life commonly found on high mountains at the equator. Compare it with the diagram above. Note that a mountain climber at the equator passes through the same zones as a traveler going northward toward the pole.

The base of the Hawaiian mountain shown here is in a dry, grassland zone. The base of Mt. Rainier in Washington is in the coniferous tree zone. In what zone do you think the base of Mt. McKinley in Alaska might be? Refer to an atlas to check your answer.

Equatorial Climates. Why is the equatorial region warm? How does this warm belt shift during the year? The diagram shows the movement of air at the equator. Why does the air rise?

Incoming air is usually moist because it has crossed oceans. What happens to this air as it rises? What happens to much of the water vapor in the air?

The photograph at the top of the page shows a forest located at the equator. Describe the climate in the forest. Discuss possible seasonal changes that might occur in the forest.

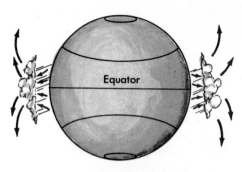

High Pressure Belts. Air at the equator rises to the stratosphere and divides, some flowing north and some flowing south. What is the approximate temperature of this air? Why does the air contain very little water vapor?

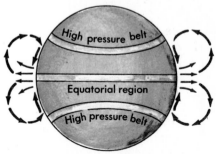

This high level air settles back to earth along two belts located about 30° on either side of the equator. What happens to the temperature of the air as it descends? What happens to the relative humidity? Why does the air along these belts have a slightly higher pressure than the air on either side?

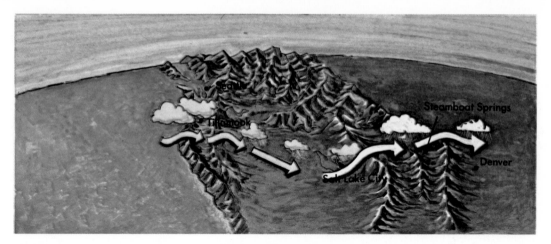

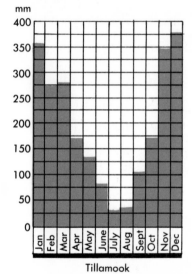

Tillamook
2 390 mm per year

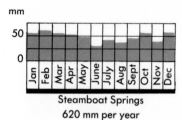

Steamboat Springs
620 mm per year

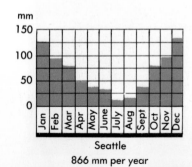

Seattle
866 mm per year

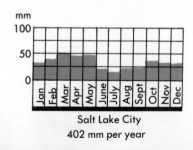

Salt Lake City
402 mm per year

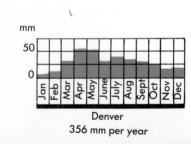

Denver
356 mm per year

The lower photograph on the previous page shows a region that is 30° north of the equator. Describe and explain the climate. Why can such a region have a rainy season?

Influence of Mountain Ranges. The diagram above describes some changes that take place in air moving across western North America. What is the prevailing wind in western North America?

Explain the changes in temperature, relative humidity, and moisture content of air rising up the mountain slopes. Explain changes in air descending on the opposite sides of the mountains. Where should rainfall be greatest? Where might there be dry regions?

The bar graphs show the rainfall of some places in western states. Describe and explain the differences in these graphs.

Locate mountain ranges on a world map that shows the prevailing winds. Where might you find heavy rainfall and little rainfall? Discuss effects of seasonal movement of the wind belts.

Climates in the United States. Climates are classified as *arid* (dry) or *humid* (moist). They may also be classified as *polar*, *subpolar*, *tropical*, or *subtropical*. Climates are sometimes classified as *marine* (influenced by air masses over nearby oceans) or *continental* (influenced by air masses over large land areas).

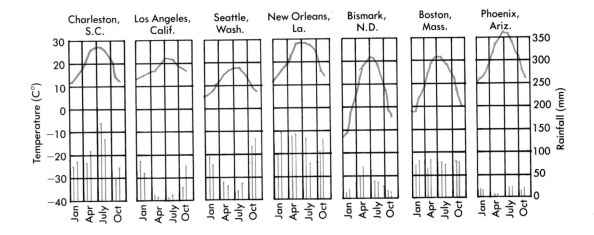

Continental climates tend to differ greatly from summer to winter, since land heats and cools rapidly. Marine climates differ comparatively little from season to season.

The climate may change from season to season as temperature and pressure belts shift north and south. A large section of the United States is influenced by dry, cold, continental air in winter, and moist, warm, marine air in summer.

The graphs above give monthly average temperatures and rainfall for a few cities. Which cities have an arid climate? A humid climate? Which can be classified as marine? Continental?

Which have a dry season and a wet season? Which have a dry season in summer? In winter?

Try to explain the differences in the climates of these cities in terms of prevailing winds, pressure belts, latitude, mountain ranges, and seasonal shifts of temperature belts.

Obtain information about monthly temperatures and rainfall of your region. Explain the climate.

GENERALIZED INFORMATION. *Atlases, encyclopedias, and geography books provide much information about climate, usually for large areas. However, within a large area there may be many small regions each having a special climate. For example, conditions on two sides of a mountain may be very unlike. A spring in a desert produces a climate not typical of the desert. Keep in mind that most reference books provide only generalized information that represents average conditions.*

CITY CLIMATES

It is impossible to cover an area of the earth with buildings, pavements, and parking areas without severely changing the local climate. The materials of which cities are made reflect, absorb, and radiate energy at different rates from soil and vegetation. Factories, home heating systems, and automobiles pour out large quantities of heat, dust, and gases into the atmosphere. Even rainfall, instead of being absorbed by the little soil there might be in the city, flows into ditches and sewers.

Solar Heating in Cities. Cities tend to absorb more solar energy than the surrounding countryside. Many materials of cities, such as asphalt, are excellent absorbers. Materials that are good reflectors when new soon change. What may happen to light-colored building materials?

The dryness of surfaces in the city also is important. There is little evaporation to remove heat. How else does the presence of water help keep temperatures low?

Study the pictures at the left. Explain how vertical surfaces of neighboring buildings trap solar radiation.

Heat Load from Fuels. An enormous amount of fuel is used in a city each day. Some is burned in heating devices. Some is burned in engines. List as many heat sources as you can. How does the nature of the weather affect the importance of each source? How might each source affect the weather?

Cities as "Heat Islands." A weather report often sounds like this: "The temperature outside our studio is 34° and at the airport it is 28°. Tonight the low will be about 24° in the city, and about 20° in the suburbs."

What time of day does this report probably represent? Keep a record of weather reports. Under what conditions are there large differences between urban and suburban temperatures? Under what conditions are the differences small?

Why do rural areas tend to absorb less solar energy than urban areas? What is the effect of chlorophyll on absorption of radiation? What is the effect of streams and ponds on the heating of the countryside? What is the effect of moisture in the soil and in leaves? Why does heat not penetrate soil deeply? Why are breezes more effective in removing heat in the country than in the city?

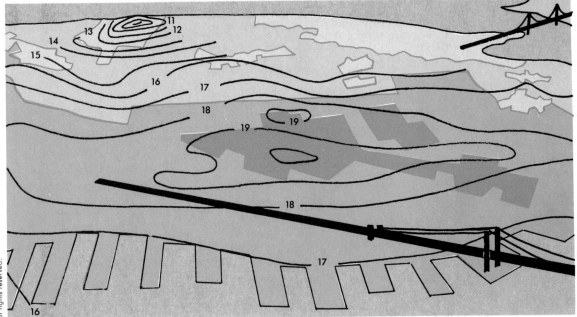

Dark blue—built-up areas; medium blue—less dense areas; light blue—open country.

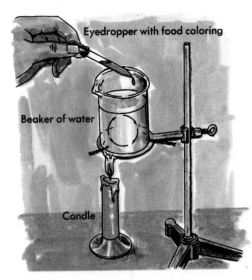

Heating causes water to rise and form a convection cell. Put a few drops of dye into water, heat, and note the circulation of the water.

The above map shows the isotherms for the San Francisco area during a spring evening. Where are the highest temperatures? Where are the lowest? How do the isotherms suggest a "heat island"?

Explain why temperature differences between a city's center and its outskirts remain large after the sun goes down. Explain why the differences may decrease toward morning. How may a strong wind affect the differences?

In regions where winters are very cold, the differences between urban and suburban temperatures may be even greater. Give some reasons why this is true. What happens to snow in cities, and how does this affect temperatures?

Urban Dust Domes. Chimneys, exhaust pipes, and many other sources pour enormous numbers of fine particles into the atmosphere. An idling automobile engine may emit more than a billion such particles per second. These particles are trapped in a convection cell, circulating through it as long as the cell is maintained, and increasing every hour.

The diagram on the next page shows a convection cell over an urban area. Dust makes such a convection cell visible.

The development of a dust dome is best observed from a distance. At first, the city can be seen distinctly. Soon, however, a haze begins to

411

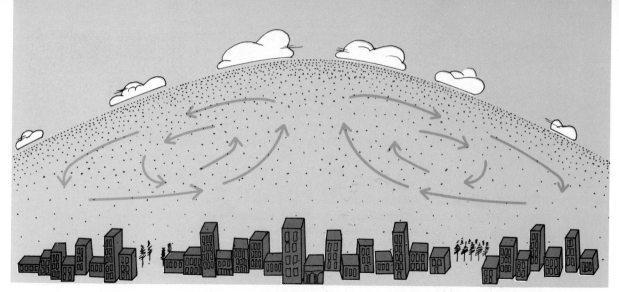

develop. This thickens and spreads both outward and upward. Within a few hours the city is blurred.

Fog and Smog. Dust-filled convection cells over cities often develop fog. Water condenses on the dust particles when the relative humidity approaches 100%. Condensation takes place during the early evening at the top of the dust dome.

The fog in the dust dome grows downward as the night advances. Sometimes it reaches the ground. This extra covering of water droplets reduces loss of heat, and keeps the city warmer at night. The fog also prevents suspended dust particles from escaping from the dust dome. The dust present one day is added on to the next day's contribution.

Obviously, the fog in an urban dust dome is no ordinary fog. It has been given the name *smog* to indicate that it is a combination of smoke and fog.

REVIEW QUESTIONS

1. What is an isotherm?
2. How do the types of trees reflect the change in climate on the slopes of a mountain in the tropics?
3. How does a mountain range influence climate?

THOUGHT QUESTIONS

1. Why does air circulate worldwide?
2. If you were a city planner, what would you do to control the amount of heat built up in a city?

Monitoring the Atmosphere

The problem of predicting tomorrow's weather involves many factors. One factor is the inventing and building of instruments to keep track of weather and to report conditions. Such instruments are called monitors. Many monitors are being positioned over our planet. Some are satellites in orbit, others are on land, in the oceans, and in the atmosphere. A second factor is maintaining a communications network to send the monitors' information to a central location. Another factor is designing a computer to handle the information at the central location as it arrives, and to draw accurate weather maps quickly.

Among the most useful monitors of the weather are the satellites. From their position in space they are able to monitor conditions over an enormous area of the earth. They are also able to keep track of conditions in areas where there are no weather stations, ships, planes, or anything else to report conditions. Weather satellites give meteorologists the chance to see world-wide weather patterns.

TIROS

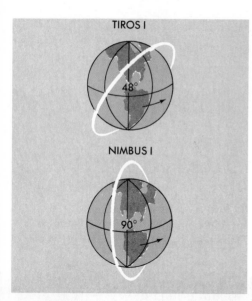

TIROS I

48°

NIMBUS I

90°

Nimbus

The First Weather Satellite. On April 1, 1960, the first satellite to monitor the earth's weather from space was launched. It was called TIROS for the initials of its scientific name, Television and Infra-Red Observation Satellite. At 670 km above the earth, each orbit took 100 minutes. It was put into an orbit from which it could watch the area where hurricanes usually develop.

Heat is the form of energy which brings about many atmospheric changes. For this reason, meteorologists are very interested in monitoring heat all over the earth. In Chapter 2 of this unit you learned that heat energy is carried in the form of infrared radiation. Radiometers aboard weather satellites measure the amount of heat being radiated from the earth.

Improvements in Weather Satellites. Satellites must have a system for controlling their position in space. TIROS I was made to spin slowly, and it kept its axis pointed always at the same star. Although this technique worked, it meant that during half of its orbit the cameras pointed away from the earth. And most of the time it took pictures at a slant to the earth's surface. Later satellites used more advanced techniques for positioning. Nimbus, a more recent satellite, was made to spin once each revolution. The cameras always pointed directly down at the earth's surface.

As a satellite orbits, the earth is also spinning at the rate of 15 degrees eastward each hour. (It takes the earth 24 hours to make one full revolution of 360 degrees.) If a satellite takes 2 hours to complete one orbit, how far will the earth have turned eastward?

TIROS I was launched into an orbit at a 48 degree angle to the equator. Study the illustration in the margin. What portions of the earth did TIROS I never pass over? At what angle to the equator must a satellite orbit to pass over the entire earth?

Other satellites have been put in orbits at 35 700 km above the surface where they take 24 hours to orbit once. They are always over the same location, and their cameras can see almost half of the planet. They are equipped to receive information from widely separated monitors on the earth's surface. From their lofty position, the satellites relay information from rain and river gauges, drifting buoys, ships, balloons, and aircraft to a central location.

Infrared Radiation. Most of the sun's energy comes as radiation with a wavelength longer than visible light. Some of the energy is absorbed at the earth's surface, and some by water vapor, carbon dioxide, and ozone in the atmosphere. The remaining energy is reflected back into space. The energy absorbed or reflected varies from place to place and from season to season. Radiometers on satellites are designed to measure the radiation from the earth's surface and its atmosphere.

The information from the radiometers can be made into photographs similar to those made by television cameras. Since the radiometers are sensitive to both infrared and visible light, they can produce useful pictures of the night side as well as the day side of earth.

Reflected Energy. Energy which is reflected from the earth's surface and atmosphere includes radiations of different wavelengths. Meteorologists use the amount of energy of shorter wavelength to calculate the land temperature of that region.

Some surfaces on the earth reflect more energy than others. In general, light colored areas reflect the most. Those which reflect the most are warmed the least. Surfaces of ice or snow and clouds may reflect as much as 80% of the energy falling on them. Desert areas reflect about 25%. Most land surfaces reflect only about 3%, while water surfaces reflect about 15% of the sun's energy.

The amount of surface covered with snow and ice has a noticeable effect on the earth's weather systems. It is estimated that the frozen area added to Antarctica in the winter is 10 – 12 million square kilometers. This greatly increases the area in which cold air masses develop. By knowing the extent of the polar ice caps, meteorologists can predict the zones where polar frontal systems will develop.

ANALYZING THE NEW DATA. When new information becomes available, it is often necessary to think of ways to organize and analyze it. Clouds seen from space do not look the same as clouds seen from the earth. A completely new classification system had to be developed for examining and analyzing the clouds in the photographs from the satellites. Typical cloud arrangements in well-known situations, such as hurricanes, had to be identified and described.

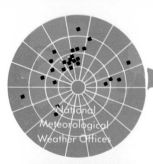

National Meteorological Weather Offices

Airplanes

Surface stations (manned and unmanned)

Ships at sea

Weather balloons

Satellites

National Meteorological Center, Maryland

prints map of Northern Hemisphere weather

Radar at weather stations helps local forecasters

Data stored for research

Satellite photos

Forecast maps sent to local forecaster

Local meteorologist prepares "Today's Forecast Map"

Using the New Data. Although there are many uses for the data gathered by satellites, one very important use is forecasting the weather. The chart above describes the steps taken in preparing a weather map from which the weather may be forecast. From what sources does the weather information come? By what means are all the data organized? In what form are the data summarized? What agency sends out the summaries?

The forecast is prepared from the national weather map, local observations, and the meteorologist's own knowledge of the regional weather trends. Weather reports are much more accurate today than in the past, thanks to a growing worldwide network of weather monitors.

The benefits of accurate forecasting are numerous. Air travel is safer; farmers can protect their crops against frost and other disasters; families can flee areas threatened by storms and floods.

OCEANOGRAPHY

The Ocean Basins

Judging from this photograph, our planet has been
misnamed. It should be called Ocean instead of Earth. Earth
scientists have estimated that over 70% of the earth's surface is
water. Also the water is deep. The average depth of the oceans is
about 3900 meters, while the average height of the continents is
only about 800 meters.

Our astronauts, exploring outer space, took this photograph.
We have learned a great deal from their explorations. The oceans
have been left unexplored until recently. As aquanauts explore
this "inner space," we are finding that it holds enormous
possibilities for providing food, minerals, and energy.

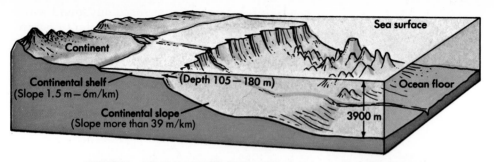

Sea surface

Continent

Continental shelf
(Slope 1.5 m — 6m/km)

(Depth 105 — 180 m)

Ocean floor

Continental slope
(Slope more than 39 m/km)

3900 m

THE CONTINENTAL MARGINS

For as long as people have sailed the oceans, they have been concerned about the depth of the water, especially where it is shallow. A measurement of the depth of water is called a sounding. *From the many soundings taken near shore, precise maps of our shores have been made.*

Only in the last 100 years have people been interested in the depth of the water far from shore. In recent years, a great many soundings made across the oceans have revealed remarkable features. Canyons larger than the Grand Canyon, and mountain chains longer, taller, and wider than any on the earth's surface have been discovered. How is it possible to discover such features by using soundings?

Using Soundings. In the bottom of a cardboard box have someone place some modeling clay formed into an interesting shape. The shape might be a mountain or a canyon. Tape a piece of graph paper over the top of the box. Mark off a knitting needle in centimeters and use it to make soundings.

Make a series of soundings along one line across the box. Write on the graph paper the depth of each sounding you make. Prepare a profile of the object inside by plotting the soundings on another piece of graph paper. Make soundings along other parallel lines and prepare their profiles.

Cut out the profiles, arrange them in order, and tape them down on cardboard so they stand up. Open the box and compare the shape of the object inside with the shape your profiles make. Oceanographers must rely on their soundings, since they cannot check their maps by opening a box.

Oceanographers Use Soundings. Soundings were once made in a similiar way to the one you used. The ship came to a stop. A weight was tied to a line which was knotted every 6 feet. The weight was let down into the water and the knots counted until the weight hit the bottom. The distance of six feet between knots was called a *fathom*.

Sounding with weights is slow and expensive. The invention of *sonar* provides a rapid method for

Taking a sounding

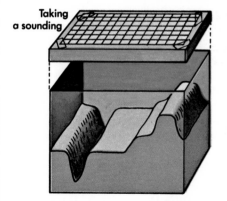

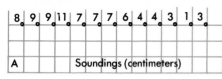

| 8 | 9 | 9 | 11 | 7 | 7 | 7 | 6 | 4 | 4 | 3 | 1 | 3 |

A | Soundings (centimeters)

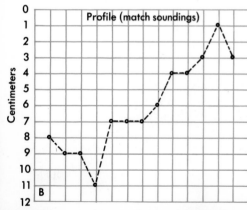

Profile (match soundings)

Centimeters

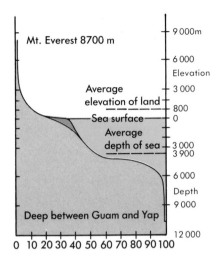

Percentage of earth's surface
above and below sea level

NORTH AMERICA

Atlantic Ocean

Pacific Ocean

SOUTH AMERICA

determining depths. Sonar sends out sound waves and measures the time it takes the echo to return. Soundings are made as the ship is moving. Today the ocean bottom is mapped using sonar.

The graph at the left sums up the results of thousands of measurements. The graph shows that about 30% of the earth's surface lies above sea level. Compare the average height of the land above sea level with the average depth of the ocean. Compare the highest and lowest points on the earth's surface. Note that the graph dips sharply just below sea level. The land masses stand above the ocean bottom like high, steep-sided plateaus.

Continental Masses. Profiles of the edges of the continents show that these are like high plateaus. Portions along the edges may lie just below sea level; these are called *continental shelves*. The steeply slanting edges are called the *continental slopes*. Note the diagram at the top of page 420.

If the ocean level were 300 meters lower, a world map would show the true shape and extent of the continental masses. Study the map below of the continental shelves. What change do you see in the shape of North and South America?

What large bodies of water lie above the continental shelves? Which coast of the Americas has a broad continental shelf? Describe the western edge of the North American continental mass.

The Continental Shelf. On the average, the continental shelf of the North American Atlantic coast is about 64 kilometers wide and 60–80 meters below the surface. The nearly flat surface of the shelf contains sediments which have come from the land and from erosion of the shelf. Oceanographers have gathered evidence from the shelf which may help to explain its history.

Submarines designed for underwater exploration have photographed and gathered samples of the material of the continental shelf. Fossil pollen grains occur in patterns indicating that trees once grew on the shelf. Ripple marks of the kind found on beaches are preserved in the flat sedimentary rock beneath the loose sediments. On the New England shelf can be found gouges and scratches apparently made by moving glaciers. Fossils of organisms which normally live in shallow water no more than 10 meters deep have been found up to 100 meters deep. What theories can you propose to account for these discoveries?

421

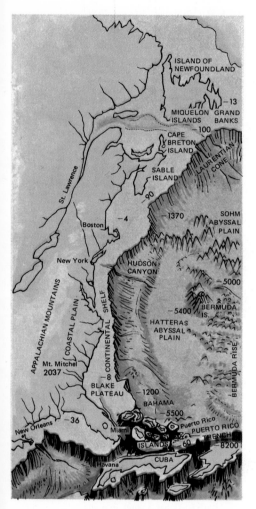

What would happen to the level of the oceans if the water evaporating from them fell on the land as snow and did not flow back into the oceans? This is apparently what happened during the ice ages within the last million years. Evidence shows that the water piled up on the land as snow. Continental glaciers formed causing the level of the oceans to drop by as much as 100 meters. On parts of the continental shelf where fish are plentiful today, birds once flew through trees.

Glaciers seem to have scoured the exposed New England shelf. As the glaciers melted, the level of the ocean began to rise. The shoreline moved up the continental shelf and left behind clues such as ripple marks, banks cut by waves, and fossils of shallow water organisms.

Continental Slopes. Soundings made farther and farther away from shore show that the water quite suddenly becomes deeper. This increase in depth marks the beginning of the continental slope. Along most continental margins the slope is very steep. Notice the margins of the east coast of North America. Describe the continental shelf. How deep is the water at the beginning of the slope? At the bottom of the slope?

Notice that there are canyons cut into the slope. Some of these canyons are larger than any similiar feature on the continent. The Hudson Canyon east of New York City is about 300 kilometers long and reaches a depth of over 1300 meters. In relation to mountains and rivers, where are the canyons located? What theory can you propose to explain how the canyons were formed?

Oceanographers are not certain how the canyons formed. Some believe the canyons formed when the oceans were at their lowest level. At this time water running off the continental shelf eroded the canyons. The canyons were flooded as the water level rose.

Others think that the canyons were formed by underwater currents and landslides. Sediments may pile up at the mouth of a canyon. An earthquake or some disturbance may start them moving. As the sediments flow through the canyon, they gain speed and force. One such current caused a number of undersea telephone cables to break. The speed of the current was determined to be about 80 km/hr.

Study the depth and form of the canyons. What objections can you think of to these two theories?

THE SEA FLOOR

Aquanauts in special submarines have taken photographs like those below. Ships on the surface have taken borings in the ocean bottom, and made soundings. The sea floor is now known to have as many interesting landforms as the continents. However, the sea floor mountains and plains are bigger and the canyons are deeper than those on the continents. In addition, many interesting features are found on the sea floor which are unlike any features found on land.

All these features on the deep sea floor are covered with sediments which seem to rain down constantly. Some of the sediments come from the continents. Others are the undissolved shells of microscopic organisms which live at the surface. Other sources of sediments include melting icebergs, volcanic eruptions, and meteorites from space. It is estimated that 35 000 to 1 million metric tons of matter from space enter the ocean each year.

Settling Rate for Sediments. Collect a liter of sand. Separate the sand into four grain sizes by sifting it through three pieces of wire screening, each with a different mesh. Drop a pinch of the largest grains into a tall jar. Time how long the particles take to reach the bottom. Time how long it takes the grains of the other sizes to settle.

Determine the average diameter of the grains of each type of sand. Measure the average diameter by lining up the grains so that they touch. Count the number of grains in 2 cm. The average diameter is 2 cm divided by the number of grains. Plot a graph of the diameter of the grains against the time it takes each to settle.

Why are the sediments with finer grains more likely to be found far from shore? What other factors might affect the rate at which particles settle in water?

Marine Sediments. Much of the ocean floor is covered with sediments. On the continental shelves, sediments are chiefly mud, sand, and gravel. Farther away on the continental slopes, the sediments consist of small particles of clay and silt.

Deep-sea sediments in mid-ocean come from two sources, the continents and the sea itself. Most sediments from the continents are very fine particles of mud and clay that have stayed in suspension long enough to be carried far from shore. In some places on the deep sea floor, very coarse sediments may be found. Oceanographers believe that these sediments are deposited by very strong

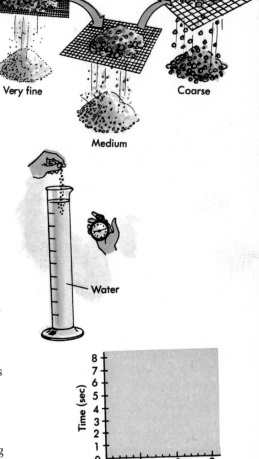

Very coarse

Very fine

Medium

Coarse

Water

Time (sec)

Particle size (mm)

currents set up when an underwater landslide occurs along the continental slopes. Other very coarse sediments may be carried by icebergs which melt far out in the ocean.

Those sediments which come from the sea itself are formed from the hard shells of small animals. After an animal dies, its shell settles to the ocean floor. Such sediments are called *oozes*. Some oozes are calcareous; others are siliceous.

The waters of the Pacific Ocean differ slightly from the Atlantic. There is more silica dissolved in Pacific Ocean water. In recent times, more siliceous oozes have formed in the Pacific than in the Atlantic. However, a siliceous ooze has been found buried under more recent sediments covering a large area of the Caribbean Sea and the Pacific Ocean. Apparently water of the Caribbean Sea dissolved more silica in the past. Where might this additional silica have come? How might water of the Pacific have mixed with water of the Atlantic at this place?

Studying Sediments. Oceanographers study sediments by taking core samples. These are obtained by sinking a tube into the sediments and bringing up the material trapped in the tube. The material is usually found in layers which often contain volcanic ash, clay, mud, and different types of oozes.

Geologists know that certain kinds of animals lived only during a specific time. Once the shells in the oozes are identified, oceanographers can determine when each ooze was deposited.

Over the last 30 years research ships have obtained core samples from many sites in all the oceans. In the Atlantic, an interesting pattern has been noted. The oldest sediments are found on the eastern and western shores. Cores taken closer and closer to the middle are composed of younger and younger sediments.

Some types of organisms are known to live in warm water while other types favor cold. This knowledge has enabled oceanographers to describe past environments in the oceans.

Islands. In some oceans, far from the shores of continents, islands can be found. By studying the map on pages 426–427, identify such islands as the Hawaiian Islands, Azores, Bermuda, and St. Helena. Geologists have researched how these islands formed. Some of the islands are composed of volcanic rocks, while others near the equator are

CALCAREOUS: Anything containing calcium, usually in the form of calcium carbonate.

SILICEOUS: Anything containing silica, the main mineral found in quartz.

Siliceous ooze magnified 4 000 x by a scanning electron microscope. Notice the variety of shells.

Specimens of core samples prepared for microscopic examination.

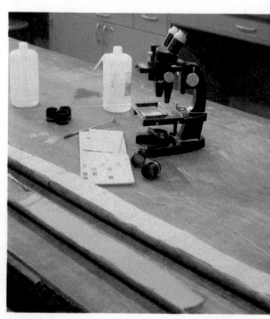

made of coral. A number of islands have active volcanoes which help to explain their origin. After each eruption of a volcano on the ocean floor, a layer of rock forms. With each succeeding eruption, more layers are added until a mountain of volcanic rock is built up which emerges above the water. The Hawaiian Islands are examples of this.

Seamounts and Guyots. Soundings made throughout the oceans reveal that there are many volcanic mountains on the ocean floor that do not rise above the water. These are called *seamounts*. Some have flat tops and are called *guyots*. On the top of some guyots are found rounded stones similar to stones on a stony beach. Ripple marks and fossils of organisms which live on coasts are found on some of these flat tops. What does this evidence show about the history of some guyots?

The tops of the guyots are not found at the same depth. If the tops had once been at the surface of the ocean as the evidence suggests, how can the tops now be as deep as several hundred meters?

There are several possible solutions to the puzzle of the guyots. One is that the sea level was once several hundred meters lower. Another is that the guyots have sunk. A third possibility is that the sea level has changed and the guyots have sunk. How might these possibilities be tested?

There were times when the earth's climate was cold so that water evaporating from the ocean fell on the land as snow. The snow did not completely melt during the warm seasons, so more and more accumulated until the northern regions were covered with glaciers hundreds of meters thick. How would these conditions affect sea level?

Coral Atolls. Some low, circular islands near the equator are composed mostly of coral with a lagoon in the middle. Scientists have drilled through the coral, as much as 1400 meters of coral before coming upon volcanic rock.

Coral animals cannot live in deep water because they require sunlight. How could coral form 1400 meters deep? One theory about the development of coral atolls is shown on this page.

The bedrock of the ocean floor is much thinner than the bedrock of the continents. As a mountain becomes bigger and heavier, it may begin to sink as the bedrock gives way beneath it. The corals grow at a rate which matches the rate of sinking. In this way they have always remained near the surface.

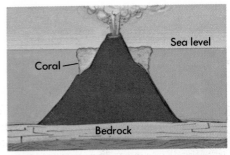

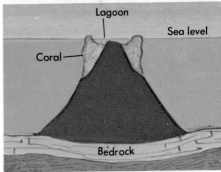

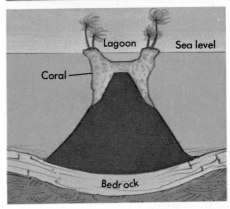

Aleutian Islands

Hawaiian Islands

Transform Fault

Transform Fault

Fracture Zone

Galapagos Islands

East Pacific Rise

Peru-Chile Trench

Puerto Rico Trench

Lesser Antilles

Bermuda Rise

Abyssal Plain

Cape Verde Abyssal Plain

Cape Verde Islands

Canary Islands

Abyssal Plain

Submarine Ridges and Rises. One striking feature on the ocean floor is the mid-ocean ridge. This continuous mountain chain runs through all the oceans. Trace the ridge on the map above. What is the name of the ridge in the Pacific? In the Atlantic? Note the islands that formed where some of the peaks of the ridge rise above sea level. Where does the ridge rise above sea level?

The Mid-Atlantic Ridge has a large split running through the middle of it called a rift. It is up to 30 km wide and hundreds of meters deep. Core samples taken at the bottom of either side of the ridge show remarkable similarities. Samples taken close to the ridge contained young sediments which

were not thick, and the rock underlying the sediments was recently formed. Core samples taken farther east and west of the ridge contained older sediments. In most cases the sediments were thicker, and the underlying rocks were older. Measurements of heat flowing up through the bedrock of the ocean floor indicate more heat flowing at the floor of the rift than elsewhere. What all these observations may mean is discussed in the next chapter.

Submarine Trenches. Locate the submarine trenches on the map. The greatest ocean depths are found in the trenches. Many trenches are bow-shaped and are usually close to a string of islands. What other observations can you make about the trenches, islands, and continents?

Transform Fault Lines. Find the fault lines on the map on the previous pages. The rocks on both sides of the fault have moved passed each other. Which direction do the fault lines run? Where has the ocean floor moved the most? Large fault lines off the west coast of North America form great cracks up to 40 km wide, 4500 km long, with cliffs rising 3 km high.

Abyssal Plains. Large areas under the ocean have a flat bottom. Core samples reveal sediments of coarse particles below layers of finer ones. Pour a mixture of fine and coarse sand into a tall glass of water. What happens?

Find the abyssal plains. What features surround most of them? What would happen if the sediments which accumulate on the continental slopes were caused by earthquakes to flow down the slope onto an abyssal plain?

The surface of the ocean bedrock has many peaks and valleys. If sediments had accumulated from the "snowing" of shells and other matter from above, the peaks and valleys would still by apparent. However, a current with sediments washing over the bedrock would fill in the valleys before covering the peaks. In this way the bottom would become flat.

REVIEW QUESTIONS

1. What is a trench?
2. Describe the mid-ocean ridge.
3. Where are the deepest parts of the ocean likely to be found?
4. Draw a profile of the edge of a continent. Label the continental shelf and slope, abyssal plain, sea level, and continent.
5. What are two theories to explain how the canyons in the continental slope were formed?

THOUGHT QUESTIONS

1. What evidence is there that the tops of all seamounts were not flattened at the same time?
2. Which techniques do oceanographers use to determine the thickness of sedimentary layers on the ocean floor?

Drifting Continents

In each area of science, just as in your classroom, there are people who are interested in small details and other people who are interested in the big picture. In geology, some scientists are interested in the forces which hold atoms together to form crystals of different shapes. Other geologists study big ideas, such as why ocean trenches usually have chains of islands or mountain ranges running next to the trenches. In this chapter you will learn how these "big idea" geologists have developed some new, different ideas about the earth and its history.

Once about every hundred years, scientists are presented with a new idea so different that they are forced to think about their subject in quite a different way. One example was when astronomers had to give up the idea that the earth was the center of the universe and use the sun as the center. Such a dramatic change is called a scientific revolution. In this chapter you will study some recent observations and their effects which caused geology to undergo a scientific revolution.

SOME EARLY IDEAS ABOUT DRIFTING CONTINENTS

Few students study the world map without noticing the amazing similarity of the coast lines of Africa and South America. Perhaps the first person to notice this idea was the person who first made a map of these coasts. Unfortunately, we do not know who that person was. The earliest recorded notice of this idea was made in 1620, soon after maps of this region became available. The oldest known observation was made by Francis Bacon, an English scientist and philosopher.

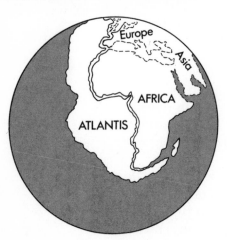

Moving the Continents Together. Locate a large map or globe. Carefully trace the shape of either Africa or South America. Place your tracing next to the other continent on the map. How closely do the two shapes fit together?

Look carefully where the two shapes overlap. Are there any large rivers on either continent in this overlapped region? How might a river alter the shape of a continent?

An Early Theory. The serious study of land forms began in the middle 1800's. At that time, features such as mountains with ocean fossils thousands of meters above sea level were explained in terms of a giant flood. The great flood in the Bible is one theory which explains how fossils got there. The map here is a copy of one drawn in 1858 by Antonio Snider-Pellegrini. He proposed that this model of the earth existed before the biblical flood. With all the land on one side, the earth was unstable. When the floodwaters were piled on the land, the spinning earth became so unstable that the land region ripped apart into its present shape.

In the late 1800's, an idea called uniformitarianism was catching on with geologists. According to this idea, the changes produced by erosion, volcanoes, and earthquakes that are happening around us today can explain the land forms that developed long ago. These land forms are still changing today. How would the acceptance of this idea affect Snider-Pellegrini's theory of the continents splitting?

Alfred Wegener's Hypothesis. In 1912, Alfred Wegener, a German physicist and meteorologist, began talking and writing about his idea that the continents were once joined as shown on the next page. In this arrangement, two different problems facing geology could be solved. One idea involved

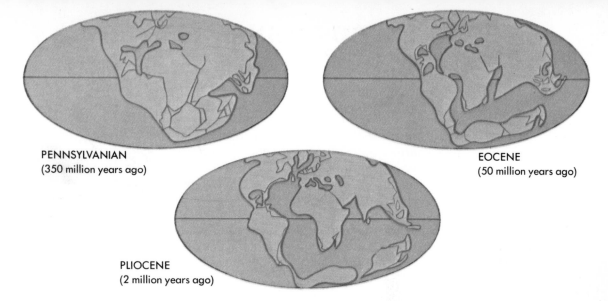

PENNSYLVANIAN
(350 million years ago)

EOCENE
(50 million years ago)

PLIOCENE
(2 million years ago)

the natural evolution of many different plants and animals. The ones in South America, Africa, India, and Australia seem to have evolved in the same place. More recently, the species have continued to evolve apart from each other. Also, these same regions were all covered by a glacier in the late Paleozoic Era . . . and no other parts of the world were. Both ideas could be explained if the continents were arranged as Wegener suggested.

Wegener referred to his idea as a *hypothesis*, an idea which needed to be tested by further observation. Most geologists, however, were not interested in testing this new hypothesis. There were many reasons for this disinterest: (1) Wegener wasn't even a geologist, so how could his ideas be correct? (2) Wegener published his ideas in German books and magazines which were not read by most geologists in other countries: (3) Wegener couldn't think of any force strong enough to pull continents apart, and (4) most biologists preferred to believe that living things evolved similarly because animals in South America, Africa, and Australia might have traveled along narrow land bridges which connected these continents. Why are animals in North and South America more similar today than animals in South America and Africa?

For the next 50 years nothing was done about this hypothesis and it only had a few believers. Wegener died in 1930 while on an expedition to Greenland. He was trying to measure the rate at which Greenland was moving away from Europe.

OBSERVATIONS SINCE WORLD WAR II

Until the end of World War II, there was generally little progress toward the acceptance of Wegener's idea. Most scientists continued to believe in land bridges to explain the similarities of plants and animals on different continents 300 million years ago and the differences in life forms on these continents today. Almost nothing was known about the ocean floor to prove or disprove whether the land bridges ever existed.

As often happens, war produced new instruments which were useful in peacetime. Submarine detectors, with slight changes, could be used to study the ocean floor. As a result, charts like the one on page 422 became available. Surplus navy ships were converted to ocean research vessels which criss-crossed the oceans. Soon maps like the one on pages 426 and 427 were developed, and the last unexplored part of the earth's surface was becoming known.

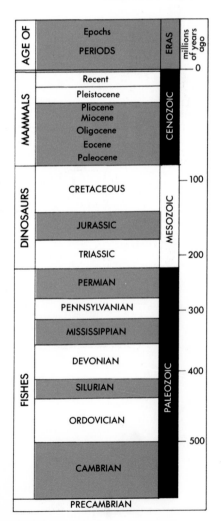

The Shape of the Mid-Atlantic Ridge. People who believed in land bridges thought that the similar shapes of the South American and African coastlines were just a coincidence. Study the Mid-Atlantic Ridge on the map on page 426. Describe its shape. Do you think its shape, similar to the South American and African coastlines is just a coincidence? Why is it more difficult to accept as coincidence three similar shapes than two similar shapes? Does the discovery of the Mid-Atlantic Ridge prove, disprove, support or not support either the land bridge or continental drift theories? Explain your answer.

Fossil Compass Needles. Half fill two small test tubes with iron filings. Place a cork or rubber stopper in the top of each test tube. Lay the test tubes on a level book cover, pointing in the same direction. Bring a strong magnet under each test tube. The magnets should be pointing in different directions, as shown here. Remove the magnets and examine the filings with a hand lens without disturbing the tubes. How do the filings in each tube differ?

A piece of the continental drift puzzle comes from the study of the magnetic iron oxide found in some rocks. Tiny particles of this mineral occur in red sandstones (sedimentary rocks) as well as basalts (igneous rocks). As these rocks form, the tiny magnetic particles line up like thousands of microscopic compass needles. Millions of years later, when the rock is studied, the weak magnetism of the rock still points to the position of the pole at the time the rock formed.

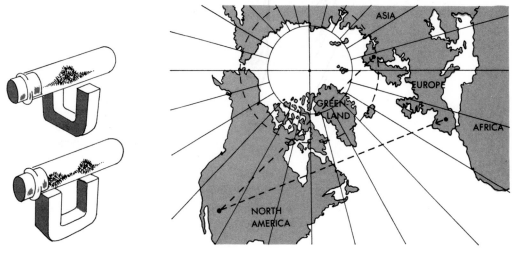

By 1955, geologists had studied fossil compasses in European rocks of many different ages. These fossil compasses did not point in the same direction. Using triangulation, geologists were able to locate the magnetic poles' position throughout the earth's history.

The drawing above shows how Precambrian rocks from two places in Europe point to the location of the magnetic pole in Precambrian times.

The solid line on the map at the right shows the position of the pole as viewed from Europe. The dashed line shows the pole's changing positions in the past as determined from rocks of different ages found in North America. Note that since the beginning of the Cenozoic Era, fossil compasses on both continents pointed to about the same place. Note that before the Mesozoic Era, the two paths remained about equally distant from each other. Imagine traveling backward in time. How might North America have moved during the Mesozoic Era to account for the different paths?

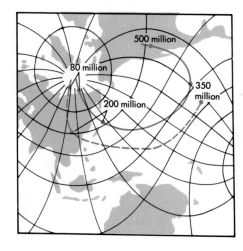

Magnetic Pole and Geographic Pole. Although the magnetic pole seems to move, geologists believe it is usually close to the geographic pole. If this is true, the paths shown here are probably close to the path of the geographic pole in the past. From this knowledge, use a globe to locate where both poles were 300 000 000 years ago, as viewed from North America during the coal forming period. Locate the equator. How does this information help to explain the warm jungle-like conditions in eastern North America which are often pictured during the coal forming period?

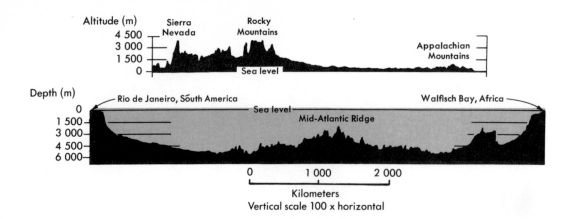

Altitude (m)
4 500
3 000
1 500
0

Sierra Nevada · Rocky Mountains · Appalachian Mountains

Sea level

Depth (m)
0
1 500
3 000
4 500
6 000

Rio de Janeiro, South America · Walfisch Bay, Africa

Sea level · Mid-Atlantic Ridge

0   1 000   2 000
Kilometers
Vertical scale 100 x horizontal

The Ocean Floor. The diagrams above show two profiles of the earth's surface. The upper profile shows the land's surface across the middle of the U.S. Locate your region on this profile. The lower profile shows the sea floor from South America to Africa. Both profiles are drawn with the vertical scale 100 times larger than the horizontal scale. How would these profiles look if the vertical scale were 100 times smaller, making both scales the same? Why are different scales used? How would the bottom profile differ if it were drawn from North America to Europe, passing through Iceland?

The Age of Ocean Floor Sediments. The photograph at the left shows two oceanographers studying a core sample of ocean sediments. The method of obtaining such samples was described in the previous chapter. The thickness of these sediments on the ocean floor varies from less than a meter to four kilometers. Near the mid-ocean ridges the sediments are thinnest. Moving away from each ridge, the sediments get thicker. Note the differing thicknesses of sediments in the drawing below.

Oceanographers have been surprised at how little sediment there is on the ocean floor. The

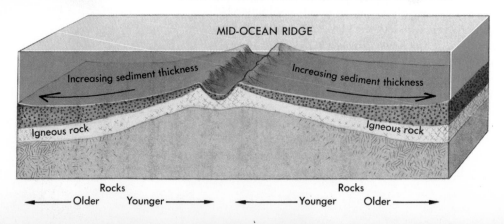

MID-OCEAN RIDGE

Increasing sediment thickness · Increasing sediment thickness

Igneous rock · Igneous rock

Rocks · Rocks
←—Older   Younger—→   ←—Younger   Older—→

average rate at which sediments are deposited is estimated to be about 2 cm every 100 years. How many years would be needed to deposit 4 km (400 000 cm) of sediments? How could this calculation be used to estimate the age of the oceans?

Geologists estimate the oldest rocks found on continents to be more than 3 billion years old. Which appears to be older, the continents or ocean floor? Discuss this difference.

Ocean Floor Rocks. Core samples of mid-ocean ridges indicate that igneous rocks are under the sediments. These rocks form from molten material below the earth's crust. Oceanographers have compared the age of the sediments and rocks of both sides of the ridges. They have made two discoveries from these comparisons. Sediments and rocks near the ridges are younger than those farther away. At about equal distances on either side of the ridge, the sediments and rocks are the same age. This idea is shown across the bottom of these pages.

Evidence From Magnetic Properties. The igneous rocks in the mid-ocean ridges contain particles of magnetic iron oxide. These particles point to the magnetic pole as described on page 433. In addition, geophysicists soon learned to which pole the magnetic particles were pointing. They discovered that bands of newer rock near the mid-ocean ridges show that the earth's magnetic poles often reversed. The north magnetic pole became the south magnetic pole.

Study the magnetic data map shown here. Note the mid-Atlantic Ridge. Look for a pattern on either side of the ridge.

The lower figure shows a simplified model of what geologists believe happens in mid-ocean ridges. How does this model fit the known data about the thickness of ocean sediments and ages of rocks in the ocean floor?

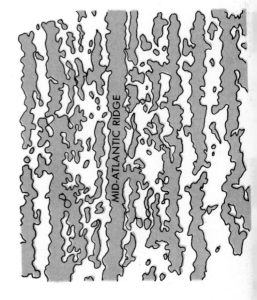

MID-ATLANTIC RIDGE

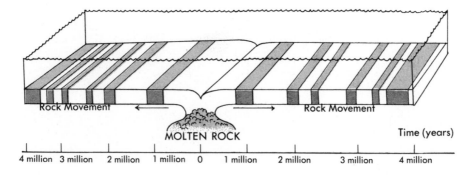

Rock Movement ← → Rock Movement

MOLTEN ROCK

Time (years)

4 million   3 million   2 million   1 million   0   1 million   2 million   3 million   4 million

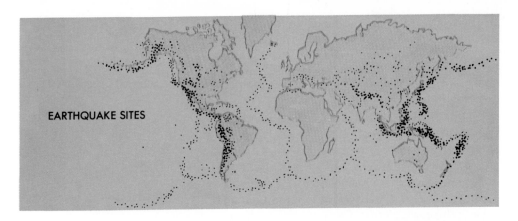

EARTHQUAKE SITES

What Goes Up Must Go Down. In the previous section, geologists described places where the sea floor was spreading apart. If this is happening, the earth's surface may be getting larger. Or there may be places on the earth where some of the surface is disappearing. Think about these two choices. Which one seems more likely? Why?

The world map above shows where earthquakes are most common. Compare these places with the map on the next page. Do all mid-ocean ridges (the sites of sea floor spreading) occur where there are earthquakes? Do all earthquakes occur where there is sea floor spreading? Explain.

A detective might reason that if new surface rocks are appearing in some earthquake regions, perhaps old rocks might be disappearing at other earthquake regions. Geologists have been fascinated since they discovered earthquake regions in the deep ocean trenches next to volcanic mountains. Note these regions on the map on the next page.

Geologists can determine the location of an earthquake by studying the shock waves which are received at two or more earthquake recording stations. The data from such recorders can also be used to calculate the depth of the earthquake. The diagram at the left shows the location and depth of many earthquakes at one ocean trench. Compare these data with the geologists' interpretation at the right. Suggest how trenches and volcanoes might occur.

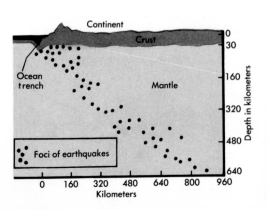

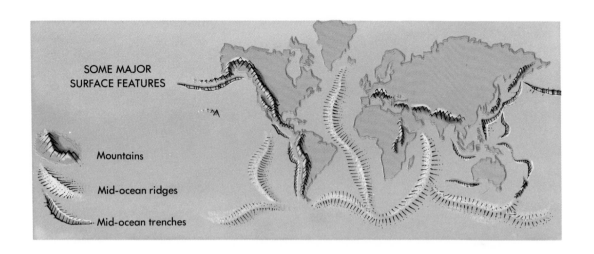

SOME MAJOR
SURFACE FEATURES

Mountains

Mid-ocean ridges

Mid-ocean trenches

What Goes Up Doesn't Necessarily Go Down.
The idea of sea floor spreading is hard to ignore. If
the earth's surface is spreading apart in some places,
perhaps there are other regions where the earth's
surface is coming together. Two common types of
mountains appear to have formed by rock layers
being compressed. These mountains are called
folded and thrust-faulted mountains. An example of
folded mountains is the Appalachian Mountains of
the eastern U.S. Examples of thrust-faulted
mountains include the Alps, Himalayas, and some
western U.S. mountain ranges.

Folded and faulted mountains have been
identified and studied for over 100 years. Until
recently, geologists explained the formation of these
mountains by thinking of the earth as similar to an
apple. As an apple dries (earth cools), the volume
decreases and the surface becomes wrinkled. Suggest
a different explanation based on the idea of sea floor
spreading.

The drawing at the left shows an artist's idea of
how ocean rocks, pushing against land rocks, might
form folded and faulted mountains. Study the map
above. Locate regions where such mountain building
may have occurred or is now occurring.

Earthquakes and Sideward Motion. Study the
two maps across the tops of these pages. Locate some
regions of earthquake activity where there is no
evidence of sea floor spreading, no evidence of deep
ocean trenches, and no evidence of folded or faulted
mountains. In these areas there is often sideward
movement of the land on either side of the
earthquake.

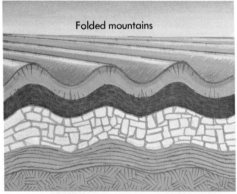

Folded mountains

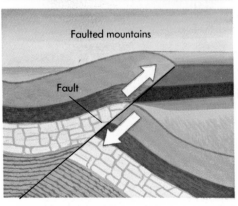

Faulted mountains

Fault

437

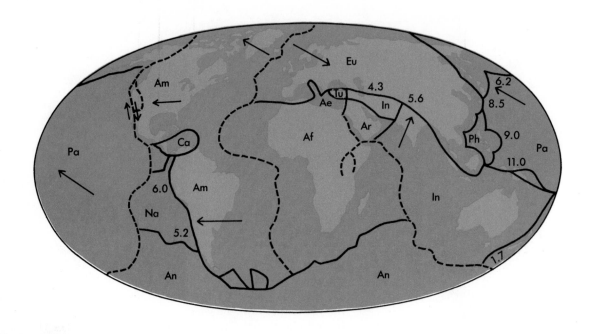

Major Plates

Af	African	Eu	Eurasian
Am	American	In	Indian-Australian
An	Antarctic	Pa	Pacific

Minor Plates

Ae	Aegean	Na	Nasca
Ar	Arabian	Ph	Philippine
Ca	Caribbean	Tu	Turkish

Juan de Fuca Plate.

– – – – – – – – – Plate boundaries of sea floor spreading

Numbers indicate cm/yr movement toward each other on adjacent plates.

THEORY OF PLATE TECTONICS

Today, geologists generally agree on how mountains, ridges, and trenches form. The development of these features is explained by the theory of plate tectonics. According to this theory, the earth's surface is divided into regions called plates. As these plates move, forces are exerted on adjoining plates, ridges, mountains, and trenches.

Note the plates on the map on page 438. Geologists sometimes simplify this map and think in terms of six major plates. Locate these six plates.

Movement Along Plate Boundaries. Where one plate touches another, one of four things may occur: (1) the plates may be moving apart, (2) the plates may be pushing together, (3) one plate may be sliding past the other, (4) the force may not be strong enough to produce any noticeable effects.

Note the arrows on North America and South America, showing the American plate colliding against the plates to the west. What land features are found along the western edges of these continents? Suggest how these features formed.

Note the arrows in the North Atlantic which show that the American plate and Eurasian plate are moving apart. What feature is found here? Suggest what direction Africa is moving relative to South America for the Mid-Atlantic Ridge to form at the edge of these two plates.

The arrow in northern India indicates that this plate is bumping into the Eurasian plate. What land feature is found here?

The earth's crust is disappearing fastest in the Pacific Ocean. Points on adjacent plates in this region are moving closer together at a speed of about 10 cm/yr. What ocean features are found where two plates collide?

Note the two arrows along the San Andreas fault in California. How does the photograph here support the idea of plate movement?

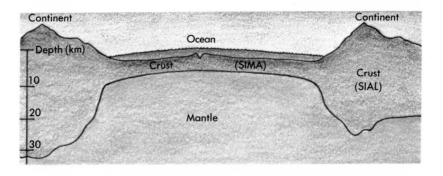

Plate Thickness. Geologists have long noted that the rocks found on continents are distinctly different from the rocks on the ocean floor. Generally, rocks found on continents are less dense and lighter colored than rocks on the ocean floor. The less dense continental rocks, rich in *silicon* and *aluminum*, are called *sial* rocks. The more dense ocean floor rocks, rich in *silicon* and *magnesium*, are called *sima* rocks.

Earthquake wave studies show that the lighter sial rocks extend deeper below the earth's surface than do the heavier sima rocks. The lower limits of the sial and sima rocks mark the boundary between the earth's crust and the mantle below. The diagram above shows some estimates of the thickness of the earth's crust. Scientists have no direct observations of the earth's mantle. Why might the first observations of the mantle be made from drilling platforms in the ocean?

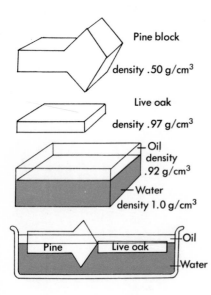

The model at the left shows the idea of thick sial rock (pine wood) and thin sima rock (much denser live oak wood). Note in this model that the sial "rocks" extend to a greater depth under mountains than in other parts of the continent. According to this model, the earth's crust (consisting here of two different kinds of wood) is floating on a denser mantle (shown here as water). The continents float higher because they consist of less dense material than do ocean rocks. This model is consistent with many observations about the earth's crust.

The depth of earthquakes at the edge of moving plates gives an idea of the thickness of each plate. Each plate consists of the crust plus the mantle to an average depth of 100 km. This thickness agrees with other calculations about the strength required for the plates to withstand all the forces acting on them. The outer layer of the earth which makes up the plates is called the *lithosphere*.

PERMIAN — 225 million years ago

TRIASSIC — 200 million years ago

JURASSIC — 135 million years ago

CRETACEOUS — 65 million years ago

CENOZOIC — Present

In the Beginning. The diagrams above show an idea of what geologists believe is the past history of the lithospheric plates. According to this idea, the continents of South America, Africa, Australia, Antarctica, and the subcontinent of India were joined together about 200 million years ago. This idea is supported by common fossils, sedimentary rock layers, and evidence of glaciation. For the last 100 million years, the fossils and present life forms have been evolving on each continent along different paths. This evidence suggests that a continuous land mass spanned these five separate regions. About 100 million years ago this land area split apart. This huge region is now given the name of *Gondwanaland*.

A second land mass, called *Laurasia*, included what today is North America, Europe, and much of Asia. Laurasia was believed to be joined until about 50–60 million years ago. Suggest why European and North American plants and animals are more similar today those of South America and Africa.

The Missing Force. Wegener's idea of drifting continents was not readily accepted. The biggest reason seems to have been Wegener's inability to describe a force which would split apart the original land mass. Today the idea of drifting continents has been changed to moving lithospheric plates (each plate contains 2, 1 or no continents). This idea is generally accepted but not because a force has been found to move the plates. Instead, an overwhelming amount of new evidence has been found which is best explained by the idea of moving plates.

At this time, many forces are being proposed and discussed. Some include gravitational forces, the forces associated with convection currents within the earth, and the inertia of plates in motion. Agreement about the forces may only occur when other ways of getting information about the earth's interior are developed, or when direct observation of the mantle is possible.

Looking Back on a Scientific Revolution. A hundred years from now, the 1960's and 1970's may be viewed as the most exciting time in the history of geology. During this time the last unexplored region, the sea floor, received its first serious study. New instruments made possible new observations, which were combined to form a completely new picture of how the earth's surface changes.

Based on past history in other sciences, probably no such similar large change, or scientific revolution, will occur in geology for 100 years or more. Instead, new observations will change only slightly our understanding of plate tectonics. The basic idea of plate tectonics, unknown to your parents when they were in school, will probably be taught to your children and grandchildren.

REVIEW QUESTIONS

1. Use the idea of plate tectonics to explain the formation of the Himalaya Mountains of northern India.
2. What are two ways that World War II speeded up ocean research?
3. Why is the earlier term, continental drift, an inexact term to describe the theory of plate tectonics?
4. What is some evidence for sea floor spreading?

THOUGHT QUESTIONS

1. If the Pacific Ocean floor is disappearing into a trench in the western Pacific at the rate of 10 cm/yr, how many years will it take for the rocks along the California coast, 10 000 km away, to disappear down this trench?
2. Suppose you, like Rip Van Winkle, slept for 25 years and awoke to find that your geography book listed Eurasia and India as two separate continents. Propose a possible explanation for such a change.

3

Ocean Water

For millions of years the water on our planet has followed a cycle. From the surface of the ocean an enormous amount of water evaporates each year to become part of the atmosphere as water vapor. As conditions in the atmosphere change, the water vapor condenses, precipitates as rain or snow, and becomes part of the land. Rain and snow are nearly pure water because when water evaporates it leaves behind whatever impurities it may have contained.

As the rain and melting snow flow over the land, erosion begins. By dissolving some materials and carrying others in suspension, rivers and streams transport large amounts of land to the ocean's shores. What happens to these materials washed into the oceans, and how these materials affect the ocean water are important concerns for oceanographers.

COMPOSITION OF SEAWATER

Seawater shares many characteristics with fresh water. The presence of dissolved salts changes certain properties and adds others. The great surface of the oceans exposes the water to conditions such as winds, temperatures, and rainfall. Consequently, ocean water varies slightly from place to place and from time to time. The behavior of ocean water and the materials dissolved in it depend partly upon these variations.

Using a magnifying glass to observe crystals growing in a dish

Evaporation of Seawater. Collect some ocean water. If you live too far from the ocean, make up some ocean water using the salts sold by pet stores dealing in tropical marine fish. Pour the ocean water into a shallow dish, and put a light over it.

Remove a few drops from the dish with an eye dropper. Put the drops on a microscope slide, and pass the slide over a flame to warm it slightly. Looking through the microscope, observe the formation of crystals. What is the shape of the crystals at the edge of the drops? Why are the crystals forming? How many different shapes of crystals can you distinguish? Which of the crystals are found in the greatest amount?

Gently boiling seawater to dryness is one way to find out how much dissolved material it contains. Measure the weight of a clean dry beaker,

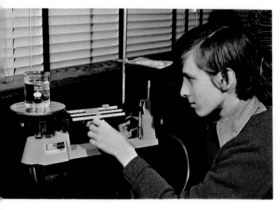

and pour 100 ml of seawater into it. Heat the water until it boils gently. Too vigorous boiling will lose some of the dissolved material in the drops that splash out. After the water has boiled away and the beaker has cooled, weigh the beaker with the remaining salts. What is the weight of the salts in the 100 ml sample of seawater? What would be the weight of the salts in a liter of seawater?

Salts in Seawater. One of the most notable characteristics of seawater is its load of dissolved materials. The table below gives the composition of one sample of seawater. What percentage does 35 000 parts per 1 000 000 parts represent? What percent of seawater is dissolved minerals? How might this value differ at the mouth of the Amazon River? How might this value differ in the Red Sea where evaporation is rapid?

What is the most plentiful mineral in seawater? What percent of seawater is composed of this mineral? How does the quantity of this mineral compare with the total of all the other minerals?

Solutions. Water dissolves many minerals. The amount of any one mineral that will dissolve in a given amount of water depends on the temperature and the pressure. Further, more of some minerals will dissolve in a certain amount of water than of others. The amount of a mineral which will dissolve in a certain volume of water is called its *solubility*. The graph at the left compares the solubility of several minerals at various temperatures. How does the solubility of each mineral change as the temperature increases? How does the solubility of sodium chloride compare to other minerals?

Saturated Solutions. Add salt to some cold water, a little at a time. Stir the mixture while adding salt. Notice that after a certain amount has been added, no more will dissolve. The solution is said to be *saturated*. What happens to the undissolved salt when you heat the solution? Try it. What happens when you let the solution cool?

The mineral, calcium carbonate, will come out of solution when the temperature drops or when the pressure rises. Seawater is nearly saturated with calcium carbonate. At the bottom of the ocean the water is colder than at the surface, and the pressure is much greater. Under these conditions calcium carbonate comes out of solution. It falls to the ocean floor, and after millions of years forms limestone.

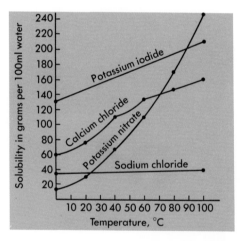

COMPOSITION OF SEAWATER

Dissolved Substance	Parts per Million Parts of Seawater
Sodium chloride (NaCl)	27 213
Magnesium chloride (MgCl₂)	3 807
Magnesium sulfate (MgSO₄)	1 658
Calcium sulfate (CaSO₄)	1 260
Potassium sulfate (K₂SO₄)	863
Calcium carbonate (CaCO₃)	123
Magnesium bromide (MgBr₂)	76
Others — traces in varying amounts	
Total	**35 000**

Analysis of Seawater. In the late 1800's, a Scottish chemist analyzed a large number of samples collected from several oceans by the *HMS Challenger* on its round-the-world voyage. A careful analysis revealed two important properties of seawater. First, the total weight of salts per liter varies only slightly in the oceans. Each liter of seawater contains from 32 to 38 grams of salt no matter where it is collected. Second, the relative amount of each of the most plentiful salts is always the same. The total amount of salts in a liter of seawater may vary, but the ratio of the amount of one salt compared to another remains the same.

The Balance of Salts in Seawater. For millions of years water has been evaporating from the sea and falling on the land as rain. The rain water flows over the land and picks up many materials. It is estimated that the rivers of the world carry more than a thousand million metric tons of dissolved salts to the ocean each year. Once in the sea the materials may dissolve or drop to the bottom.

Some of the salts stay in solution. Some become attached to very fine solid particles and sink. Some are removed by living organisms that use the salts for making shells, bones, and flesh. Other salts are blown into the air from sea spray. All these possible paths appear to be in balance. As certain salts increase, more animals grow more shells and use up the excess salts, thus restoring the balance. Scientists studying the chemistry of seawater are mostly concerned with describing the paths of the elements in the sea.

Dissolved Gases. In addition to salts, there are gases dissolved in seawater. Oxygen and carbon dioxide are found in seawater. The temperature of the water and the pressure on it greatly affect the amount of gas which will dissolve in the water.

Collect a glass of cold water, note the temperature of the water, and let it stand for several minutes. Notice the bubbles collecting along the sides and the bottom of the glass. Where do you think the bubbles come from? Compare the temperature of the water with what it was before.

Heat some water and pour it into a clear plastic tumbler. Note its temperature and let it stand. What do you observe after several minutes? How is this glass of water different from the one you set up before? How does temperature affect the amount of gas that will remain dissolved in water?

PROPERTIES OF SEAWATER

The conditions of temperature and pressure have a great effect on the amount and kind of material that will dissolve in seawater. More solid material will dissolve in warm seawater than in cold. However, a smaller volume of gas will dissolve in warm seawater than in cold.

As the character of seawater changes, so do many of its properties. Sound is conducted more quickly through cold water than through warm. Ships using sonar must take the temperature of the water into account in order to make accurate soundings.

Freezing Temperature of Seawater. Observe the effect of dissolved salts on the freezing temperature of water. Make a mixture of one cup of crushed ice with one tablespoonful of table salt. Stir the mixture and take its temperature. Add one tablespoonful of salt after another up to six. Stir and record the temperature each time more salt is added. What effect does adding salt have on the temperature of the mixture? What do you think will happen if you keep adding salt indefinitely?

Half fill one test tube with seawater and another with fresh water. Place them in the salt and crushed ice mixture. Insert a thermometer into each test tube and record the temperature as each begins to freeze. Water with a salinity of 35 000 parts per million freezes at about −2°C. How does the freezing point of seawater compare to the freezing point of fresh water?

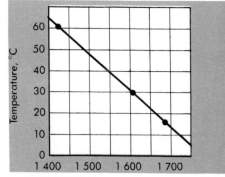

Speed of sound in water, m/s

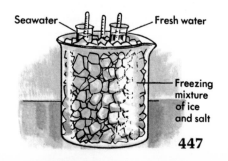

447

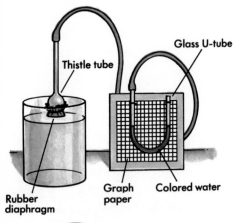

Thistle tube

Glass U-tube

Graph paper

Colored water

Rubber diaphragm

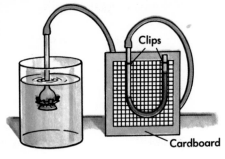

Clips

Cardboard

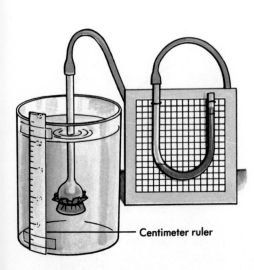

Centimeter ruler

Density of Fresh Water and Seawater. To find out how much one liter of fresh water weighs, weigh a container, pour in a liter of water, and weigh it again. How much does a liter of fresh water weigh? *Density* can be expressed as weight divided by volume. What is the density of fresh water in grams per milliliter? Obtain some seawater and calculate its density.

Underwater Pressures. Fill a pail with water. Put your hand into a plastic bag. Place your hand wrapped watertight just under the surface of the water. Slowly move your hand deeper into the water. How does the pressure on your hand change?

Depth of Water and Pressure. Cut a piece of rubber from a balloon. Stretch it tightly over the mouth of a thistle tube. Bend a piece of glass tubing into the shape of a U. Mount the tube on a piece of cardboard covered with graph paper. Fill the U-tube half full with colored water. Connect it to the thistle tube with a rubber tubing about 75 cm long.

Note the level of the water in the open end of the U-tube. Place the thistle tube under water. How has the level of the water in the U-tube changed?

Water pressure can be calculated by measuring the weight of water on a square centimeter. One cubic centimeter of fresh water weighs one gram. The pressure under one centimeter of fresh water is one gram per square centimeter. What would the pressure be under 5 cm of fresh water?

Place a ruler along the side of a container. Record the level of the water in the U-tube as the thistle tube is submerged 5 cm deeper. Calibrate the U-tube in grams per square centimeter.

Undersea Pressure. Imagine a cubic container one meter on a side filled with seawater. What is the pressure on each square centimeter of the bottom? This is the pressure under a meter of seawater. What is the pressure under 2 meters of water? What is the pressure at the ocean's greatest depth (about 12 000 meters)?

Underwater Explorations. Pressures in the ocean depths are too great for humans. Divers are limited to depths of about a hundred meters.

For greater depths, strongly constructed vessels must be used. Many of these are small, because the total force on a large vessel is enormous. Sometimes the crew and instruments are carried by a small sphere, and everything that is not damaged by pressure is carried outside.

The Specific Gravity of Ice. *Specific gravity* is a comparison of the weight of a material with the weight of an equal volume of fresh water. Suppose a piece of wood has a volume of 10 cubic centimeters and weighs 5 grams. As you know, 10 cubic centimeters of water weighs 10 grams. The specific gravity of the wood, therefore, is 0.5.

Make a hole near the top of a plastic container. Fill it up to the hole with ice water. Wipe a large ice cube dry with paper towels and weigh it. Put the ice cube in the plastic container, hold it beneath the surface with two sticks, and collect the overflow. Weigh the overflow. Does the ice cube weigh more or less than the same volume of water? What is the specific gravity of the ice cube?

The piece of wood with a specific gravity of 0.5 will float in water with five-tenths (one-half) of its volume under water. Another piece of wood having a specific gravity of 0.2 will float with two-tenths of its volume under water. Use your calculation of the specific gravity of ice to predict the volume of an ice cube under water. Test your prediction.

Ice Cubes in Water. Float an ice cube in a glass of warm water. Estimate how much of the ice is above the water and how much is below. Observe the ice cube as it melts. What difference do you notice in the rate of melting in the portion under water compared to that above water? Did your ice cube roll over as it melted? Why might it do that?

Mix fine clay or mud with water and freeze the mixture. Place a frozen cube of this mixture in a glass of water, and observe the cube as it melts. In what way is this cube more like an iceberg than an ice cube of pure water? Describe how the ice cube looks underwater and above the water. What happens to the material frozen in the ice cube?

Icebergs. A glacier that flows into the sea breaks up, and the pieces float away. These pieces are *icebergs*.

Icebergs may be several kilometers long and 300 meters thick. Most icebergs come from the Greenland Icecap from which more than 100 glaciers reach the sea. The largest icebergs come from the shelf ice of Antarctica.

Ice has a density slightly less than water. Therefore, icebergs float with most of their volume below water. An iceberg may extend below water eight or nine times its height above water.

Icebergs carried by ocean currents into shipping

449

lanes become a major menace. Avoiding them is especially difficult when fog forms in the region.

Specific Heat. A property of water that distinguishes it from other liquids is the amount of heat necessary to raise its temperature. This property is called the *specific heat*. Specific heat is measured by the number of calories needed to raise one gram of water one Celsius degree.

Weigh out 100 grams of water into one beaker. Weigh out 100 grams of mineral oil and pour it into another. Measure the temperature of the water and the oil. Heat each beaker on the same hot plate, and measure the temperature of the two liquids every two minutes for ten minutes. Which liquid heats up more quickly? Which liquid has more heat energy added to it?

Allow the two beakers to cool together. Read the thermometer every 2 minutes in each beaker. Plot a graph of temperature against time. Which liquid gives up heat faster?

The capacity for holding heat is an important property of water. The fact that water has a large capacity for heat, and that it releases heat slowly, has a significant effect on the climate of land areas near water. Why do seacoasts have milder winters than inland areas?

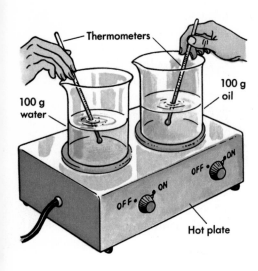

Thermometers

100 g oil

100 g water

OFF ON
OFF ON

Hot plate

REVIEW QUESTIONS

1. Name four salts found dissolved in seawater.
2. How does a change in temperature affect the solubility of salts? The density of seawater?
3. Describe a procedure for determining the total amount of dissolved material in seawater.
4. What does solubility mean?
5. What is a saturated solution?
6. In what two ways is seawater the same in every ocean?

THOUGHT QUESTIONS

1. Under what conditions does seawater become more dense in the ocean?
2. How does limestone form on the ocean floor?
3. Name three paths that dissolved material in rivers may take as they are washed into the oceans.

Seawater in Motion

Ocean water is never still, because it is acted upon by many forces. The sun and moon exert a force of gravity on the ocean. The constant blowing of the wind moves the surface of the oceans, resulting in waves and currents. The earth's gravity is another force. It causes the more dense water to flow under less dense water.

The motion of the water brings about changes. Waves and currents shape the edges of the continents both below and above the water. The heat energy of the currents affects the temperature of coastal climates.

Within the ocean are some of the largest, longest, and widest flowing bodies of water in the world. Oceanographers study the currents in an effort to find out why they flow as they do.

WAVES AND WAVE ACTION

Waves are the most obvious feature of the open ocean. Probably there is no moment when the surface of the sea is not in motion. During storms, waves can be awesome.

All waves, except certain unusual ones, obtain their energy from the wind. This energy builds up as long as the wind blows. After the wind ceases, the energy lessens because of friction. Meanwhile the waves may spread outward hundreds of kilometers from their point of origin.

Description of an Ocean Wave. Study the diagram on this page. The top of a wave is called the *crest* while the lowest part is called the *trough*. The distance between the crest and trough is the *height* of the wave. The distance between two crests is the *wavelength*. Waves not only disturb the surface of the water, but they also disturb the water under the surface. The depth of a wave is a measure of how deep the water is being moved by the surface wave.

A careful observer might notice that as a wave travels along the surface, the water does not flow along with the wave. Instead, the water moves up and down as well as side to side. A cork placed on the water moves in a circle, as shown by the circles in the diagram. Most likely the water molecules are moving in the same way. The paths which the water molecules follow are called *orbits*.

Oceanographers and sailors are interested in the time it takes for a boat to rise up with the wave, fall into the trough, and rise again. This time is called the *period* of the wave.

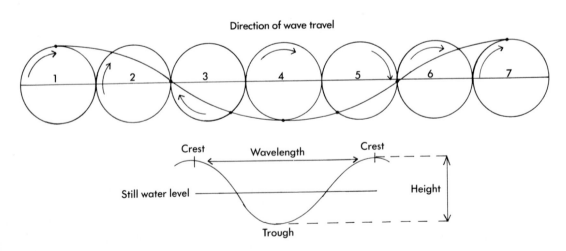

Direction of wave travel

Crest   Wavelength   Crest

Still water level

Height

Trough

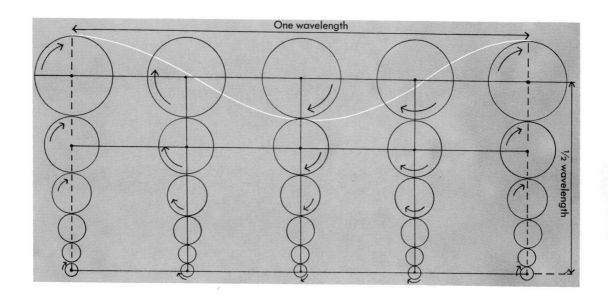

One wavelength

½ wavelength

Origin of Ocean Waves. There is friction between moving air and the surface of ocean water. The water molecules at the surface are struck by the air molecules. This causes the water molecules to pile up until the force of the air molecules is balanced by the gravitational force (mass) of the wave crest. The uppermost water molecules tumble forward and move back to the surface molecules being pushed by the wind. In this manner the wind forces the surface molecules into circular orbits.

Study the diagram on this page. In relation to the wind, which way is a water molecule on the surface of a crest moving? Which way is a molecule on the surface of a trough moving? If you were in a boat when a wave passed under you, what motions would you feel?

The energy of motion of the surface molecules is transferred to molecules under the surface. Some of the energy is lost to friction, so the orbital motion of molecules below the surface is less. Notice in the diagram that the orbits of water molecules decrease in size as the molecules are deeper in the water. At some depth (often about half the wavelength) the molecules are no longer affected by the surface wave. Why do fish tend to swim at greater depths during a storm?

Period (seconds)	Wave-length (meters)	Speed (m/s)	Wave depth (meters)
6	56.1	9.3	28.0
8	99.4	12.4	49.7
10	156.5	15.6	78.2
12	223.2	18.6	111.6
14	304.8	21.8	152.4
16	399.3	25.0	199.6

The Speed of a Wave. As you might suppose, not all waves travel at the same speed. Study the chart which shows measurements made on different-sized waves. Are there any relationships which exist among speed, period, and wavelength? What characteristics do waves moving at high speeds have compared to waves moving at slower speeds? If you measured the period of a wave, how could you calculate its speed? What relationship do you see between the depth of the wave and its wavelength? When a wave moves into shallow water, its characteristics will begin to change because the ocean bottom has an underwater effect on the wave. As the molecules strike the bottom, their orbits are flattened and the molecules are slowed down. In a sense, the wave is running aground.

Swells. As anyone knows who has been on a beach and observed waves, they tend to come in groups. After a storm has passed, the waves made by the storm's winds continue to move across the ocean. As the waves move through areas of calm, their crests become rounded and the waves tend to move in groups. Such a group is called a *swell*. In this form the waves can travel thousands of kilometers across deep water with very little loss of energy.

As a swell moves into water that is undisturbed, the first few waves lose their energy as they cause the undisturbed water to begin moving. As a result, the leading waves of a swell continuously disappear

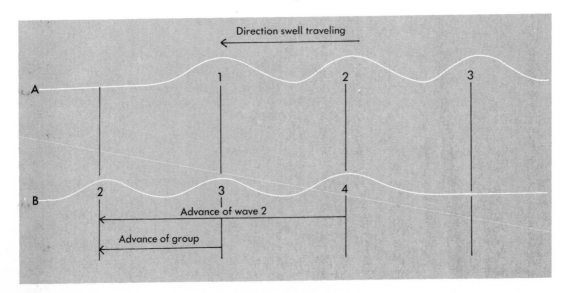

but new waves form at the end of the group. Thus, the number of waves in a swell remains constant and the total energy remains nearly constant. However, the speed of the group becomes less than the speed of each wave. Measurements made on the speed of a swell show that the swell moves at about half the speed of each individual wave.

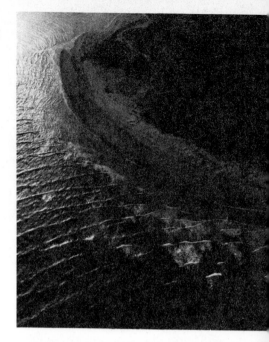

Waves on Beaches. As waves travel in water that is deeper than one half their wavelength, the ocean bottom has no effect on them. However, as waves approach the shore and enter shallow water, the bottom does have an effect. Study the photograph and the diagram. How do waves change as they move into shallow water?

Notice in the photograph that the wavelength of this group of waves is shorter nearer the beach. Notice also that the waves are bending. The part of the wave nearest the beach must not be traveling as fast as the part in deeper water.

Note in the diagram that the height of a wave increases as it nears the beach. Those who have been swimming in the surf have seen how quickly the height of a wave increases. Swimmers also may have noticed that the shape of the crest changes from rounded to peaked. This occurs as the orbits of the water molecules are changed from circular to elliptical as they strike the bottom. Eventually the wave reaches a height when it becomes unstable and the crest plunges forward to form a breaker. In this manner the waves hit the beach with much force, and the energy gained from some storm far out to sea is spent on the beach.

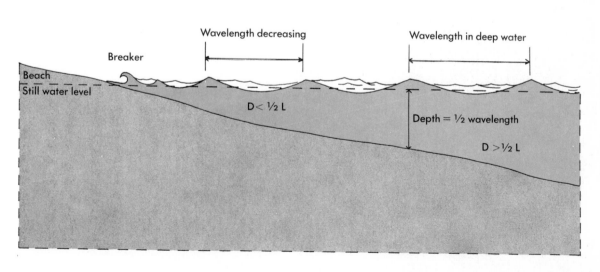

Sea Cliffs and Terraces. Direct effects of wave action are limited to regions between high and low tides. However, as waves attack the base of a rocky shore the higher parts are undermined and tumble down. A sea cliff is formed.

Waves do not erode much below low-tide level. They tend to leave a rocky terrace, as shown in the photograph on the next page. Explain the rounded boulders on this terrace.

Loose materials may be dragged seaward by receding waves. These extend the width of the terrace as they drop over the edge. What happens to the rate of erosion as a cliff recedes and the terrace broadens. Why?

Erosive Power of Waves. The energy of waves is expended in several ways. Much is changed to heat by friction. Some energy may set up currents parallel to the coast. The remainder causes erosion.

Large waves strike cliffs with enough force to shake the ground. This pounding loosens the rock along the bedding and the joint cracks. Loose blocks fall into the waves and become agents of erosion.

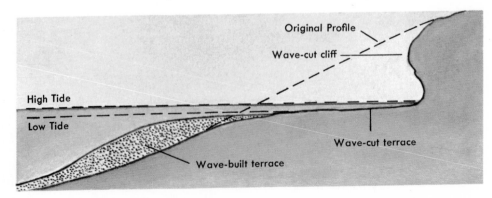

They are dashed against other rocks and dragged along the bottom as waves advance and recede.

Sea Caves, Arches, and Stacks. The rocky cliffs of a shoreline may be weaker in some portions than in others. Waves pounding against these cliffs produce picturesque features as they erode faster in some places.

The waves may hollow out cavities and form *sea caves*. If erosion continues until it breaks through the back of a cave, a *sea arch* is produced.

An arch may be widened by wave action until the top collapses. A column, called a *stack*, remains. What do you think will eventually happen to the stack?

Stack

Sea arch

Notch

Sea cave

Beach

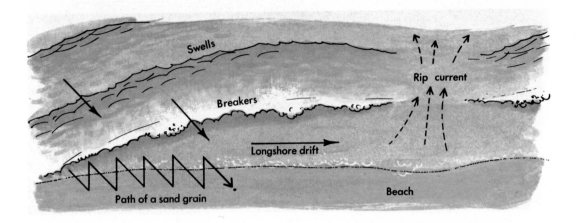

Swells

Breakers

Rip current

Longshore drift

Path of a sand grain

Beach

Wave Load. Over thousands of years the seas have gradually worn away the rocky coastlines to leave beaches of pebbles and sand. In some instances the sources of the beach materials are the eroded cliffs nearby. But in most cases the pebbles and sand have come from far up the coastline.

The water at the shore carries sand, pebbles, and bits of shells. This material carried by the waves is known as the *load*. The bigger the wave, the greater is the load the wave can carry.

When waves advance up the beach they deposit much of their load. They also drag small pebbles and sand back when they retreat. If this back and forth motion were straight in and out, the beach material would stay on the same beach. Many beaches are composed of materials which do not exist in the immediate area. Why?

Longshore Drift. It is unusual for waves to wash in parallel to the beach because the wind which produces the waves seldom blows directly toward shore. Waves more often come in at an angle to the beach. The flow of water, therefore, is not straight in and out, but is more of a zig-zag movement. The waves wash up the beach at an angle and retreat straight back. Thus as the waves wash in and out along the shore, they take with them a load of pebbles and sand. This is known as *longshore drift*

Longshore drift carries an enormous amount of sand down the coast. In order to slow down this erosion of their beaches, people build piers. How do the piers in the photo cause the sand to pile up and retard erosion? In which direction is the longshore drift moving?

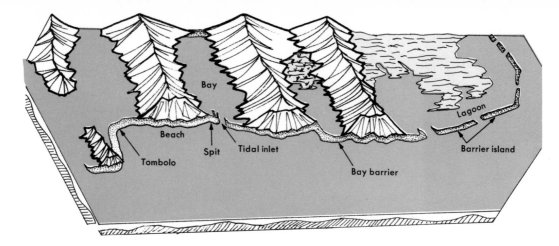

Bay

Beach

Tombolo

Spit

Tidal inlet

Bay barrier

Lagoon

Barrier island

Rip Currents. The water from a wave that has washed up on the beach must flow back down the beach to the sea. Since the waves are constantly washing up, the water flows back under the waves. Sometimes this water forms a fast flowing current called a *rip current*. A rip current can be very dangerous to bathers, because it will drag them out away from the beach. The rip current will end when it has reached a point outside the zone where the breakers form. What would you advise a bather to do if caught in a rip current?

Wave Deposits. Sediments deposited on beaches come partly from the erosion of headlands and partly from streams discharging into the ocean. In some situations, the latter source is far more important.

Wave-induced currents sweep sediments away from headlands where erosion is taking place. The sediments are carried along the shoreline until quieter water, such as a bay, is encountered. There the sediments are dropped. Explain the structures shown in the diagram above.

Bays and lagoons behind bars and barrier islands often become marshes. What is the probable source of the sediments that fill them? How do plants assist in the process of filling? Why may the water in the marsh be less salty than seawater?

SUMMARY QUESTIONS

1. What natural conditions may be responsible for setting water in motion?
2. What is the motion of a water molecule in a wave?
3. Explain how a wave breaks on a beach.
4. How does longshore drift develop?

The Gulf Stream as drawn by Timothy Folger of Nantucket for Benjamin Franklin

CURRENTS

Years ago captains of sailing ships knew of some ocean currents. They took advantage of these currents to shorten the time of their trips. The best known current, the Gulf Stream, was mapped by Timothy Folger, a whaling captain. He was asked by Benjamin Franklin, then deputy postmaster of the American colonies, to draw a map for publication.

Since the map of the Gulf Stream was drawn, scientific expeditions have discovered many other currents. They have found other surface currents as well as currents below the surface. Oceanographers want to know what causes currents in the ocean, and what effects the currents have.

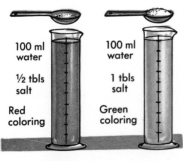

100 ml water

½ tbls salt

Red coloring

100 ml water

1 tbls salt

Green coloring

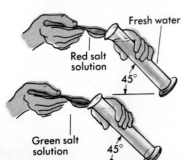

Fresh water

Red salt solution

45°

Green salt solution

45°

The Effect of Salinity Differences. Prepare the following two solutions of salt water. To 100 ml of fresh water add a half tablespoonful of salt plus red food coloring. To another 100 ml of fresh water add one tablespoonful of salt plus green food coloring. Which of these solutions is more dense? You will not be able to determine the difference in density by comparing the weight of equal volumes because the difference will be so small.

Fill a cylinder almost to the top with fresh water. Set the cylinder at a slant of 45 degrees. Pour a small amount of one salt solution into the cylinder. Describe the motion of the salt water.

Set another cylinder of fresh water at the same slant of 45 degrees, and pour in a small amount of the other salt solution. What happens in this case? Did one of the solutions move down the cylinder faster than the other?

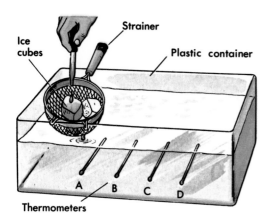

Ice cubes

Strainer

Plastic container

Thermometers

A B C D

TEMPERATURE CHANGE

Time	A	B	C	D
Start				
2 min				
4 min				
6 min				
8 min				
10 min				

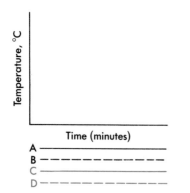

Temperature, °C

Time (minutes)

A ————————————
B – – – – – – – –
C ————————————
D – – – – – – – –

Set up a race between the two solutions. Note the time it takes each solution to travel the length of the cylinder. Each time use fresh water and slant the cylinder at the same angle. Why must you always begin with fresh water? What general statement can you make about what happens when solutions of different densities flow together?

The Effects of Temperature Differences. Fill a clear pan with lukewarm water. Place four thermometers evenly spaced along the bottom of the pan. At one end put several ice cubes in a strainer. Put a drop of food dye over each thermometer. These will color the water and let you observe the currents in the water. Read the temperature of each thermometer every two minutes for ten minutes. How does the temperature change at each thermometer? Why does the temperature change?

Make a graph of the temperature change at each thermometer every two minutes. How could an oceanographer, studying only the temperatures you recorded, predict the direction of the currents in the pan? How could you calculate the speed of the currents?

Look at the currents through the side of the pan. Which way do the currents move? Put a drop of food coloring on the ice, and describe the motion of the water under the ice. Make a diagram of the currents as seen from the side of the pan.

When water flows away from one end of the pan, water from elsewhere flows in to take its place. You probably noticed two currents in the pan, one flowing in the opposite direction to the other. Such a current is called a *counter current*. Oceanographers have found that wherever a surface current is found, there is a counter current associated with it.

Density Differences. You have seen what happens when waters with different salinities are mixed, and when waters with different temperatures are mixed. How does the salinity of water affect its density? Does the temperature of water affect its density?

Set up an experiment to test the last question. Fill a cylinder almost to the top with hot, fresh water. Set the cylinder at a 45-degree slant. Add a drop of food coloring to some cold water. Pour the cold water into the slanted cylinder. Describe and explain what happens. What effect does the temperature of water have on its density?

461

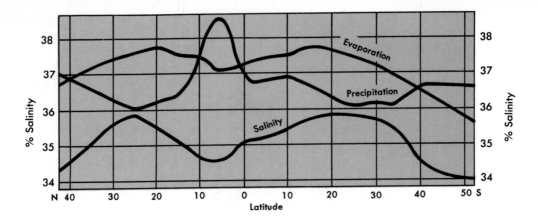

Density Currents. Oceanographers have discovered currents deep in the sea. These currents at the bottom of the oceans are too far below the surface to have been caused by winds. Oceanographers think that these currents are waters with different densities flowing past each other. If it is true that there are waters in the oceans with varying densities, what are the natural processes that could cause the different densities?

Find out what happens to the density of seawater as it evaporates. Pour 100 ml of seawater into a beaker and heat it until there are only 75 ml left. Compare the density of this water with natural seawater. Partly fill a cylinder with natural seawater and set the cylinder on a 45-degree slant. Pour a little of the evaporated seawater into it. Describe what happens. Which is denser, the evaporated or the natural seawater? Why?

Evaporation is a natural occurrence that changes the density of seawater. Another natural occurrence is rain. How might rain change the density of seawater?

Study the graph above which compares evaporation, precipitation, and salinity of surface waters with latitude. If precipitation is greater than evaporation, what effect would this have on the salinity of water? How could this condition affect the density of the seawater? At which latitudes does more precipitation occur than evaporation? In which direction would water flow based on these data?

There are other occurrences which change the density of seawater. How would the salinity of surface water be affected by freezing? At what latitudes is the density of surface water likely to be greatest?

The density of seawater depends on the temperature and salinity of the water. Freezing, like evaporation, removes fresh water from the sea. The salts are left in the liquid seawater, increasing the salinity. The water under the ice is very cold. These two factors account for the most dense water being found around Antarctica and Greenland. These waters sink and flow toward the equator as deep currents. The rate at which these currents flow has been estimated at 20 kilometers per year. At that rate how long would it take water from Greenland to reach the equator?

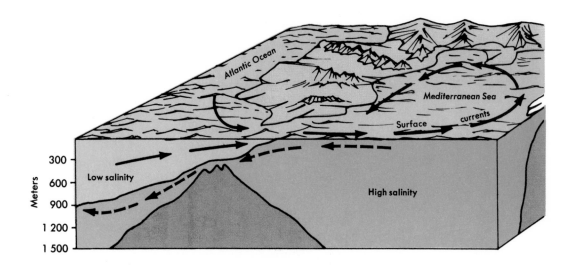

The Straits of Gibraltar. A strong current flows through the Straits of Gibraltar into the Mediterranean Sea. It was long believed that there was some hidden exit for the water in the Eastern Mediterranean. It is now known that there is a strong subsurface counter current flowing westward through the strait.

Rapid evaporation into the dry air of the Mediterranean basin increases the salinity of the water. This denser water flows out through the bottom of the strait, and less saline water enters at the top from the Atlantic.

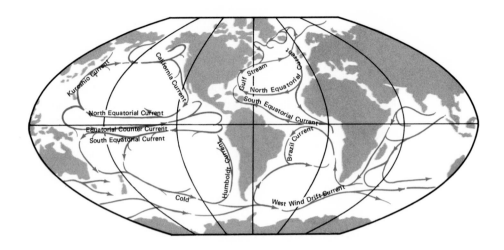

Major Surface Currents. The map above shows the general pattern of surface currents in the oceans. Many minor currents are not shown. The pattern of major currents is remarkably persistent, changing from season to season, but only slightly over the years.

These surface currents move rather slowly, usually less than two kilometers per hour. In some places they move five or six kilometers per hour.

Explaining the Surface Currents. Scientists have assumed that winds cause the major surface currents. Compare this map with a map of the prevailing winds on the next page. What evidence supports this hypothesis?

A less popular hypothesis assumes that the earth's rotation piles up water on the oceans' western shores in the tropics. This water then slides along the continents and returns eastward near the poles.

Another hypothesis suggests that major currents are due to temperature differences. The water near the poles is generally less than 2°C; that near the equator is usually above 21°C. The water at the poles settles and flows below the surface toward the equator. This cold water displaces the warmer water and drives it along the surface toward the poles. All the currents curve because of the Coriolis effect due to the earth's rotation.

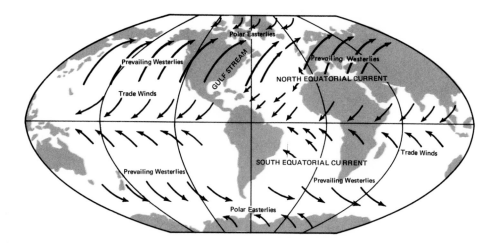

Ocean Currents and Climate. The map on the previous page shows the major surface currents of the oceans. In which part of the oceans does the water flow generally westward? In which part does the water flow generally eastward?

Where does the water in these currents receive more heat daily than it loses? In which part of the world does the water lose more heat daily than it gains? How does the water lose most of the heat?

How may a current flowing toward a pole affect the lands that it passes? How may a current flowing toward the equator affect the lands it passes?

Find the Gulf Stream on the map. Is this current made up of warm or cold water? What European lands are touched by the Gulf Stream? What is their climate? Compare their climate with that of Labrador. Compare and explain the differences.

Discuss the effect of the California Current on the climate of western North America. Compare the climates of coastal regions of the same latitude. Explain the differences by nearby ocean currents.

SUMMARY QUESTIONS

1. Under what conditions does the density of water become greater?
2. Name two conditions that cause ocean currents.

TIDES

The photograph shown here was taken at low tide in the Bay of Fundy on the Atlantic coast. What clues can you find to indicate the water level at high tide? Anyone visiting the seashore finds the tide to be a major influence on the lives of the animals and people living there. Fishing, swimming, boating, and beachcombing are a few of the many activities that must be planned around the tides.

Gravity and Tides. Ocean tides are often described as resulting from gravitational attraction between the sun, moon, and earth. Some observations about tides can be accounted for on the basis of gravitational attraction. Other tidal observations cannot be explained so simply.

If the moon and sun influence the tides, there should be some relation between the tides and their motions. Study the times of high and low tide. How much time is there between one high tide and the next? How many minutes later each day does the tide come in?

Notice the times of moonrise. How many minutes later each day does the moon rise? What relation can you see between the times of the tides and the times of the moon's rising and setting?

Using the chart on page 467, make a graph comparing the height of high tide with the day. Also on your graph draw the appearance of the moon on the dates when it is new, first quarter, full, and last quarter. How do the heights of the tides relate to the phases of the moon? Explain why the tides are highest during a full moon.

Tidal Range. Half fill a bowl with water. Move a spoon back and forth in the water, slowly at first, then faster until the water and spoon are moving together. The time it takes for the water to slosh back and forth is the natural period for that bowl. Try the same activity with containers having different shapes. How does the shape of a container affect its natural period?

The tidal range in bays depends on several factors. One is the shape of the bottom and sides of the bay. If the time between high and low tide is nearly the same as the natural period for the bay, very high and very low tides result. This situation exists in the Bay of Fundy between Nova Scotia and New Brunswick, Canada, which experiences the highest tides in the world.

NEW BRUNSWICK

Bay of Fundy

NOVA SCOTIA

Atlantic Ocean

Day	Moon Rise	Set	Time of High Tides AM	PM	Height of Tide (meters above sea level) AM	PM	Phases of Moon
1	5:44 PM	7:00 AM	12:14	12:28	3.2	3.4	
2	6:18 PM	8:05 AM	12:52	1:05	3.2	3.4	○ Full moon
3	6:58 PM	9:09 AM	1:32	1:46	3.1	3.4	
4	7:44 PM	10:13 AM	2:15	2:28	3.1	3.4	
5	8:40 PM	11:12 AM	3:01	3:15	3.0	3.3	
6	9:42 PM	12:07 PM	3:52	4:09	2.9	3.3	
7	10:50 PM	12:55 PM	4:47	5:06	2.9	3.3	
8	– –	1:37 PM	5:48	6:10	2.9	3.2	◖ 3rd quarter
9	12:02 AM	2:13 PM	6:51	7:13	3.0	3.3	
10	1:15 AM	2:46 PM	7:52	8:17	3.1	3.4	
11	2:29 AM	3:17 PM	8:52	9:18	3.3	3.4	
12	3:43 AM	3:48 PM	9:47	10:14	3.5	3.5	
13	4:57 AM	4:19 PM	10:38	11:07	3.7	3.5	
14	6:10 AM	4:53 PM	11:28	11:58	3.8	3.5	
15	7:22 AM	5:30 PM	–	12:15	–	3.8	
16	8:30 AM	6:13 PM	12:47	1:01	3.4	3.7	● New moon
17	9:34 AM	7:00 PM	1:35	1:47	3.4	3.7	
18	10:30 AM	7:52 PM	2:22	2:35	3.2	3.5	
19	11:19 AM	8:48 PM	3:10	3:24	3.0	3.3	
20	12:01 PM	9:46 PM	4:00	4:15	2.9	3.1	
21	12:36 PM	10:46 PM	4:53	5:09	2.8	3.0	
22	1:07 PM	11:45 PM	5:47	6:06	2.7	2.9	◗ 1st quarter
23	1:34 PM	–	6:43	7:02	2.7	2.8	
24	2:00 PM	12:44 AM	7:38	7:56	2.7	2.8	
25	2:25 PM	1:43 AM	8:25	8:47	2.8	2.9	
26	2:50 PM	2:43 AM	9:13	9:34	3.0	3.0	
27	3:16 PM	3:45 AM	9:55	10:19	3.1	3.0	
28	3:44 PM	4:48 AM	10:35	11:03	3.2	3.1	
29	4:17 PM	5:52 AM	11:16	11:44	3.4	3.1	○ Full moon
30	4:56 PM	6:58 AM	11:55	–	3.5	–	

REVIEW QUESTIONS

1. What is the name of the rock formation just off shore in the top photograph on page 466?
2. How many tides do most coasts experience each 24 hour period?

THOUGHT QUESTIONS

1. Where do pebbles, stones, and sand grains on beaches come from?
2. Why is the time of high tide a little later each day?

5

Investigating on Your Own

Although people have sailed across the surface of the sea for hundreds of years, only recently have they begun exploring beneath the surface. Even the most imaginative sailors would not have dreamed that the world's tallest mountains, and deepest and longest canyons were passing under their ship.

The challenge of studying the oceans is to use the ideas and theories from all of the fields of science. A description of the oceans must include at least the chemistry of the water, the physics of the water's motion, the biology of living things, and the geology of ocean basins.

Now that scientists have developed skills and know-how to explore the deep sea, oceanography is becoming a most important activity. There is excitement in solving the mysteries of the ocean. There is satisfaction in finding ways of using the oceans for the benefit of people.

INVESTIGATING A BODY OF WATER

The intent of this chapter is to introduce you to some of the activities of oceanographers. Select a small body of water that you can study throughout a year. Any small pond, natural or artificial, will serve as a miniature ocean. You should plan to study certain aspects of the pond in each season of the year. You will find that some very interesting and revealing changes take place in a pond during a year.

Physical Features. An accurate description of the shape of the pond is one of the first activities to plan. Knowing the physical features of the pond will help you to plan other investigations. When you sample the water, the bottom, or the living things, you will want to record the places from which you took the samples.

You may want to make a map of the pond using the base line method. To use this method, pound two stakes, A and B, into the ground at least 20 meters apart. Measure the distance between the stakes as accurately as you can. Choose a scale such that the map will fit onto the piece of paper you plan to use. Pin the paper onto a drawing board. Draw the base line to scale on the paper labeling one end of the line A, the other end B. Place a pin at each end of the base line.

Mapping the pond

Put the drawing board on stake A so that point A is on top of the stake and point B is aligned with stake B. Select an obvious feature on the shore at each end of the pond. Align the pin at A with the feature and mark an X, noting the direction of the feature on the paper. Sight on several other features. Move the drawing board to stake B, and position the paper so that point B is over stake B and point A is aligned with stake A. Again sight on the same features from point B. Note the direction of these features on the paper by marking Y. Locate these features on your map by extending the sighting lines until pairs of lines pointing to the same feature intersect. Pick out several other features and locate them on the map by sighting from stakes A and B. Use these points to sketch your map.

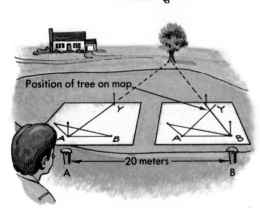

Position of tree on map

20 meters

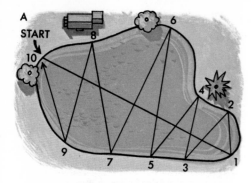

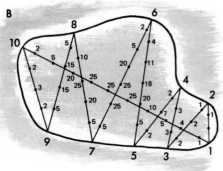

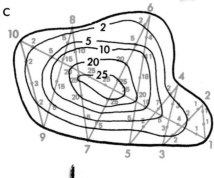

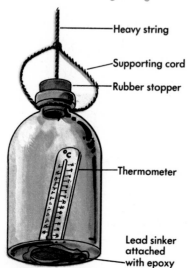

Heavy string

Supporting cord

Rubber stopper

Thermometer

Lead sinker
attached
with epoxy

The Bottom of the Pond. Determine the shape of the bottom of the pond. Where is the deepest point? What is the general shape of the pond's basin?

Line up one site, such as a tree or a rock at one end of the pond, with another site at the other end. Stretch a string with corks tied every 10 meters from one site to the other. Measure the depth of the pond at each cork using another string with knots tied every meter and a weight to pull it down.

Select several sites on each side of the pond. Locate these sites on your map. Stretch the string with the corks across the pond, and take soundings at each cork. Record the soundings on your map. Connect the places of equal depth by a line, thus making a contour map of the bottom of your pond.

Collecting Water Samples. Once you have found the deepest spot in your pond, collect water samples from the top and bottom. If the pond has a stream flowing into it, collect water at the mouth of the stream. Also sample the water at the beginning of a stream which flows out of the pond. Note on the map the locations where water is sampled.

To collect water samples use a bottle rigged as shown below. The bottle is lowered to the depth desired with the stopper in position. A quick jerk will remove the stopper and water will flow into the bottle. Keep the water collected in small jars labeled with the place and depth at which the water was sampled.

Temperature. Place a thermometer in the bottle so that you can also note the temperature of the water. After jerking the stopper free, hold the bottle still for the thermometer to record the temperature. Then raise the bottle quickly and read the temperature.

Measure the temperature of the water at every meter of depth. Record the temperatures and depths on a chart. Plot a graph comparing temperature with depth. Repeat these temperature measurements during each season of the year. At what depth does the temperature suddenly become colder? Does this sudden change occur at different depths throughout the year?

Acidity. Prepare a solution of red cabbage juice to test the acidity of the water. Pour some of the water collected from various parts of the pond into separate test tubes. Add red cabbage juice to each

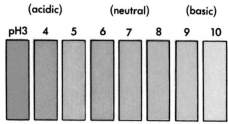

(acidic)　　　(neutral)　　　(basic)

pH3　4　5　6　7　8　9　10

Color range of purple cabbage juice indicator

PURPLE CABBAGE JUICE

Boil several purple cabbage
leaves in two cupfuls of
water. Pour off the colored
water and store it in a
refrigerator until needed.

Testing conductivity

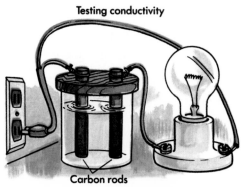

Carbon rods

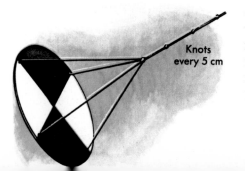

Knots
every 5 cm

sample and note the color of the mixture. What
differences do you notice among the samples? Note
the acidity of the water samples on your map. Do
you see any pattern to the water conditions in the
pond?

Electrical Conductivity. Remove the central
carbon posts from two D-cells. Drill two holes 5 cm
apart through a wooden board to hold the carbon
posts. This board should be big enough to rest on
top of a beaker. Connect wires from the carbon rods
to a wall plug and to the light as shown below.
Add some water from a faucet to the beaker, and
place the rods in the water. (**CAUTION:** The carbon
rods may get very hot.) What happens to the light?
Add salt to the water and test again. How does
adding salt affect the light? Try adding substances
other than salt to tap water. Test the samples of
water you took from your pond and enter the
results on your map.

Observing Underwater. Devise a way to see
clearly what is under the water. If you are to look
down into the water you need to eliminate the glare
and distortion caused by the water's surface. A
piece of glass or clear plastic fitted watertight into
the bottom of a pail will allow you to see clearly.

Cut a hole in the bottom of the pail. Make a
frame of wood to hold the glass or plastic in place.
Make the window watertight by using waterproof
putty or aquarium cement. Paint the inside of the
pail black to further reduce glare.

The pail can be used at night as well as during
the day. Light the darkness underwater with a
flashlight sealed in a jar. Observe the living
organisms in the pond.

Transparency. One measure of the amount of
material suspended in the water is its transparency.
To make this measurement oceanographers use a
Secchi disk. A Secchi disk is a metal disk 20 cm in
diameter. The disk is divided into quarters with two
opposite quarters painted white, the other two
painted black.

In a shady part of the pond, lower the Secchi
disk into the water until it just disappears. Record
the depth. Then raise the disk until it just
reappears. Record this depth, and average the two
measurements. Use the glass-bottom pail for
measuring the transparency of the water at sunny
parts of the pond.

471

THE LIFE IN A BODY OF WATER

When you investigate the living things of the pond, you will probably find a great many different kinds. You will find some kinds in greater numbers in some places than in others. Often the presence of certain organisms serves as an indicator of the conditions at that place. If you continue your investigations throughout a year, you will find fascinating changes occurring in the life forms of the pond.

Collecting Live Specimens. A useful tool for catching animals is a long-handled dip net. By sweeping along the bottom or along the plants near shore you may catch fish, amphibians, reptiles, insects, and other creatures.

You can make your own dip net by securely tying a kitchen strainer to the end of a long broom handle. Once you have made a sweep through the pond, empty your catch into a pan. A white pan will allow you to see more easily what you have caught. In addition, an eye dropper, a kitchen baster, several collecting jars, and a magnifying glass are handy items for sorting.

For collecting samples from the bottom, cut a can into a scoop as shown. Bolt the can to a long pole. Use this device with the same motion you use in raking leaves. When the can has scooped up some bottom material, dump the contents into the white pan.

If you decide to keep some organisms for study, put them in collection bottles. Label each bottle with the date, time of day, and location where the animal was found. You can study their behavior and development if you follow these rules for keeping aquariums:

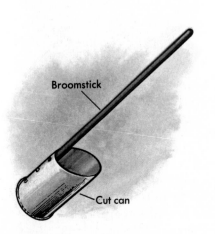

Broomstick

Cut can

1.   Use very little bottom material. About 90% of the volume of the aquarium should be water.

2.   Place the aquarium in a northern exposure. Direct sunlight causes too rapid growth of algae.

3.   Cover the aquarium with glass cut to size. The glass will reduce evaporation and prevent organisms from entering or leaving the aquarium.

4.   Keep the water temperature as close to the temperature of the natural environment as possible.

EXPERIMENTAL RESEARCH

1.   Pour a little oil on the surface of some water in a pan. Try various techniques for removing the oil. Use something to soak it up. Try pushing it together and siphoning it off. What effects do detergents have on the oil? Look up the latest techniques used to clean up oil spills.

2.   Compare air and water as heat conductors. Put one ice cube into a glass of water, another of the same size on a wire screen across the top of a glass. Which ice cube melts faster? Why?

3.   Tear a paper match out of a match book. Put the match in a jar full of water with a metal cover. Tightly screw the cover down. Press down on the cover and observe the match. What happens when you let up on the cover? How can you explain the action of the match in terms of pressure?

4.   Purchase some brine shrimp eggs from a pet store. Raise the shrimp and keep a notebook of your observations on their development. Test their reactions to light, heat, and sound.

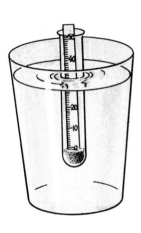

5.   Test the effect of dissolved materials on the density of water. Cut a strip of paper to fit inside a test tube. Mark off the edge of the paper in millimeters, numbering every tenth one, and insert it into the test tube. Nearly fill a glass with water and float the test tube in the glass by adding just enough sand to cause the test tube to float upright. Note the level on the strip of paper at which the test tube floats. Add salt to the water one spoonful at a time. How does the salt affect the floating test tube? This device is called a *hydrometer*

A

Warm water

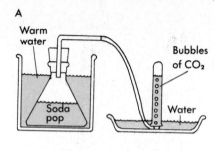

Bubbles of CO₂

Soda pop

Water

B

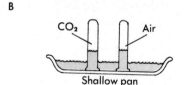

CO₂

Air

Shallow pan

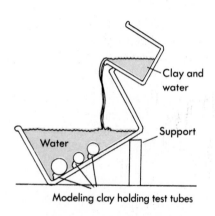

Clay and water

Support

Water

Modeling clay holding test tubes

6. How soluble is carbon dioxide in water? Set up your apparatus as shown at the left. Collect a test tube full of carbon dioxide and allow it to stand upside down in a shallow pan of water. Set up a clean, dry identical test tube filled with air. Cover the mouth of the test tube and place it upside down next to the test tube containing carbon dioxide. After one hour, note any difference in the water levels within the test tubes.

Does the temperature of the water have an effect on the solubility of carbon dioxide? Experiment by placing the test tubes in the shallow pan and refrigerating them. What are your results?

7. Find a world map with the continental shelves indicated. Use thin paper and trace the outline of the continental shelves. Cut out the tracings and try fitting them together into a picture puzzle. How well do they fit together? Which continents seem to fit together best? If all the continents were once together, in which direction has each moved? How far has each moved?

8. Investigate how sediments are distributed over an uneven surface. Place test tubes of varying sizes along the bottom of an aquarium. Secure them in place with bits of modeling clay. Fill the aquarium about half full and tip it so one end is higher than the other.

While the water in the aquarium calms down, make up a mixture of water and clay. Use enough clay so there will be a visible layer when the clay has settled to the bottom of the aquarium. Pour the water and clay mixture into the raised end of the aquarium and allow time for the clay to settle. Again add water and clay, only this time color the mixture with food coloring to distinguish this layer from the last. Every other time the mixture is poured in, add food coloring. Photograph the results and discuss the implications for oceanographers studying the ocean bottom.

INVESTIGATIONS AND PROJECTS

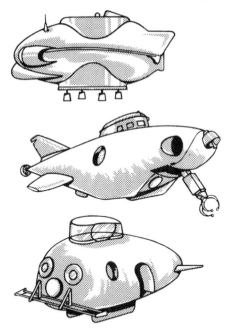

1. Prepare a report on the types of submersibles being used to explore the depths of the oceans, such as Trieste, Deepstar, Aluminaut, Sea Lab, and Ben Franklin.

2. Place an ice cube in ocean water and measure the volume above and below water. Calculate the percent of the ice cube that is submerged.

3. Visit a shore at high and low tides. If the shore is rocky, make diagrams of sea cliffs, wave-cut terraces, and wave deposits. If the shore is sandy, make a study of the shapes of sand bars.

4. Collect pictures of marine life and prepare an exhibit that shows the types of organisms that live at different depths.

5. Read about the theory of continental drift and how it explains the nature of the continental blocks, submarine ridges, and ocean depths.

6. Obtain a map of the seacoast near your home. Label the geological features shown on the map, using a geology book to find the proper terms.

7. Make a salt solution with the density of ocean water (30 grams of table salt to the liter). Put some of this solution into a freezer until it is partly frozen. Separate the ice, melt it, and compare its taste with that of the original solution.

8. Read *Twenty Thousand Leagues Under the Sea* by Jules Verne. Compare this fiction with what is actually known about the oceans.

9. Study the fish resources of the oceans. Which species are most valuable, which are being overfished, and what is being done to protect them.

10. Find out how seaweeds are used for foods and for other purposes. Report on methods of harvesting.

11. At the seashore, when waves are small, observe the distances waves advance up the beach. Find out why some go higher than others.

12. Make a chart that shows the depths to which different explorers have gone in the ocean, both with and without submersibles. Calculate and show the pressure at each depth.

13. Study the career opportunities for oceanographers and make a report to the class.

14. Look up Gaspard G. de Coriolis to see how he discovered the effect our spinning earth has on the paths of winds and ocean currents.

15. Demonstrate the Coriolis effect by filling a glass pie plate half full of water and placing it on a Lazy Susan. Float a small piece of wood as a boat. Push the boat straight across the water. Mark where the boat landed. Put the boat back where it was and push it toward the mark. As the boat moves toward the mark, turn the plate one quarter of a turn. Where did the boat land this time? Why?

16. Write a report and make a poster about the techniques being used to study the layers of sediments on the ocean floor.

17. Report on the voyages of the oceanographic research ship, Glomar Challenger.

18. There are how many techniques for catching fish? Make a chart with pictures or drawings of some techniques. Include the types of fish caught by each technique and where it is used.

19. Study the diagram at the left. Use it to show convection currents in water. Why does the hot water come out of the small bottle?

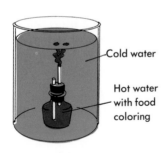

Cold water

Hot water with food coloring

20. Report to the class about ocean currents and their role in history. (Read about the voyage of the Kon Tiki by Thor Heyerdahl.)

21. Trace a world map and draw in the major ocean currents and the average speed of each.

22. Read about scuba diving. Report to the class about the effects of water pressure on the body.

23. Find out how tidal waves (tsunamis) are formed and how they can do great damage.

24. Use a road atlas to find states bordered by an ocean. Study the coastlines. Are they submerging or emerging? Along which coasts would you expect to find sand bars, sand spits, and tombolos? Where might you find stacks and arches?

25. How are the heights of tides measured? Find out and use a large diagram to show the class.

26. Look up aquaculture and mariculture. Find out how some people raise sea animals.

The Oceans as a Resource

Human history on earth can be described partly as a search for new farmland, new forests to cut or hunt, and new mineral wealth to mine. Today, almost no land area remains to be explored and settled. Still our population increases, demanding that more people be fed, clothed, and provided with books, houses, cars, and television sets!

Where can the resources be found to meet the increasing demands of the world's growing population? One answer has always been to use the resources of the sea. In this chapter you will learn about the sea's resources. Only from such knowledge can people decide whether the oceans can solve our resource problems.

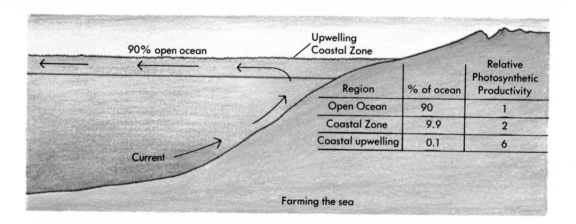

Region	% of ocean	Relative Photosynthetic Productivity
Open Ocean	90	1
Coastal Zone	9.9	2
Coastal upwelling	0.1	6

Farming the sea

FARMING THE SEA

Soil scientists say that only about 25% of the earth's land surface is usable for farming. This farmland must supply all the food for the world's population. These farming regions must not be too steep, too dry, or too cold. The soil must contain sufficient minerals. There is, however, a great untapped part of the earth which is not too steep and not too dry. This region might also be used for supplying plant food for the world's people. This region is the oceans of the world.

Food Energy in the Ocean. The diagram at the right shows a typical food chain in the ocean. Most such chains begin with the many varieties of microscopic plants, called *phytoplankton* . Phytoplankton grow if the light, minerals, and temperature are right for them. They are the source of food for microscopic animals (called *zooplankton*)which in turn are eaten by other animals. Thus, the number of fish an ocean region can support is determined by the amount of phytoplankton which exists there.

Fertility of the Oceans. The diagram above shows how marine biologists describe the fertility of the oceans. The green regions show where conditions favor growth of phytoplankton. What prevents plants from growing in the deep oceans?

The top layer of the open ocean does not contain enough of the right types of minerals for phytoplankton to grow well. Thus, the open ocean resembles a desert in its plant growth. About twice as much plant life occurs in each square meter along the coastal zone. What source provides minerals for this region?

The most productive phytoplankton growth occurs along the coastal regions where deep ocean

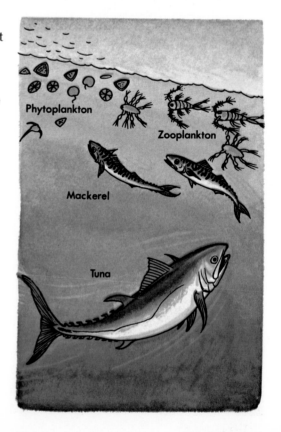

Region	Annual Fish Production (metric tons)
Open ocean	160 000
Coastal Zone	120 000 000
Coastal Upwelling	120 000 000

(Note: 1 metric ton equals 1000 kg)

Baleen whales have no teeth. They have plates called baleen, which hang like curtains, and strain food out of the water.

currents rise to the surface. These upwellings bring to the surface vast quantities of minerals. How would these minerals affect phytoplankton growth? How would abundant phytoplankton growth affect zooplankton? How would abundant zooplankton affect the growth of fish?

Long and Short Food Chains. The open ocean has the smallest supply of nutrients. Very small phytoplankton and zooplankton grow here. These plankton do not live close together, so the small fish which feed there must travel far to get enough food. In this type of food chain, plant food is eaten by larger and larger animals an average of five times before humans catch and eat fish from the chain. More than 100 kilograms of phytoplankton are eaten by zooplankton and other members of this food chain for every gram of fish we eat.

Along the continental shelf, particularly in upwelling regions, minerals fertilize the ocean. Phytoplankton here are much larger than in the open ocean. They also grow much closer together. As a result, large animals such as whales feed directly on plankton. Here a much greater percentage of the weight of plankton ends up as weight of edible meat. Marine biologists estimate that 100 grams of phytoplankton in these regions produce 10 grams of edible meat. Compare the economic value of phytoplankton in the open ocean and in upwelling regions. Study the table. Where do most fish grow? Explain.

Algae Farming. Over 90% of the food in the ocean consists of plants, mostly algae. However, 99% of the food we take from the ocean is animal, mostly fish. Today, Japan leads all other countries in the amount of plant food gathered from the sea.

The photograph at the left shows a "farmer" gathering kelp, a giant sea alga. Kelp is farmed in many countries. This seaweed contains a chemical called algin which keeps oil and water from separating. As a result, it is used in salad dressings, ice cream, and cake icings, as well as paints, drugs, and cosmetics. In Japan, other types of algae are used in soups, salads, and garnishes. Many countries have experimented with growing algae as a profitable food plant. One alga, chlorella, can be grown easily. Dried chlorella can be ground and added to flour, increasing the flour's food value. This alga has been considered as a source of food and oxygen for astronauts.

479

THE OCEAN AS A HUNTING GROUND

One of the oldest used resources in the ocean has been its animal life. No one knows when people first started to fish for food. Today, however, in some countries, such as Japan, over 50% of the protein each day comes from fish. Less than 5% of the daily protein of Americans comes from fish. What are some possible reasons for this difference?

Whale Hunting. The history of whaling has been traced back a thousand years to a small town in Spain. After the people in that village killed all the whales near their coastal town, they had to travel out to sea. Later, people from this village traveled as far as Greenland in search of whales.

In the mid-1800's about 1000 whaling ships sailed from New England in search of whales. Only the oil was saved; all else was thrown back.

In the late 1800's, whale stations on shore were established. Whalers hunted at the time of year when whales were near their regions. Killed whales were towed to these land stations where not only the oil was extracted, but the meat and bones were also processed. By 1915, Newfoundland had 18 whaling stations. By the 1930's, many of these stations were closing and new ones were opening on the Pacific coast. Today, all the whaling stations in North America are closed because whalers killed too many whales each year. Not enough adult whales were left to produce young which would insure whales in the future.

Only two or three countries continue to hunt the few remaining whales. In addition, at least one country has begun netting the small plants and animals which the few remaining whales eat. How will this affect the whale population?

Underfishing and Overfishing. The map on the next page was published in 1949 by the United Nations to help nations feed themselves. At that time, the 30 fish shown on the map were listed as being underfished. There were more fish in these regions than could be taken by the local fish catchers. More people began fishing these regions.

In 1968, these sites were restudied. Fourteen fish sites on the original list (shown in blue) were now classified as *overfished*. This means that more fish were being caught each year than could be replenished. What will happen if the fishing continues at this rate?

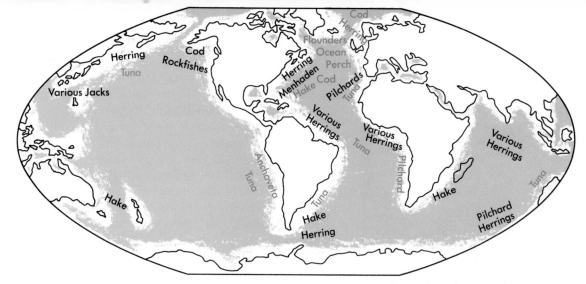

Maximum Sustainable Yield. The population of tuna in the oceans consists of many small tuna, a smaller number of average size tuna, and a still smaller number of very large tuna. If people decide to catch and eat the larger members of any animal population, such as tuna, one might ask, "What is going to happen to the number of large tuna if you start removing them from the population?" The answer from ecologists is plain and simple. Any amount of fishing for large tuna will reduce the proportion of large tuna in the population.

Perhaps a more important question is: "What happens to the total population of tuna (or any other wild animal), if you catch more and more of them every year?" Population ecologists have studied this question in hundreds of cases. The number of fish caught each year can increase, up to a point. Beyond this amount, there will be fewer fish available to catch the following year. As a result, the catch each following year will be less if the same people catch as much fish.

The greatest amount of any kind of animal which can be caught each year and still have the same amount available to catch the following year is called the *maximum sustainable yield*. Many fish in the ocean are now being caught in numbers greater than the maximum sustainable yield. What will happen to the numbers of this kind of fish caught in future years? Discuss the problems involved in getting fish catchers, pet food company presidents, other fish users, and bankers of different countries to agree to catch only the maximum sustainable yield each year.

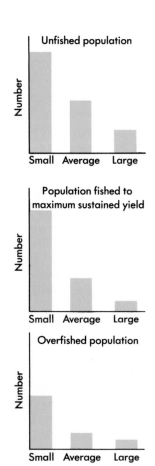

481

A desalinization plant

Manganese nodules
on the ocean floor

The Fish Potential of the Ocean. In 1974, about 160 million kilograms of fish were caught in the entire world each day. If every bit of these fish were eaten by the world's four billion people, how much fish would each person eat each day?

Today, most of the fish which are cheapest to catch are being caught at a rate at or above the maximum sustainable yield. Some fish are hard to catch because they live at greater depths, travel in smaller numbers, or live in distant places. Why is the Indian Ocean not overfished today?

Experts predict that the ocean could produce a maximum sustainable yield of about 300 million kilograms of fish a day if all the fish in all the oceans were hunted. To catch this many fish would require new technology which is not available. How much fish will each person then be able to eat each day? (The world population in the year 2000 will be about 7 billion.) Discuss these predictions. What assumptions may affect them?

Food from the Sea in the Future. The present use of the sea as a food source is similar to the early American colonists' use of woodlands as a source of food. They relied on wild deer for meat. By changing the woodlands to grazing areas, cattle could be raised, yielding more meat per hectare. Perhaps the same idea is possible for the oceans.

The photograph above shows oysters being grown in tidal waters. This experimental effort results in the production of nearly 100 million oysters a year from this location in Australia.

The problem of raising and "herding" fish is difficult. When ranchers raise herds of cattle, they get the meat to repay their efforts. Laws and sheriffs protect their herds. But who would protect "herds" of fish raised in the open ocean?

Experts point out that large fish eat about ten times their weight in smaller organisms. If this is true, couldn't we get ten times as much food from the sea by harvesting the smaller organisms that large fish feed on? In a few cases, this idea has been tried successfully. However, in most cases the smaller organisms are so widely spaced that it costs more to catch them than they are worth. Large fish help us by catching the smaller organisms. When the large fish collect in numbers, it is easier and less expensive to catch them. However, in the future, with better technology, we may be more successful in hunting "back down the food chain."

MINING THE OCEAN

The ocean is a vast storehouse of dissolved minerals. Each cubic kilometer of seawater contains nearly ¼ million dollars of dissolved metals, and there are 1½ billion cubic kilometers of seawater in the oceans! Unfortunately, with today's equipment, it usually costs more to remove the metals from seawater than they are worth. Better equipment must be invented before we can use the sea's minerals.

ELEMENTS IN A CUBIC KILOMETER OF SEA WATER

Element	Metric Tons
Chlorine	19 000 000
Sodium	10 000 000
Magnesium	1 300 000
Sulfur	900 000
Calcium	410 000
Potassium	380 000
Bromine	65 000
Carbon	28 000
Strontium	8 100
Boron	4 700
Silicon	3 000
Fluorine	1 300
Argon	600
Nitrogen	510
Lithium	170
Phosphorus	71
Iodine	60
Barium	30
Zinc	10
Iron	10
Aluminum	10
Molybdenum	10
Tin	3.0
Copper	3.0
Arsenic	3.0
Uranium	3.0
Nickel	1.9
Manganese	1.9
Titanium	1.9
Antimony	0.4
Cesium	0.4
Cobalt	0.4
Silver	0.2
Gold	.004

Dissolved Minerals. The table here lists some elements in seawater. Engineers know that the most common elements in seawater may not be the best ones to try to remove. For example, the elements sodium and chlorine, which make up table salt, are very common in seawater. However, these materials can be obtained cheaply on land. Thus, the value of an element, as well as its cost to produce, has to be considered before deciding to separate it from seawater. Today, only salt, magnesium, and bromine are produced profitably from seawater. Gold sells for about $5 a gram. Calculate the value of the 4000 grams of gold in each cubic kilometer of seawater.

Fresh Water from Seawater. Most cities in the United States figure the cost of their drinking water at about 10 cents to find, transport, and purify every 1000 liters. However, in some areas where fresh, pure water is scarce, this cost is much higher. Using today's methods, salt water can be converted to fresh water for about 20 cents for every 1000 liters. Thus, if a region is near the sea and fresh water is scarce, the cheapest source of fresh water may be the sea. Such facilities which remove the minerals from seawater are called *desalinization plants*. In the future, as populations increase, more people will probably be drinking desalinized water.

Mining the Sea Floor. Much of the sand and gravel used for construction in seacoast cities comes from the sea floor. Giant dredges scoop up the sea floor. Sand, gravel, and shells are separated. The shells are used as a source of lime.

At greater depths on the continental shelf, phosphorus-rich nodules have been discovered. One large deposit has been found off the southern California coast. Other nodules of manganese have been found on the ocean floor. Such nodules are shown in the photograph on page 482. New techniques will permit mining these resources.

Mining the Crust Below the Sea. Nearly 20% of the world's supply of oil comes from offshore oil wells like the one at the left. This percentage is increasing rapidly. As oil beneath the land areas is used up, we must continue to develop better technology to get at the oil under the sea. About 90% of the mineral wealth mined each year from below the sea floor consists of gas and oil.

Many land deposits of minerals extend out under the ocean. As a result, many mines also extend under the ocean. These mines are as deep as 2 kilometers below sea level and as far as 25 kilometers from shore. Today, over 100 mines exist under the sea floor removing coal, limestone, and the ores of nickel, copper, and tin.

REVIEW QUESTIONS

1. Is the sea a likely source of food for great numbers of people in the near future? Explain.
2. Is the sea a likely source of minerals in the future?
3. What are some inventions which would improve our chances of getting minerals from the sea?
4. What is maximum sustainable yield?
5. What are phytoplankton? Why are they important?
6. Why are fish scarce in some regions of the ocean, but plentiful in other regions?
7. What conditions limit the growth of phytoplankton?
8. Why aren't more minerals mined from seawater?

THOUGHT QUESTIONS

1. In 1933, 29 000 whales were caught which produced over 2½ million barrels of whale oil. In 1966, exactly twice as many whales were caught, but they produced only 1½ million barrels of oil. How would a population ecologist explain this?
2. What happens to the cost of fish which are being caught in numbers greater than maximum sustainable yield? Why would it save money to support the international enforcement of fishing only to the maximum sustainable yield?

7

Ocean Pollution

Once upon a time there were ten families in a distant village. Two of the families owned the factory in the village. The other families worked in the factory. Waste from the factory was emptied into the river. As a result, the poor people in the village who drank the river water often became sick. In bad weather, the smoke from the factory made the people ill who lived near the factory. The people did not complain, however, because if the factory didn't dump wastes into the river and give off smoke, it would mean there was no work. If the factory closed, most of the people of the village would starve.

The people who owned the factory lived several kilometers from the factory. They drank water from wells, and they happened to live upwind from the factory so that the smoke almost never blew in their direction. These people felt sorry for the families who got sick, but the poor families continued to suffer.

The manager of this factory set the price for the products. The price was determined by adding up the cost of the material, labor, light, heat, taxes, and so on, plus a profit. What cost involved in making the product was not passed on to the buyer? Who benefited most from this lower cost? Who suffered most?

CONSUMERS ARE POLLUTERS

About 15–20% of the world's population, including much of the population of North America and Western Europe, are the major consumers of the world. Unfortunately, these consumers do not pay the full price of the products they use up.

Every time a mining company cuts costs by leaving huge piles of debris and polluted air and streams, the consumer of those minerals saves. But many nonconsumers of those minerals pay the difference by having their environment affected. Every time a paper mill saves money by putting waste into a river, drive-in restaurant customers save on the price of paper-wrapped food. The unfortunate nonconsumer of paper-wrapped food who lives downstream from the paper mill has to "pay" for this saving.

Many ecologists believe that the total cost of consuming our resources should be paid by the consumer. Instead, today we divide up part of the cost and spread it among consumers and nonconsumers alike. What do you think?

Polluting Our Waterways. For the past 350 years, Americans lived their personal and business lives as described in the fable. To keep taxes low, sewers dumped waste from homes directly into rivers, lakes, and oceans. To keep the cost of manufacturing low, wastes from smokestacks, solid wastes, and liquid wastes have all been spewed out into the environment. Although the owners of homes and businesses reaped the benefits of lower costs and taxes, the actual costs were shared by everyone in the environment. Today, everyone lives with polluted lakes and streams, and with air which is growing worse to breathe.

By the year 1960, almost no stream or river which flowed through any small town was fit for drinking. After flowing through any large town or small city, it was not even fit for swimming. And after flowing through any large city, there was almost no life in the river at all.

A few people in the early twentieth century were aware of what was happening to our streams. However, almost no one believed that water pollution was any more widespread than just pollution of lakes and streams. But today there is abundant evidence that years of river pollution have brought about seriously polluted oceans. Of course, most of our largest cities did not contribute to polluting our upland streams. These seacoast cities have been polluting the oceans directly for centuries.

Pollution in Estuaries. The photograph here shows a region where a waterway slows down as it enters the ocean. As the river loses velocity, the particles it carries fall to the bottom. These particles block the flow of water so that the river is constantly being pushed aside. These wide marshy areas are called *estuaries*. Estuaries flood and drain twice a day as the tide flows in and out.

In past years, estuaries were considered to have little value. Although they were at important places where rivers meet oceans, nothing could be built there. The land was too wet. One use for such areas was as a dump. After many years, the solid waste would be piled high enough to build something, such as a factory or shopping center.

Soon, estuary markings began to disappear from many maps. Solid land with buildings took their place. At the same time, another phenomenon was noticed. Many different forms of life were becoming scarce in the ocean. After much study, it was learned that many sea animals such as menhaden, anchovies, and striped bass spend part of their lives in estuaries. Without estuaries, not only these animals, but other animals which are part of the food web with these fish, will also die.

Food Chains Concentrate Pollutants. Rivers bring dissolved chemicals to the sea. Sea animals such as clams remove calcium carbonate from seawater. Calcium carbonate is used by sea animals to make their shells. Recently, however, several new chemicals have been added to the sea. Some harmful examples are the insecticides DDD, DDT, and dieldrin.

Since World War II, people have used insecticides like DDT to kill insects. Molecules of DDT are not completely *biodegradable*. That is, they are not readily broken down into smaller harmless molecules the way molecules in a dead tree are broken down. As a result, much of the DDT which has been sprayed in the world is still with us. By now, much of it has washed down to the sea. Today, only a tiny percent of seawater consists of DDT, but its effects are very noticeable.

Algae and other phytoplankton take in DDT from the sea. Zooplankton feed on these plants containing DDT. As more plants are eaten, more DDT is taken in. Animals do not readily get rid of DDT in their bodies, so the amount of DDT in them gradually increases. These small animals are eaten

Concentration of DDD insecticide
(in parts per million)

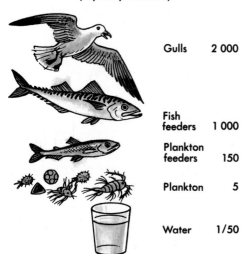

Gulls	2 000
Fish feeders	1 000
Plankton feeders	150
Plankton	5
Water	1/50

ZOOPLANKTON — Microscopic animals that live at the surface of the oceans.

487

by larger animals and the DDT levels are even higher. As a result, the larger animals in this food chain have the highest concentration and the most serious effects. For example, some types of salmon can no longer be sold, because their DDT level is too high to be eaten safely. The bald eagle is almost extinct because DDT affects the egg shells so that they are easily broken.

The Continental Shelf: Polluting Our Food Supply. Not all areas of the oceans are equally polluted. The most polluted regions are the continental shelves which are close to industrial countries which do most of the polluting. Unfortunately, these same continental shelf regions provide nearly all the fish and other seafood we eat. Industrial countries generally are the greatest polluters of the sea.

Recently, some biologists have noted that some types of algae and other phytoplankton are affected by very tiny amounts of DDT. These plants produce less food and oxygen when as few as one or two molecules of DDT are present for every billion molecules of water! This amount of DDT is often found along the continental shelf. Since we don't eat algae, why would scientists worry about how well algae grow?

Heat Pollution Along the Coasts. Not all pollution of the seas and rivers consists of chemicals. Many large factories and power plants are located along rivers and seacoasts. These plants often use water to cool their equipment. As a result, the water which is returned to the river or ocean is warmer than when it left. Note the picture at the left taken with infrared film. With this film, the warmer the water, the brighter the picture. Where did this warm water enter the river?

How will this change affect the amount of gas dissolved in the water? How might this change affect the organisms which live in this river, and the ocean into which it flows? How will the amount of thermal pollution change as increasing demand for electricity causes more power plants to be built?

Oil Pollution at Sea. Oil is one of the most widely used earth resources. Unfortunately, oil is not often found where it is needed. Thus, oil is always being transported from where it was found to where it will be used. As you read this, hundreds of oil tankers at sea are transporting millions of barrels of oil.

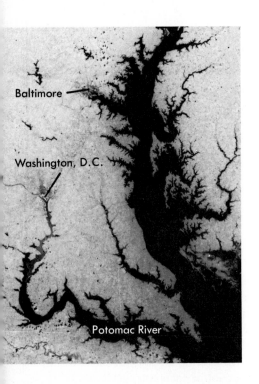

Baltimore

Washington, D.C.

Potomac River

There are many chances for spilling oil. Accidents at offshore wells like the one at the right sometimes cause oil slicks many square kilometers in area. Oil may spill into the ocean as it is being loaded on or off an oil tanker. Ships at sea sometimes collide, break apart in storms, or run aground. What would happen if such an accident involved an oil tanker?

In the past, oil tankers have carried oil to refineries. If the ship returned empty, it floated much higher in the water and was unsafe. For this reason, seawater was pumped into the nearly empty tanker for the return trip. When the tanker approached the port where more oil was to be taken on, the seawater (containing ½–1% oil) was pumped out into the sea. Today, methods and equipment have been developed to greatly reduce the amount of oil which enters the sea by this method. What are some reasons why oil tanker captains might or might not use such equipment?

Countries have agreed that offshore mineral rights extend only 5 kilometers from shore. The problem of oil pollution at sea shows the problem of making and enforcing laws in a region which no country owns. Who might fine the captain of an oil tanker which flushed its tanks 3 kilometers from shore? Six kilometers from shore?

The World's Largest Dump. For centuries, we have considered the rivers, lakes, and seas as convenient places for dumping wastes. All nations do not pollute the sea equally. Most observers agree that the countries which are the wealthiest are the ones which pollute most. Which countries should be the most able to pay for cleaning up their wastes to reduce ocean pollution?

Today, seacoast cities are debating the problems involved in developing other methods of treating garbage, trash, and sewer waste besides dumping them a few kilometers from shore. At the same time, people who are developing atomic power plants have a problem. How can they dispose of very radioactive material which is a waste product of using atomic energy to make electricity? Some people are now surrounding these wastes with concrete, and dropping them into the deepest part of the ocean. To be safe, they must not affect man in any way for 100 000 years. After that time, they will be less radioactive. Do you think this method of disposing of radioactive wastes is a good one? Propose a better solution.

REVIEW QUESTIONS

1. Generally, what types of countries do the most polluting?
2. What is an estuary? Why are they important?
3. Although ocean waters contain only very small amounts of mercury, the level of mercury in some ocean fish is too high to eat. How does this happen?
4. Why should there be laws prohibiting people from filling in marshes and estuaries?
5. Very few people eat algae, yet scientists worry about its growth. Why?
6. What is thermal pollution? How does it affect river and ocean water?

THOUGHT QUESTIONS

1. Why do most towns which use rivers for drinking water take in their water upstream from the town?
2. What are some reasons for changing the boundaries of our country so that the border extends to 300 kilometers offshore?

8

Careers

Since 1969 when astronauts first photographed the earth from the moon, people have become painfully aware of our planet as a spaceship with limited resources. Seeing the whole earth seemingly stopped and alone in space makes the earth seem small and fragile. The empty, desert-like terrain of the moon stands in stark contrast to the shining, blue-green planet we call home.

The astronauts on that epic voyage had to budget the use of their materials. So too must we on earth use the earth's resources with intelligence. There will be an increasing need for people who understand the materials of our earth and how they interact. This chapter introduces you to the wide variety of careers open to you in the earth sciences.

CAREERS IN EARTH SCIENCE

The study of the earth, from the lightless ocean floor to the upper atmosphere, offers a great many career opportunities. The science of studying the earth has been divided into several fields of interest. Some of these fields are listed in the chart below.

There are careers for individuals with interests and talents of all kinds. Technicians are needed to service, operate, and maintain the devices used in scientific activities all over the world. Engineers are needed to design and build apparatus to accomplish tasks never imagined a few years ago. Scientists are needed to conduct investigations, propose hypotheses, and counsel people about the effects of their activities on their environment.

The space missions and other explorations reported by television tend to glamorize the few individuals who must perform while the mission is in progress. These people have exciting tasks, but they represent a very small fraction of the total number who are involved. As a mission progresses, these less-publicized people take great pride in observing their success.

Even more important than these glamorous missions is the mission which concerns us all. That mission is to maintain and use the limited resources of our planet wisely. To accomplish this mission, people with knowledge of the earth sciences are needed.

FIELDS OF INTEREST IN EARTH SCIENCE

Historical geology	Interpret the history of the earth
Physical geology	Discover the nature of the earth
Paleontology	Study ancient life forms
Petrology	Investigate rocks and how they form
Meteorology	Study the atmosphere and its weather
Oceanography	Explore the oceans, earth's inner space
Astronomy	Explore planets, stars, and outer space
Cosmogony	Relate the earth to the solar system and the universe

Geologists. Geologists concern themselves with the earth and its history. In much the same way that detectives search for clues, geologists look for evidence to reconstruct the conditions of our planet millions of years ago.

In addition to advancing knowledge and gaining personal satisfaction from solving problems, there are practical advantages to a study of the earth's history. Many valuable materials, such as ores, gas, coal, and oil, are in constant demand because of the ever-increasing needs of a growing population. Knowledge of the earth's history is a necessary requirement if we are to find these materials.

Not all geologists devote their time to reconstructing the past. Most geologists are employed in mining and petroleum fields. A geologist's knowledge is vitally important to engineers who build dams, highways, bridges, and buildings. They need the geologist's knowledge of the nature of the underlying soils and rocks. The chemical composition and structure of the soil are factors which engineers must take into account if they are to build long-lasting structures which are environmentally sound.

Several specialized fields are recognized in geology. *Paleontology* is the study of the earth in ancient times. This study includes a concern for the living things as well as landforms. *Geochronologists* are mainly concerned with organizing geologic events into a time sequence. In many cases, they become involved in finding methods for establishing the age of a land formation or the fossil of a once living thing.

Geomorphologists. The changing shape of the earth's crust is of particular interest to the *geomorphologists*. They study the energy and forces which shape the ocean bottoms and the continents. The work of these scientists has contributed to the development of a theory to explain the shape of the earth's crust. They are trying to explain the position of the continents, the origin of the ocean basins, and the occurrence of mountain chains and other surface features.

A geomorphologist may be hired by any business interested in developing underground resources. Geomorphologists can survey land to determine the existence and extent of minerals underground.

493

The techniques and equipment used by geomorphologists often require the skill of trained technicians. Qualified technicians assist with field surveys as well as in the laboratory testing programs.

Meteorologists. This field of study covers all the interactions in the atmosphere. People involved in meteorology include not only the familiar weather forecaster, but also physicists, chemists, geographers, and mathematicians. On a day-to-day basis, many meteorologists concern themselves with understanding the weather well enough to forecast the atmospheric conditions for days, and even weeks ahead. Some develop mathematical models for use with computers. As computers receive data from hundreds of weather stations, the data are arranged according to mathematical models. When the arrangement is completed by the computer, certain weather conditions become obvious. Long-term forecasts can be made from this information.

Another function of the computer is to use data to draw a map of the weather as it exists across the country. These maps may be adapted for special purposes as needed by airlines and the coast guard. Farmers who might lose their crops unless they make preparations are interested in accurate weather maps.

Other meteorologists are involved in learning more about how the air is polluted. They want to know how the air can be kept clean. They also study the natural processes by which pollutants are removed from the air.

Still another interesting aspect of meteorology is controlling the weather. The damage brought on by hurricanes, tornadoes, floods, and droughts are a matter of history. If meteorologists can find out how to prevent the formation of these serious conditions, a great service will have been performed for everyone.

There are some questions about long-range changes in climates which meteorologists are researching. Some of these questions center around the ice ages. What factors are responsible for bringing about such drastic climatic changes as an ice age? Do ice ages occur in cycles? Are we now just coming out of an ice age or just about to enter one?

CAREERS IN OCEANOGRAPHY

The field of oceanography is rapidly becoming one of the most important fields among the earth sciences. Our increasing human population will make more and more use of the oceans. The oceans are a potential source for recreation, food, and minerals. In addition, the oceans are being used to dispose of wastes. These uses will require knowledgeable people to manage the oceans so that they are not ruined.

Oceanographers study all aspects of the oceans. They study the physical properties of ocean water, and relate changing conditions of the water to seasons, changing currents, pollution, or other causes. They study the living organisms in the oceans, finding out where and how they reproduce, what they eat, and what eats them. They study the shapes of the ocean basins and the processes which work to change the basins.

The results of the studies by oceanographers have greatly extended our knowledge of our planet. The studies have also contributed to the development of practical methods for weather forecasting, fishing, and mining the ocean and its floor.

The Work of an Oceanographer. As an oceanographer, you might plan and carry out an experiment in some part of the ocean. Some of these programs might be carried out aboard an oceanographic ship. Most oceanographers spend some of their time at sea collecting data. The remainder of the time is used for analyzing data, planning more studies, and so on. Other activities might include writing, teaching, or being a consultant for government or industry.

Biological Oceanographers. Most oceanographers specialize in one of the branches of oceanography. Biological oceanographers give their attention to the living things in the ocean. From their studies, many important discoveries have been made to the benefit of the fishing industry, the medical world, and our understanding of the ocean.

The migratory paths of many valuable marine animals have been discovered. Estimates of the populations of fish, whales, and other animals have helped individual countries and worldwide organizations to decide how best to control the harvest of these sources of food.

Thousands of sea animals are poisonous. The poisons within these animals may be used in the development of new drugs. Research with these poisons has resulted in the development of new ways to treat heart disease and cancer.

Field	Industry		Government		Universities	
	1976	1986	1976	1986	1976	1986
Geologists	26 250	30 000	3 150	4 000	10 600	12 000
Oceanographers	5 200	6 500	2 200	2 800	1 400	2 000
Meteorologists	2 000	3 000	2 000	2 500	1 300	1 800

Deep-diving submersible

Physical Oceanographers. The circulation of ocean water is one of the main concerns of physical oceanographers. Although they will often make observations themselves, they rely heavily on unmanned stations anchored far out at sea for data on the condition of the ocean water.

The atmosphere and the oceans have profound affects on each other. The air above the sea has properties and behaves in ways that are sometimes different from air over land. Measurements of atmospheric conditions made by aircraft have been compared with those made by ships. From these and other data, a more complete understanding is developing of how storms form over the oceans.

Geological Oceanographers. The structure and composition of the ocean basins are the major interests of scientists in this field. Drilling into the oceans floor has provided important evidence that the continents have been drifting across the surface of our planet. Geological oceanographers study sediments, measure the heat coming from within the earth, and explore the depths of the ocean in deep-diving submarines.

Ocean Engineers. Ocean engineers are responsible for designing and building the devices needed for research and equipment for harvesting the ocean's resources. Unmanned instrument packages designed to take measurements on a regular schedule have provided scientists with data over a greater area than ever before.

Ocean structures require special designs and materials to withstand the conditions in the open ocean. Such structures include the platforms from which drilling for gas and oil is done. Also, the large buoys which are anchored to the bottom of the sea and operate as robot observation stations require special designs by ocean engineers.

Feeding penguins at an aquarium

Wood's Hole Oceanographic Inst., Mass.

Education for Geologists and Oceanographers. At least 330 colleges and universities offer undergraduate courses in geology and oceanography leading to a bachelor of arts degree. There are a minimum of 160 institutions that offer advanced degrees. The outline of studies usually calls for a quarter of the academic time to be spent on courses in geology and oceanography. About one-third of the time is spent on mathematics, chemistry, and physics. The remaining time involves general academic subjects, including languages, literature, history, philosophy, and so forth.

Biography of an Oceanographer. Many situations, ideas, and experiences together seem to steer a person into one interest or another. In the case of one oceanographer, it was the gift of an aquarium with tropical fish, a newly built aquarium nearby, and extended visits to the ocean shore during vacation periods. All this occurred when this person was approaching the time to decide which courses to take in high school. This person was particularly interested in science and math, and took one science course and one math course each high school year.

During the high school years, this person's interest in aquatic organisms became stronger. Much time was spent at the nearby aquarium, raising several varieties of tropical fish at home, and visiting the ocean shore.

In the summer between junior and senior years, an application for a job at the commercial aquarium brought results. The duties included a lot of cleaning up, but this person got to know who did what and why, and how they felt about it. The associations made that summer were very helpful. Several people at the aquarium referred this person to their associates at other institutions. The succeeding summers were spent working at a sea circus, at an oceanographic institute, and cruising for six weeks on one of the institute's research vessels.

Meanwhile, this person decided to go to college to earn a bachelor of arts degree in oceanography. This decision meant taking mostly science and math courses, along with German, several courses in literature, philosophy, and art history.

Upon graduation from college, this person applied for a commission in the officer corps of the National Oceanic and Atmospheric Administration.

This is one of the United States' seven uniformed services. This person was commissioned as an ensign (the Army equivalent is a second lieutenant) in the National Ocean Survey. The ensign can expect sea duty aboard any one of several ships in the Survey's fleet. Within a few years, this person might command a small research ship. The ensign might also be selected for flight training, and be assigned to weather research and charting. There is also the possibility of working at one of the laboratories that process the data from the field.

From this biography of an oceanographer you can identify some important times when decisions had to be made. At what times did this person have to decide which courses to take in school? How were summer jobs helpful? What opportunities will this person have to choose from in the near future?

At several points in this person's career, decisions had to be made about continuing academic training. Although this person did elect to continue, the decision could have been to concentrate on technical training. If this had been the case, several different opportunities would have occurred. What might some of them have been?

REVIEW QUESTIONS

1. What role do technicians play in the scientific study of the earth?
2. What are some of the tasks of engineers and scientists in advancing our knowledge of the earth?
3. Name and describe three fields of study in geology.
4. What are scientists who study the atmosphere called?
5. Name and describe three fields of study in oceanography.

THOUGHT QUESTIONS

1. Why is it believed that the demand will increase for people who know about earth science?
2. Who supports most of the work done by earth scientists?
3. At what times during the next ten years will you have to make decisions about your career? What kinds of decisions will they be? Upon what considerations will the decisions depend?

A New Eye for the Scientist

Observations are the building blocks for all scientific understandings. As more and different types of observations are made, scientific understanding increases. Early observers used only their eyes to develop the first ideas about astronomy. Later, the invention of the telescope brought new observations which resulted in a much better understanding of the moon and planets. The large telescopes of this century gave still more observations which increased our knowledge of galaxies.

In recent years, earth scientists have used another new tool which yields the best observations yet. These observations come from satellites. Satellites allow astronomers a view of outer space unhampered by air and dust. Satellites also allow earth scientists new and better views of the continents, oceans, and weather systems.

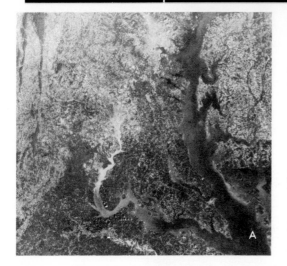

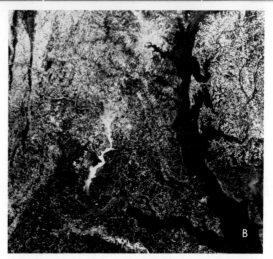

LANDSAT SATELLITE PHOTOGRAPHY

In July 1972 and January 1975, two satellites were placed in orbit. These satellites were originally ERTS (Earth Resources Technology Satellites), but are now called Landsats. The orbits are nearly circular, going from pole to pole. Each satellite travels around the earth 14 times a day at a height of 920 km. Each satellite sees 14 different strips on the lighted side of the earth each day. On the next day, 14 different strips are seen which slightly overlap the previous day's view. Every 18 days, the observed strips repeat. If there are no clouds, each region of the earth can be carefully observed each 18 days.

Because the satellites do not have people on them, the "seeing" does not depend on human eyes. Instead, more different types of light coming from the earth's surface are observed than any human eye can see. This idea is described below.

A Brief Review of Energy and Light. The diagram across the tops of these pages shows types of energy waves which are constantly traveling through the air around you. The waves from gamma through infrared can all be detected by photographic film. Which rays can you detect with your senses? Which rays can you detect with the aid of the detector shown at the left?

When energy waves strike a surface, several changes may occur. The waves may (1) pass through, (2) be absorbed by the new material, or (3) be reflected. In the diagrams on page 501, two different wavelengths of energy are pictured as being

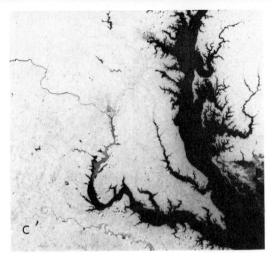

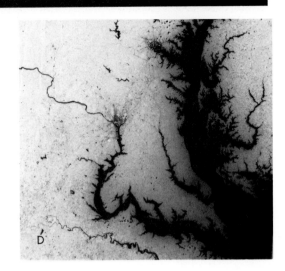

reflected. The shorter wavelength is a type of visible radiation called green light which the human eye can detect. The image which forms at the back of the eye stimulates nerve endings there and a message is sent to the brain.

At the same time a longer wavelength of energy reflects off the plant and enters the eye. This wavelength is called infrared and is shown in black. The image which forms in the eye is not detected by the nerve endings there. No information is sent to the brain from this wavelength.

Satellite Photography. The four black and white photographs on these pages were taken at the same time from a Landsat satellite. Note Washington, Baltimore, Chesapeake Bay, and the Potomac River. Each picture was taken with filtered light of a different wavelength, similar to the lower diagram. These wavelengths correspond to two visible colors (green and red) and two colors our eyes normally cannot see (very near and near infrared). Study the pictures. Which photographs best show the sediments being carried in the Potomac River? Suggest a cause of the sediments. Suggest a possible explanation of why the sediments at this time were only in a part of the river.

Which photograph received the most reflected light (the brightest image)? Which photographs contrast land and water best? Which photograph shows details on the land best?

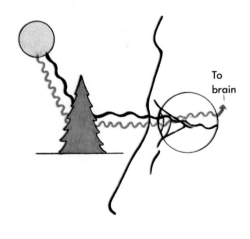

To brain

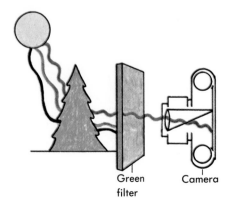

Green filter

Camera

501

Printing Color Images. Most color pictures are printed in books and magazines using a printing press with four different cylinders. The press supplies the three primary colors (red, yellow, and blue) plus black. By varying the amount of ink from each cylinder, the press can produce all different shades and colors. Note the effect of each additional cylinder in the color print here.

False Color Photographs. Two or more photographs like those shown on pages 500 and 501 can be combined to form a color photograph. The photograph on page 503 was made by combining photographs A, B, and D. If the photographs were printed as the eye normally sees them, all the information in D, the near infrared photograph, would be invisible. Instead, each photograph is printed in a different color. The green filter image (A) is printed in blue ink. The red filter image (B) is printed in green ink. The near infrared filter image (D) is printed in red ink. The resulting print is called a *false color photograph*. The photographs in this chapter are all false color Landsat photographs.

Plants reflect large amounts of light of near infrared wavelength. This wavelength is printed in red ink on false color photographs. Where are trees and other plants abundant in the false color photograph? Yellowish-white is the false color of freshly plowed or harvested fields. Where are such farmlands located in this photograph? Suggest a possible reason for the unusual red area along the left edge of the photograph.

Note Bull Run southwest of Washington. What evidence supports the idea that a recent rain in the upper Potomac River watershed did not produce much rain in the Bull Run watershed?

Look for evidence of the effects of people. How do cities look in these photographs? Can you find other evidence such as highways? What is the narrowest region you can detect on these photographs as reproduced in this book?

Recognizing Faults. The false color photograph on page 504 shows the Los Angeles region of California. Los Angeles is the light blue area along the coast at the bottom right of the photograph. Why hasn't Los Angeles grown outward at the same rate in all directions?

The Mohave Desert is the whitish region in the northeast (upper right) part of the photograph.

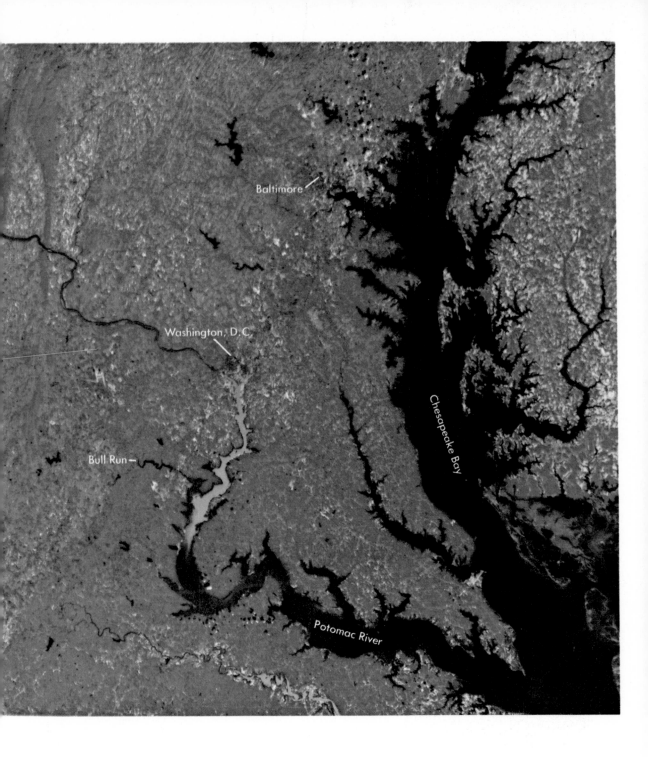

Baltimore

Washington, D.C.

Bull Run

Chesapeake Bay

Potomac River

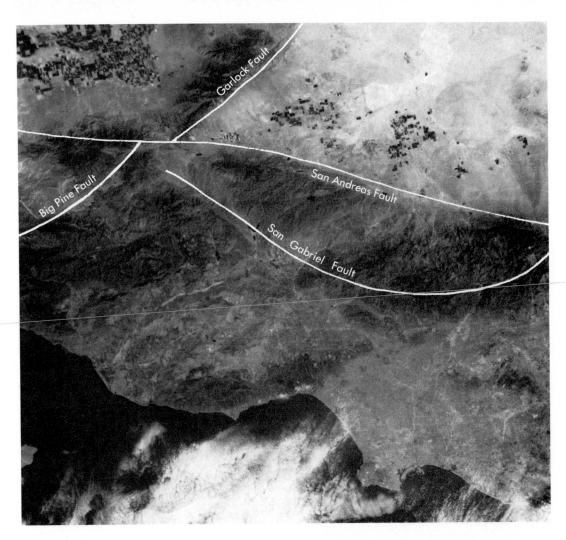

Locate some irrigated lands and an alluvial fan in this region. Propose how the fan developed.

Note the straight line-like boundaries on the west and south sides of the desert. These borders mark two important faults. Note these two faults on page 504. Geologists know that faults often mark the edge of a straight mountain range or the path of a straight valley. Look for other examples of line-like structures in the photograph. Compare these structures with the position of known faults as shown on the photograph above.

Note Big Pine and Garlock Faults. Some geologists believe that these two faults were once one continuous fault. How might the land on either side of the San Andreas Fault have moved to produce the present positions?

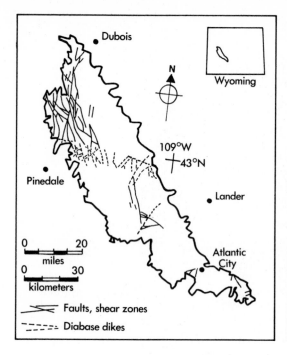 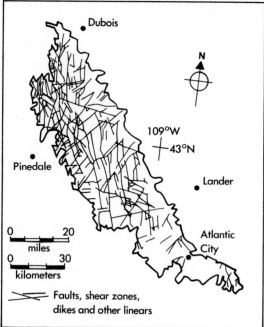

Using Photos to Map Faults. On the previous page, known faults were identified with a Landsat photograph. Of greater importance is the identification of previously unknown faults from photographs. Such information is important for at least two different reasons: (1) knowledge about faults aids geologists in understanding the geological history of a region, and (2) faults are often important mineral and mining sites.

The left map above shows the outline of the rugged Wind River Mountain Range in Wyoming. The lines within the region show the faults which were identified during five summers of geological (bedrock) mapping (much geological mapping is done by college geology teachers during their summers).

The right map shows the faults identified in eight hours by a person skilled in identifying faults from Landsat photographs. These nearly straight line structures still need to be checked by visits to the sites, but the advance knowledge in planning detailed mapping is very valuable.

Geologists also know that curved bedrock fractures often indicate igneous rock intrusions, such as the curved structures left by an old volcanic neck. Such structures also have the potential of being important sites for future mines.

506

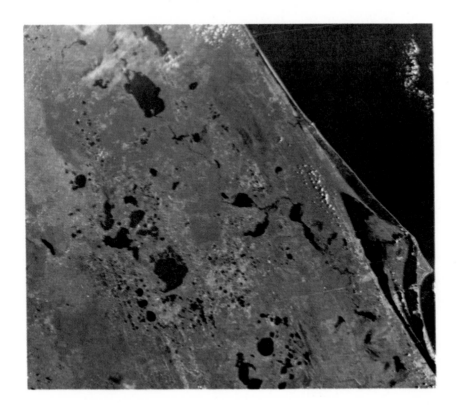

A SAMPLE OF SATELLITE PHOTOGRAPH USES

In the previous section, you learned how satellite photographs are used to study faults. In this section, a few quite different earth science uses of satellite photos are presented. Many more uses for such photos will probably be developed in the future.

Florida's Beaches, Old and New. The photograph above shows a part of eastern Florida. The straight shoreline with its offshore bar is a feature of ocean coasts where the land is slowly rising (or the sea is slowly dropping). Most geologists believe that Florida is the youngest part of the United States. Its land is still slowly rising. If this idea is correct, what will happen to the shoreline in this region in the next million years? Where would the shoreline have been a million years ago? Look for evidence of parallel beaches indicating where the shoreline was long ago.

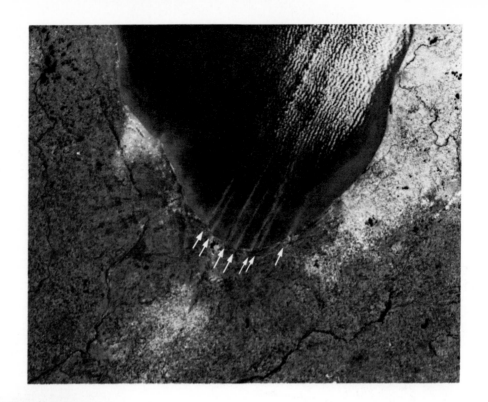

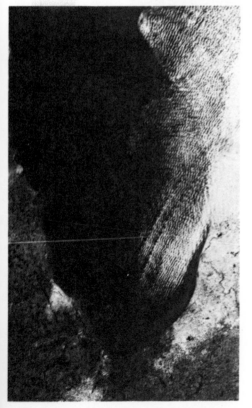

There are two other interesting land features of interior Florida. Water-filled round sinkholes occur when ground water dissolves the soft young limestone bedrock. There are hundreds of springs, many of which pour billions of liters of fresh water a day into Florida's lakes and rivers. Locate some round, water-filled sinkholes. Streams can be located by the darker vegetation which grows along the banks.

Humans Affect Weather. The pictures on this page are the first photographic proof of how humans have affected the weather. The top photograph has seven arrows pointing to the smokestacks of power plants and steel mills in the Gary, Indiana and Chicago, Illinois areas. The smokestacks add many metric tons of smoke and dust particles to the air each day. Each particle serves as a nucleus on which water vapor condenses. The water vapor is provided by evaporation from Lake Michigan. The millions of tiny droplets which result can be seen as clouds which form down-wind from the smokestacks. How do these clouds affect the weather to the northeast? Compare the ground cover to the northeast with the ground cover around Chicago.

The "Whiting" of Lake Michigan. The photograph here was taken in August, at a time when lake users were noticing that the lake water appeared milky-colored. Similar observations are becoming more common not only in this lake but in Lake Erie and Lake Ontario as well. During such "whitings" city water intakes often become clogged with thick masses of algae.

Careful study of the lake water at this time showed a surprisingly large increase in the number of algae and other plant microorganisms growing in the water. These plants remove carbon dioxide from the water, making the water less acidic. Water which is acidic can dissolve calcium carbonate; water which is not acidic cannot dissolve calcium carbonate. Thus, as the lake water became less acidic, the dissolved calcium carbonate precipitated out (formed a white powder which would not dissolve in the water). The precipitate made the water milky-colored. If the precipitated calcium carbonate settles on the lake bottom and hardens, a layer of limestone (calcium carbonate rock) will result.

Why the algae suddenly increased in numbers was a question which puzzled the lake investigators. Such increases occur in laboratories only when more nutrients are added to the algae's environment. Usually algae grow at the top of a lake and most of the nutrients stay near the bottom. For several days before this photo was taken, there had been a strong wind blowing from the west. This force would cause the water level on the lake's east side to increase slightly. Then the wind quickly changed and blew from the east. Quickly the surface waters moved to the west side of the lake. This shift brought deep nutrient-rich water to the surface on the east side of the lake. How does the photograph support this theory?

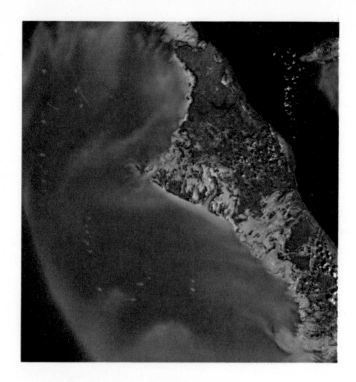

The Bahamas Platform. The photograph above shows some of the hundreds of islands which make up the Bahama Islands, located 300 km southeast of southern Florida. These islands are perched on many separate seamounts sticking up from the ocean floor. Note these islands on the diagram on page 422.

This region is one of the few places where earth scientists can study firsthand the processes which form the world's limestone, a common sedimentary rock. The main producers of insoluble calcium carbonate are sea animals (coral) and sea plants (algae). The coral, like clams and other relatives, take dissolved minerals from the seawater and form hard body coverings. Wave action breaks these structures apart and the sediments settle to the sea floor. The algae remove calcium carbonate as described on the previous page. How does warm, clear water aid algae growth?

Most coral growth is along the east coast of the islands. The sediments are carried to the shallow region to the west. Note the much deeper regions along each edge of the photograph.

The ocean floor in this region varies from 500 to 4000 meters. The limey sediments under the island reach a depth of 5000 meters. Explain this in terms of the crust under the island.

Southern Iceland. Few regions of the world interest geologists as much as Iceland. Its geological variety includes (1) the easiest place to study the mid-ocean ridge (it's above water here), (2) active volcanoes, (3) numerous glaciers, (4) bubbling molten lava surrounded by a glacier, (5) enough steam vents and hot springs to heat and power the largest city of Iceland, and (6) a new island off the south coast which formed from 1963 to 1967 as a volcano on the sea floor. The island grew so large it reached the ocean's surface. This small island, labeled S on the photograph, is called Surtsey.

Iceland's land surface includes snow and ice (white on the photograph), bare lava rock (dark brown), cultivated fields, grasslands, dwarf trees, and lichen-covered rocks (differing shades of red). Identify some snow-covered mountains and some valley glaciers coming from mountain snow fields.

The letter H points to Mt. Hekla which last erupted in 1973. A lava flow from 1973 can be seen on the northwest side between the mountains and the river. Trace the river to its outlet on the south coast. What causes the blue color in the ocean here? Why might this river produce more of this material than other rivers?

Note the line-like structures to the west of this river. These structures, resembling faults, mark the center of the Mid-Atlantic Ridge, where the earth is spreading apart. What will happen to Iceland's size in the future?

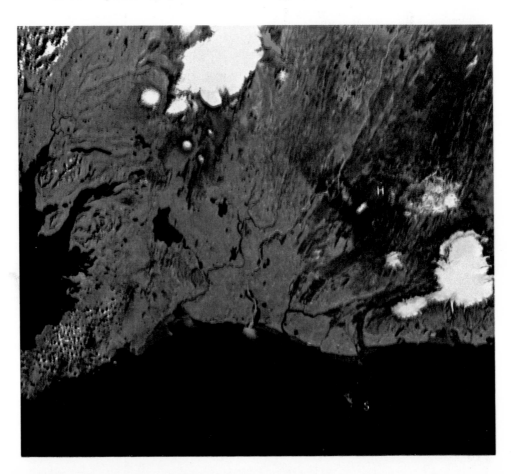

Interested Specialist	Applications/Problems	Comments
Map Maker	How to keep maps up-to-date in rapidly changing urban, shoreline or remote areas.	
Mineralogist	How to identify potential mining sites.	Rock formations and rock and soil color are useful.
Glaciologist	How to observe advances and retreats of valley glaciers in remote regions.	
Water Resource Manager	How to identify potential water well sites. How to predict water storage (snow) and spring runoff. How to map accurately the extent of flooding. How to determine the best location of water intakes.	Useful in determining need for federal disaster assistance; show areas of high and low sediments.
Archeologist	How to identify roads in earlier civilizations.	Landsat photos doubled the known length of roads built by 10th and 11th century Pueblo Indians.
Land Manager	How to record the extent of avalanches and landslides in remote areas. How to map the extent of forest fire damage. How to monitor the extent to which strip miners are restoring land to its natural state.	
Shipper	How to inform ships of the best routes when ice breaks up in the spring.	
Meteorologist	How to trace tornado paths. How to monitor effect of cloud-seeding experiments. How to trace the development of storm systems to aid in weather prediction.	Damage paths show clearly in wooded regions. Satellite pictures shown regularly on TV news.
Agriculturist	How to measure the effect of different grazing policies on grasslands. How to observe the extent of insect damage in orchards and forests.	
Road Designer	Designing roads in unmapped deserts.	

REVIEW QUESTIONS

1. What is a false color photograph? Why is it more useful than regular color photography?
2. How can faults be identified in satellite photos?
3. Why is the knowledge of faults useful?

THOUGHT QUESTION

1. Propose a use for satellite photographs not discussed in this chapter.

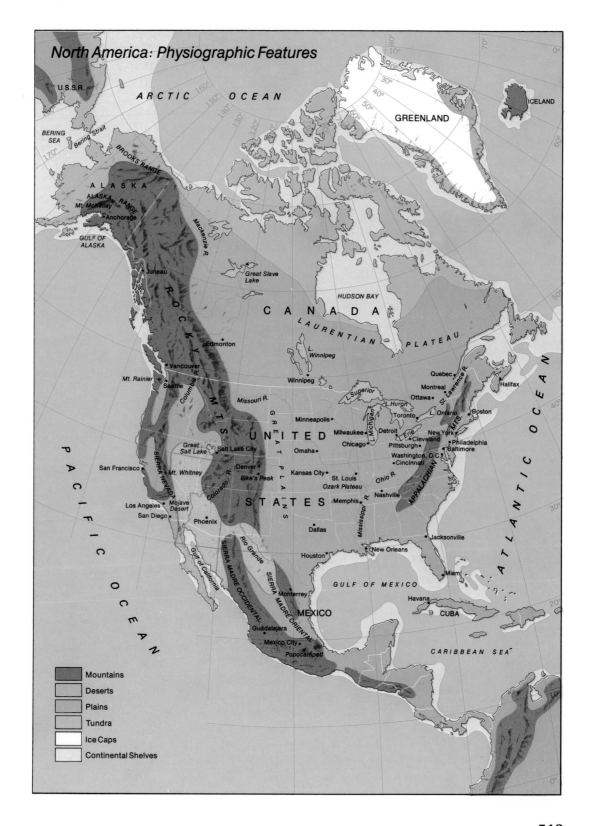

North America: Physiographic Features

Mountains
Deserts
Plains
Tundra
Ice Caps
Continental Shelves

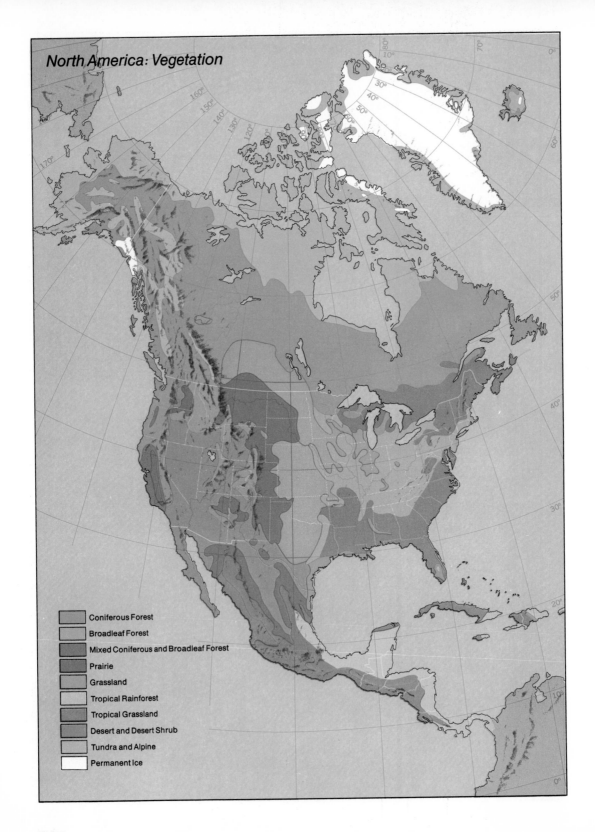

North America: Vegetation

Coniferous Forest
Broadleaf Forest
Mixed Coniferous and Broadleaf Forest
Prairie
Grassland
Tropical Rainforest
Tropical Grassland
Desert and Desert Shrub
Tundra and Alpine
Permanent Ice

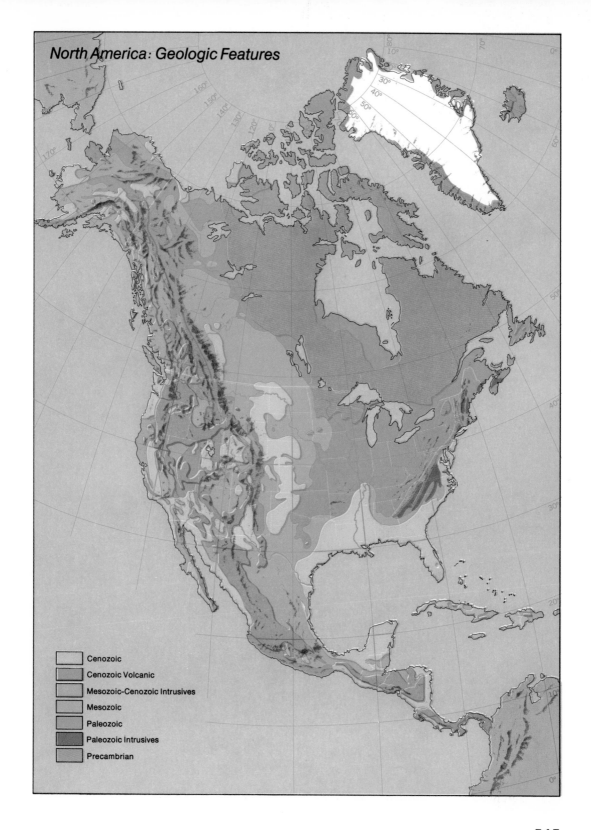

North America: Geologic Features

Cenozoic
Cenozoic Volcanic
Mesozoic-Cenozoic Intrusives
Mesozoic
Paleozoic
Paleozoic Intrusives
Precambrian

North America: Surface Rocks and
Fossil Deposits

Igneous and Metamorphic Rock

Sedimentary Rocks

Glacial Ice

F   Major Fossil Locations

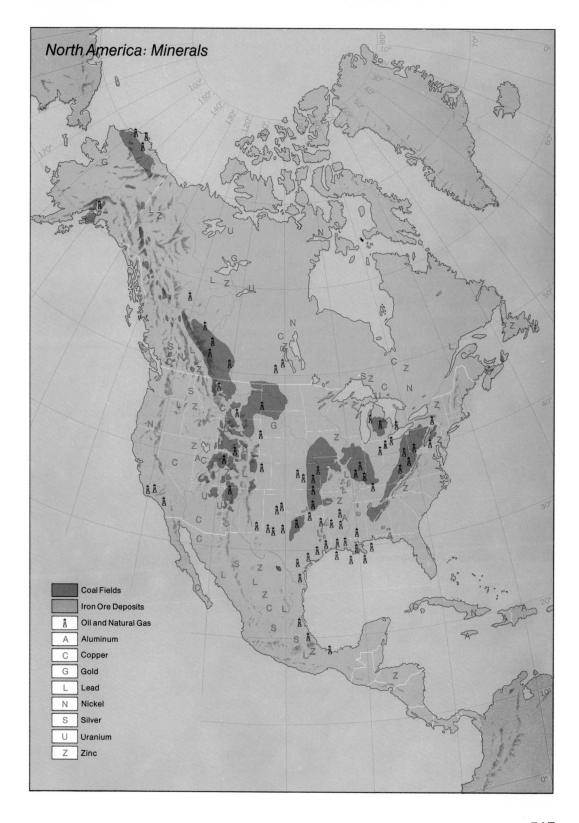

North America: Minerals

Coal Fields
Iron Ore Deposits
Oil and Natural Gas
A   Aluminum
C   Copper
G   Gold
L   Lead
N   Nickel
S   Silver
U   Uranium
Z   Zinc

ATOMS AND ELEMENTS—
MOLECULES AND COMPOUNDS

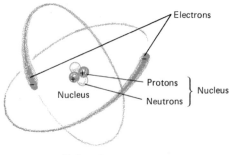

Helium atom

Atoms. Scientists believe that all matter is composed of *atoms*. Just as all the words in the English language are made up of 26 letters of the alphabet, so all matter is composed of about 100 different atoms. Some atoms are more plentiful than others. For example, in the earth's crust there are more atoms of oxygen than of any other element.

Each atom has a tiny dense core called a *nucleus*. There are two kinds of particle in the nucleus. *Protons* are nuclear particles that carry a positive charge. The other kind of particle is the *neutron* which does not carry any charge. So the nucleus of an atom has a positive charge.

Around the nucleus there are shells of *electrons* which bear a negative charge. The number of electrons in an uncharged atom equals the number of protons inside the nucleus. If there are fewer electrons than protons, the atom has a positive charge. Such a charged atom is called an *ion*.

Elements. Matter found in nature is composed of atoms having from 1 to 92 positive charges. These atoms are the smallest particles of the 92 natural *elements*. Scientists have created 13 additional elements. The table on pages 520–521 lists 105 elements that have been identified.

Most elements are found in combination with others. Only a few are found as a pure substance. Gold is one element found in pure form.

Compounds. The arrangement of electrons in an atom determines its chemical properties. Most atoms will combine chemically with other atoms to form *compounds*. The various atoms of a compound join together by the interaction of their electrons. The way the electrons interact usually determines the number of atoms of one kind that will combine with other kinds of atoms.

The illustration on the next page shows a sodium atom giving up an electron to a chlorine atom. Sodium will give up only one electron, and chlorine will attract only one. When sodium and chlorine atoms combine through electron interaction, one atom of sodium always reacts with one atom of chlorine.

When an atom of sodium unites with an atom of chlorine, a new substance is formed which does not resemble either element. The compound so formed usually has quite different properties from the elements of which it is composed.

Molecules. A group of atoms tightly bound together is called a *molecule*. If all the atoms are of the same kind, we have a molecule of an element. If the atoms are of different kinds, we have a molecule of a compound. One molecule of the compound hydrogen chloride contains one hydrogen atom and one chlorine atom.

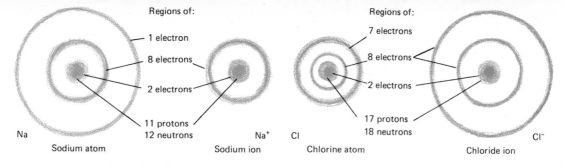

Regions of:

— 1 electron
— 8 electrons
— 2 electrons

11 protons
12 neutrons

Na
Sodium atom

Na⁺
Sodium ion

Regions of:

7 electrons
8 electrons
2 electrons

17 protons
18 neutrons

Cl
Chlorine atom

Cl⁻
Chloride ion

THE PERIODIC TABLE

Atomic Weights. Each element represents a particular kind of atom, and any quantity of the element is made up of identical atoms of this kind. What distinguishes one element from another is the nature of its atoms. The basic physical difference between atoms is their weight, or mass. For instance, sulfur atoms are twice as heavy as oxygen atoms. Chemists have measured the relative weights of different atoms and arranged them in order of increasing weight. The relative weight of each atom is called its *atomic weight*.

Atomic Number. Each kind of atom has a particular number of protons in the nucleus. The *atomic number* represents the number of protons. Oxygen, for instance, has 8 protons, fluorine has 9, and nitrogen has 7. Knowing the number of protons and the

8 electrons

Nucleus { 8 Protons +
           8 Neutrons —

Oxygen atom

atomic weight enables you to determine the number of neutrons as well. The atomic weight consists almost entirely of the sum of the protons and neutrons. (Electrons have too little mass to be significant.) By subtracting the atomic number from the atomic weight (rounded off to the nearest whole number) you can determine the number of neutrons in the nucleus of an atom.

Atomic Symbols. Chemists save time and space by writing symbols for each kind of atom. Most of the symbols are easy to remember since they contain the first letter and one other prominent letter in the name of the element. A few symbols, such as Ag for silver, are used because these elements were known centuries ago when Latin was the language used by scholars. The Latin word for silver is *argentum*.

Chemical Formulas. Elements will combine with other elements forming *chemical bonds* by the interaction of their electrons. When two or more elements combine chemically, they always combine in set proportions. For example, when hydrogen and oxygen combine to form water, they always combine in the proportion of two hydrogen atoms to one oxygen atom. Chemists write the *formula* for one molecule of water as H_2O. A subscript number to the right of a symbol indicates the number of those atoms contained in the molecule. If a chemist wants to indicate two molecules of water, the formula would be written $2 H_2O$. Molecules of water are sometimes chemically combined in a mineral such as gypsum. The formula for gypsum is written $CaSO_4 \cdot 2H_2O$.

519

THE PERIODIC TABLE OF THE ELEMENTS

1.008	← Atomic Weight
Hydrogen	← Name of element
1 (H)	← Atomic number Symbol

The blue areas of the table represent the two rare-earth series: The lanthanides and the actinides, named after their respective first members. The top number in each box indicates the atomic weight of the element. The number at the lower left is the atomic number.

6.94 Lithium 3 (Li)	9.0122 Beryllium 4 (Be)							
22.9898 Sodium 11 (Na)	24.312 Magnesium 12 (Mg)							
39.102 Potassium 19 (K)	40.08 Calcium 20 (Ca)	44.956 Scandium 21 (Sc)	47.90 Titanium 22 (Ti)	50.942 Vanadium 23 (V)	51.996 Chromium 24 (Cr)	54.938 Manganese 25 (Mn)	55.847 Iron 26 (Fe)	58.9332 Cobalt 27 (Co)
85.47 Rubidium 37 (Rb)	87.62 Strontium 38 (Sr)	88.906 Yttrium 39 (Y)	91.22 Zirconium 40 (Zr)	92.906 Niobium 41 (Nb)	95.94 Molybdenum 42 (Mo)	98.91 Technetium 43 (Tc)	101.07 Ruthenium 44 (Ru)	102.905 Rhodium 45 (Rh)
132.905 Cesium 55 (Cs)	137.34 Barium 56 (Ba)	138.91 Lanthanum 57 (La)	140.12 Cerium 58 (Ce)	140.91 Praseodymium 59 (Pr)	144.24 Neodymium 60 (Nd)	(147) Promethium 61 (Pm)	150.4 Samarium 62 (Sm)	151.96 Europium 63 (Eu)
			178.49 Hafnium 72 (Hf)	180.948 Tantalum 73 (Ta)	183.85 Tungsten 74 (W)	186.2 Rhenium 75 (Re)	190.2 Osmium 76 (Os)	192.2 Iridium 77 (Ir)
(223) Francium 87 (Fr)	226.03 Radium 88 (Ra)	(227) Actinium 89 (Ac)	232.038 Thorium 90 (Th)	(231) Protactinium 91 (Pa)	238.03 Uranium 92 (U)	(237) Neptunium 93 (Np)	(242) Plutonium 94 (Pu)	(243) Americium 95 (Am)
			(261) Kurchatovium? 104 (Ku?)	(260) Hahnium? 105 (Ha?)				

Gas at room temperature

Liquid at room temperature

						4.0026 Helium 2 (He)
10.81 Boron 5 (B)	12.011 Carbon 6 (C)	14.0067 Nitrogen 7 (N)	15.9994 Oxygen 8 (O)	18.9984 Fluorine 9 (F)	20.183 Neon 10 (Ne)	
26.9815 Aluminum 13 (Al)	28.086 Silicon 14 (Si)	30.9738 Phosphorus 15 (P)	32.06 Sulfur 16 (S)	35.453 Chlorine 17 (Cl)	39.948 Argon 18 (A)	

58.71 Nickel 28 (Ni)	63.546 Copper 29 (Cu)	65.37 Zinc 30 (Zn)	69.72 Gallium 31 (Ga)	72.59 Germanium 32 (Ge)	74.9216 Arsenic 33 (As)	78.96 Selenium 34 (Se)	79.904 Bromine 35 (Br)	83.80 Krypton 36 (Kr)
106.4 Palladium 46 (Pd)	107.868 Silver 47 (Ag)	112.40 Cadmium 48 (Cd)	114.82 Indium 49 (In)	118.69 Tin 50 (Sn)	121.75 Antimony 51 (Sb)	127.60 Tellurium 52 (Te)	126.904 Iodine 53 (I)	131.30 Xenon 54 (Xe)
157.25 Gadolinium 64 (Gd)	158.924 Terbium 65 (Tb)	162.50 Dysprosium 66 (Dy)	164.930 Holmium 67 (Ho)	167.26 Erbium 68 (Er)	168.934 Thulium 69 (Tm)	173.04 Ytterbium 70 (Yb)	174.97 Lutetium 71 (Lu)	**Actinide Series** ←
195.09 Platinum 78 (Pt)	196.967 Gold 79 (Au)	200.59 Mercury 80 (Hg)	204.37 Thallium 81 (Tl)	207.2 Lead 82 (Pb)	208.980 Bismuth 83 (Bi)	(210) Polonium 84 (Po)	(210) Astatine 85 (At)	(222) Radon 86 (Rn)
(245) Curium 96 (Cm)	(247) Berkelium 97 (Bk)	(249) Californium 98 (Cf)	(254) Einsteinium 99 (Es)	(255) Fermium 100 (Fm)	(256) Mendelevium 101 (Md)	(254) Nobelium 102 (No)	(257) Lawrencium 103 (Lr)	**Lanthanide Series** ←

CELSIUS TEMPERATURE SCALE

Scientists and most other people in the world use the Celsius temperature scale. On this scale, water freezes at 0° and boils at 100°. The space between 0 and 100 on the thermometer is divided into 100 equal parts. Spaces of the same size are added to the thermometer above 100° and below 0°. The Celsius temperatures for a number of environments and changes of state are given on the scale at the right.

THE METRIC SYSTEM

The metric system of measurement was developed by a commission of French scientists during the late 1700's. During the 1800's, it was adopted by many other European countries and Japan as the legal system of weights and measures. More recently, China, India, and Egypt, and now Great Britain, Canada, and Australia have switched over to the metric system. It is used by scientists throughout the world and by the people of most nations except the United States. The system used in the United States is called the English system.

The basis of the metric measurement of length was originally the distance between the North Pole and the Equator of the earth. Dividing this distance into ten million equal parts gave a length of one meter. This equals 39.37 inches, or 1.1 yards in the English system.

The three most commonly used metric units are: (1) meter, for length; (2) liter, for volume; and (3) gram, for mass. (Weights are often expressed in grams, since weight is used as a measure of mass.) A prefix can be added to any of these units to obtain a smaller or larger unit. These prefixes are always multiples or fractions of ten. For example, a milliliter is 1/1000 of a liter; a

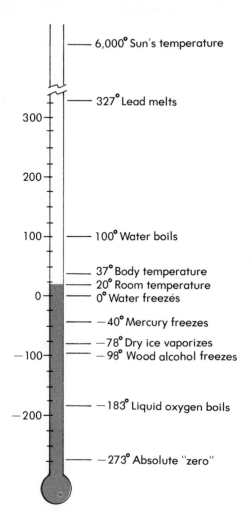

6,000° Sun's temperature

327° Lead melts

300

200

100 — 100° Water boils

37° Body temperature
20° Room temperature
0 — 0° Water freezes

−40° Mercury freezes
−78° Dry ice vaporizes
−100 — −98° Wood alcohol freezes

−183° Liquid oxygen boils
−200

−273° Absolute "zero"

kilogram is 1000 grams. These prefixes can be used with any metric unit. The adjoining table shows many of the prefixes used in the metric system. Those most commonly used are *milli-*, *centi-*, and *kilo-*.

THE METRIC SYSTEM

Unit	Length	Volume	Mass
METRIC NAME	METER	LITER	GRAM
Approximate Amount	from your waist to the floor	4 cups	a dime has a mass of 2 grams

METRIC PREFIXES

Prefix	Abr	Meaning	Commonly Used Lengths	Commonly Used Volumes	Commonly Used Masses
micro-	μ	$1/1\,000\,000$	micron	—	microgram
milli-	m	$1/1\,000$	millimeter	milliliter	milligram
centi-	c	$1/100$	centimeter	—	—
deci-	d	$1/10$	—	deciliter	—
(no prefix)		1	meter	liter	gram
deka-	da	10	—	—	—
hecto-	h	100	—	—	—
kilo-	k	1 000	kilometer	—	kilogram
mega-	M	1 000 000	megameter	—	megagram

COMPARING LENGTH TO VOLUME TO MASS (WEIGHT)

This distance |←——→| is 1 centimeter of length

   (100 of these equals 1 meter)

A cube 1 centimeter on each side has a volume of 1 milliliter

   (1 000 of these 1-milliliter cubes equals 1 liter)

One milliliter of water has a mass of one gram

OUT-OF-CLASS STUDIES
Listed by Seasons

Fall

1. Sept. 21st—Sun rises exactly east and sets exactly west. Locate these points on the horizon. Determine the hours of daylight.
2. Note the position of the Big Dipper in the evening.
3. Record air and water temperatures throughout Sept. and Oct.
4. Watch weather maps for hurricanes and frost lines.
5. Pound stakes flush into the ground exactly three feet from eroding stream banks and head ends of gulleys. Look for changes in the positions of the stakes after spring erosion.
6. After the first killing frost, look for plants that were not killed. Identify the factors which protected these plants.
7. Paint a section on a cliff. Next spring, look for fallen painted rocks.
8. Watch for meteor showers on these dates: Oct. 10 (Draco), Oct. 22 (Orion), Nov. 17 (Leo), Dec. 12 (Gemini).

Winter

1. Dec. 21st—Note the rising and setting positions (and times) of the sun. Determine the number of hours of sunlight. Estimate the angle of elevation of the noontime sun.
2. Note the position of the Big Dipper. Compare with your fall data.
3. Find the depth to which soil freezes.
4. Drill holes through the ice and determine the depth of ponds and lakes. Make a topographic map of the lake bottom.
5. Study lake temperatures at different depths monthly through spring.
6. Compare monthly temperatures of wells and springs with lakes and streams.
7. Note the change in Orion's 8:00 P.M. position during winter and spring.
8. Take photos of wind erosion in snow; angles of rest of snow.
9. Watch for meteor showers on these dates: Dec. 22 (Little Dipper), Jan. 3 (Boötes, Draco).

Spring

1. March 21st—Sun rises exactly east and sets exactly west. Record the rising and setting times. Determine the hours of sunlight at the equinox.
2. Compare the evening position of the Big Dipper with your previous data.
3. Measure a stream's velocity at high water. Compare this velocity with the velocity at low water.
4. Visit your erosion points (see no. 5, Fall) and look for evidence of erosion.
5. Visit the painted cliff section (see no. 7, Fall). Look for changes.
6. Record the daily temperatures of a pond and the air.
7. Watch for meteor showers on these dates: April 21 (Lyra), May 4 (Aquarius), July 29 (Aquarius), Aug. 12 (Perseus).

GEOLOGY

Aran, Joel, *Rocks & Minerals.* New York, N.Y.: Bantam, 1973, 145 pp.

Farb, Peter, *Living Earth.* New York, N.Y.: Pyramid, 1959, 160 pp.

Fay, Gordon, *The Rockhound's Manual.* New York, N.Y.: Harper & Row, 1972, 290 pp.

Fenton, Caroll Lane and Mildred Adams Fenton, *The Rock Book.* Garden City, N.Y.: Doubleday, 1940, 357 pp.

Hurlbut, C. S., Jr., *Dana's Minerals and How to Study Them.* New York, N.Y.: Wiley, 1949, 323 pp.

Matthews, William H., III, *Soils.* New York, N.Y.: Franklin Watts, 1970, 89 pp.

Pearl, Richard M., *How to Know the Minerals and Rocks.* New York, N.Y.: Signet, 1955, 192 pp.

Pough, Frederick H., *Field Guide to Rocks & Minerals.* Boston, Mass.: Houghton Mifflin, 1953, 333 pp.

U.S. Dept. of Agriculture, *Soil. The 1957 Yearbook of Agriculture.* Washington, D.C.: U.S. Gov't Printing Office, 1957, 784 pp.

Zim, Herbert S. and Paul R. Shaffer, *Rocks & Minerals.* New York, N.Y.: Golden Press, 1957, 160 pp.

THE EARTH'S SURFACE

Anderson, Alan, *The Drifting Continent.* New York, N.Y.: Putnam, 1971, 193 pp.

Bauer, Ernst, *Wonders of the Earth.* New York, N.Y.: Franklin Watts, 1973.

Golden, Frederic, *The Moving Continents: The Story of the New Geology.* New York, N.Y.: Scribners, 1972.

Hyde, Margaret O., *Exploring Earth and Space Science.* New York, N.Y.: McGraw-Hill, 1970, 194 pp.

Marcus, Rebecca B., *The First Book of Volcanoes and Earthquakes.* New York, N.Y.: Franklin Watts, 1972.

Neal, Harry Edward, *Of Maps and Men.* New York, N.Y.: Funk and Wagnalls, 1970, 179 pp.

Ogburn, Charlton, *The Forging of Our Continent.* New York, N.Y.: American Heritage, 1968, 160 pp.

Silverberg, Robert, *Clocks for the Ages; How Scientists Date the Past.* New York, N.Y.: Macmillan, 1971, 238 pp.

ASTRONOMY

Beet, E. A., *Mathematical Astronomy for Amateurs.* New York, N.Y.: Norton, 1972, 143 pp.

Bova, Ben, *Starflight & Other Improbabilities.* Philadelphia, Pa.: Westminster, 1973, 126 pp.

Brown, Peter Lancaster, *What Star is That?* New York, N.Y.: Viking, 1971, 224 pp.

Gardiner, Martin, *Space Puzzles: Curious Questions & Answers about the Solar System.* New York, N.Y.: Simon & Schuster, 1971, 95 pp.

Jastrow, Robert and Malcolm H. Thompson, *Astronomy: Fundamentals & Frontiers.* New York, N.Y.: Wiley, 1972, 439 pp.

Kaufmann, William J., III, *Relativity & Cosmology.* New York, N.Y.: Harper & Row, 1973, 134 pp.

Moore, Patrick (ed.), *1975 Yearbook of Astronomy.* New York, N.Y.: Norton, 1974, 250 pp.

Moore, Patrick (ed.), *The Atlas of the Universe.* Chicago, Ill.: Rand McNally, 1970, 272 pp.

Neely, Henry M., *The Stars by Clock & Fist.* New York, N.Y.: Viking, 1972, 192 pp.

Page, Lou Williams, *Ideas from Astronomy.* Reading, Mass.: Addison-Wesley, 1973, 250 pp.

Richardson, Robert S., *The Stars & Serendipity.* New York, N.Y.: Pantheon, 1971, 129 pp.

THE ATMOSPHERE

Bova, Ben, *Man Changes the Weather.* Reading, Mass.: Addison-Wesley, 1973, 159 pp.

Branley, Franklyn M. *The Earth: Planet Number Three.* Philadelphia, Pa.: Crowell-Collier, 1974.

Brindze, Ruth, *Hurricanes: Monsters from the Sea.* New York, N.Y.: Atheneum, 1973, 106 pp.

Chandler, T. J., *The Air Around Us: Man Looks at His Atmosphere.* New York, N.Y.: Natural History Press, 1969, 156 pp.

Forsdyke, A. G., *Weather and Weather Forecasting.* New York, N.Y.: Grosset, 1970, 159 pp.

Gannon, Robert, *What's Under A Rock?* New York, N.Y.: Dutton, 1974.

Helman, Hal, *Light and Electricity in the Atmosphere.* New York, N.Y.: Holiday, 1968, 223 pp.

Milgrom, Harry. *Understanding Weather.* New York, N.Y.: Macmillan, 1970, 84 pp.

Navarra, John Gabriel, *Nature Strikes Back.* New York, N.Y.: Natural History Press, 1971, 224 pp.

Ross, Frank, *Storms and Man.* New York, N.Y.: Lothrop, 1971, 191 pp.

Rubin, Louis D., *Forecasting the Weather.* Franklin Watts, 1970, 64 pp.

Stambler, Irwin, *Weather Instruments: How They Work.* New York, N.Y.: Putnam, 1978, 95 pp.

OCEANOGRAPHY

Boyd, Waldo T., *Your Career in Oceanography.* New York, N.Y.: Messner, 1968, 219 pp.

Carson, Rachel, *The Sea Around Us.* New York, N.Y.: Oxford, 1961.

Carter, Samuel, *The Gulf Stream Story.* Garden City, N.Y.: Doubleday, 1970, 181 pp.

Claiborne, Robert, *On Every Side the Sea. Man's Involvement with the Oceans.* New York, N.Y.: American Heritage, 1971, 175 pp.

Chester, Michael, *Submersibles and Undersea Laboratories of the World.* New York, N.Y.: Grosset, 1970, 147 pp.

Cousteau, Jacques-Yves, *Three Adventures: Galapagos, Titicaca, The Blue Holes.* Garden City, N.Y.:Doubleday, 1973, 304 pp.

Pennington, Howard, *The New Ocean Explorers; Into the Sea in the Space Age.* Boston, Mass.: Little Brown, 1972, 282 pp.

Rhodes, Frank H. T., *Geology.* New York, N.Y.: Golden, 1972, 160 pp.

Scott, Frances, *Exploring Ocean Frontiers; A Background Book on Who Owns the Seas.* New York, N.Y.: Parents' Magazine, 1970, 220 pp.

Taber, Robert W., *1001 Questions Answered About the Oceans and Oceanography.* New York, N.Y.: Dodd Mead, 1972, 269 pp.

Voss, Gilbert, *Oceanography.* New York, N.Y.: Golden, 1972, 160 pp.

GLOSSARY

Abrasion (uh BRAY shuhn): The wearing away of rocks as they rub against each other.

Absolute time: Relating an event by calculating how much time has passed since the event took place.

Adhesion (add HEE shuhn): The attraction of two unlike substances.

Advection fog: A fog formed by moist air being chilled below its dew point when blown across a cold surface.

Air mass: A large body of air that has become uniform throughout.

Algin: A chemical in seaweed which keeps oil and water from separating.

Alluvial (uh LOO vee uhl) **fan:** Rocks, sand, mud and other materials that are dropped by a current at the mouth of a stream or river.

Analogy (uhn AL uh jee): An explanation of something by comparing it with something that is similar.

Aneroid: Refers to a barometer which contains no liquids.

Angle of elevation: The angle between the horizon and the sun.

Angle of rest: The angle of steepest slope upon which loose material can remain.

Annular solar eclipse (ih KLIPS): Occurs when the moon's disk does not completely cover the sun's disk, leaving a ring of sunlight around the moon.

Anthracite (AN thruh syt) **coal:** A hard coal.

Anticline (ANT ih klyn): An upward bend in a rock layer.

Aphelion (ah FEEL yuhn): The point in a planet's orbit that is farthest from the sun.

Apogee (AP uh jee): The point where the moon is at its greatest distance from the earth.

Apparent motion: Refers to the sun which appears to move through the sky.

Argon: A colorless, odorless gas which makes up less than 1% of the atmosphere.

Auroras: Northern and southern lights.

Aurora (ah ROHR uh) **Borealis** (bohr ee AL uhs): Northern lights.

Axis: Imaginary line which helps to classify crystals into different shapes.

Barometer (buh RAHM uh tuhr): An instrument which measures air pressure.

Barred galaxies: Galaxies which appear like spiral galaxies but have a thick band of visible light along the diameter of the galaxies.

Basalt (buh SALT): Igneous rock made up chiefly of microscopic dark-colored minerals.

Base line mapping: A horizontal line which serves as the base on a map from which a scale model can be drawn.

Bauxite (BOX ite): The ore of aluminum.

Bedding: Sedimentary rocks which are found in layers.

Beds: Layers of course particles dropped on layers of fine particles.

Biodegradable (by oh dee GRAYD uh buhl): Molecules that can be broken down into harmless compounds.

Bituminous (by TOO muh nuhs) **coal:** A soft coal.

Brachiopods (BRAK ee uh pahds): The group of marine organisms which has two armlike organs within its shells.

Breccia (BRECH ee uh): Rock that is similar to conglomerate but contains angular pebbles.

Caldera (kal DER uh): A large cavity near the top of a volcano.

Capillary (KAP uh ler ee) **action:** Liquid rising in hairlike openings in a solid.

Carbon dioxide: A colorless gas which is used by green plants to manufacture their food supply.

Cast: Hardened plaster which has been poured into a mold. Also mud or minerals deposited and hardened where the body of an organism has decayed.

Center of mass: An object where the mass is centered at one point.

Cephalopods (SEF al uh pahds): The group to which octopuses and squids belong.

Chronometer (kron OHM ee tuh): An instrument which accurately shows longitude.

Cilia (SIL ee uh): Tiny hairlike projections that brachiopods use for locomotion.

Cinder cone volcano: Steep-walled mountainlike volcano formed from the cinders of many eruptions.

Cinders (SIN drs): Large particles of ash that fall near a volcano.

Circumpolar (SIR kum POE la) **constellations** (kahn stuh LAY shuhns): Objects in the sky that travel around the sky pole without setting.

Clay: Fine grains of small, flat flakes that feel smooth and contain water.

Clay soil: Soil which is harsh when dry and sticky when wet, and composed of particles so small that they can be seen only with an electron microscope.

Cleavage (KLEE vij): A smooth, flat break in a mineral that leaves smooth surfaces.

Coal: Sedimentary rock containing the remains of plant sediments.

Cold front: The invisible boundary between masses of warm air and cold air.

Colloidal (kuh LOY dl) **particles:** Small bits of soil that keep water cloudy, composed of clay and humus particles.

Column: A stalagmite which grows beneath and joins a stalactite.

Comet (KAHM uht): Mass of dust and ice which orbits the sun.

Compost: Plant remains which are digested by soil organisms and broken down into humus.

Conglomerate (kuhn GLAHM uh ruht): Rock which has water-worn and rounded pebble-sized pieces cemented together.

Continental drift: A theory which explains movements of the earth's crust.

Continental drift theory: The hypothesis held by geologists which strongly suggests that the continents were once joined, but broke apart and drifted away from each other.

Continental shelf: The fairly flat top of the continental slope.

Continental slope: That part of the sea floor that slopes and marks the boundary between a continent and the sea floor.

Contour (KAHN thr) **interval:** The difference in elevation between two contour lines next to each other on a topographic map.

Contour lines: Lines on a topographic map that connect all points of the same elevation.

Controlled experiment: Experiment performed where all variables are controlled but one.

Convection (kuhn VEK shuhn) **cell:** A complete circuit of moving air, caused by the heating and cooling of the atmosphere.

Core: That region of the earth that is below a depth of 2 900 kilometers.

Corona (kuh ROH nuh): The glowing outer atmosphere of the sun.

Correlation (KOR uh LAY shuhn): In geology, piecing together the sequence of rock layers from one place to another to construct geologic history.

Counter current: A current which flows in the opposite direction to another.

Crater (KRAYT uhr): The pit at the top of a volcanic cone.

Creep: The slow downward movement of loose material on a slope.

Crest: The top of a water wave.

Crinoid (CRY noyd): Marine animal whose body is at the top of a stalk and resembles a plant.

Cross-bedding: Sand dune that has slopes at different angles.

Cross profile: A graph which connects two points on either side of a stream.

Crust: The outer layer of the earth.

Cumulus: Puffy clouds that look like cotton balls.

Cyclones (SY klohns): In the Northern Hemisphere, wind spinning counterclockwise around a center of low atmospheric pressure.

Delta: A deposit of sediments at the mouth of a river or stream.

Dependent variable: The variable which responds to other variables.

Depression (dee PRESH uhn) **contours:** Lines on a topographic map which show where the surface of the land is low.

Desalination (dee sal ih NAY shun) **plants:** Facilities where minerals are removed from seawater.

Dew: Water vapor which has condensed out of the air.

Dew point: The temperature at which water vapor begins condensing out of the air.

Diorite (DI uh rite): Light-colored plutonic rock that contains little or no quartz.

Doppler (DAHP luhr) **effect:** The change in the rate of wavelengths of light.

Dormant: A volcano which hasn't shown any signs of activity for many years.

Drumlin: Underlying loose materials pushed by glaciers into long, cigar-shaped ridges.

Dune: A deposit of sand caused by winds.

Earthshine: The faint light which makes the dark part of the moon visible.

Eccentric: Not quite circular in shape or motion.

Ecliptic (ih KLIP tik): The plane in which the planets orbit the sun.

Ellipse (ih LIPS): Curved plane surface started by a point that moves from two fixed points so that the sum of its distance is constant.

Elliptical galaxies: Galaxies which are not quite circular in shape.

Epicycle: Term used to describe planetary motions; specifically, a circle whose center moves along the circumference of another, larger circle.

Erode (uh ROAD): The wearing away of earth and rock.

Estuary (ES chu wer ee): Wide marshy area which floods and drains twice a day as the tide flows in and out.

Exfoliation (eks foh lee AY shuhn): The peeling off of thin layers from the bare surface of a rock.

External (EK stur nl) **mold:** A mold of the outside of an object, such as a shell.

False color photograph: A color photograph made by assigning different bands of the electromagnetic spectrum to each of four black and white images of the same subject, then projecting each band through a different color filter and combining them.

Fathom: In the English system of measurement, a nautical distance of six feet used to measure the depth of water.

Fault: Movement of rock along both sides of a crack in the earth's crust.

Felsite: Igneous rock made up chiefly of microscopic light-colored minerals.

Firepit: Lava that remains molten in the bottom of a volcanic crater.

Flood plain: A flat region composed of silt and clay particles along the sides of a river or stream.

Flowstone: A deposit caused by a solution of calcium carbonate which trickles down a wall or slope of a cave.

Fluorescence (flur ES ens): Light which is absorbed by atoms which then give off the energy as visible light.

Focus: Either of the two fixed points used in making an ellipse.

Fog: A cloud of tiny droplets of water which has condensed on dust particles in the atmosphere.

Fold: A bend in rock layers.

Foliation (foh lee AY shuhn): Metamorphic rock which has good cleavage and splits apart easily through the bands of other minerals within it.

Fossil (FAHS uhl) **correlation** (KOR uh lay shuhn): The comparison of fossils among rock layers.

Fossil (FAHS uhl) **fuels:** Energy reserves such as natural gas, oil or coal which were formed from the remains of animals or plants.

Foucault (foo KO) **pendulum** (PEN juh luhm): A special pendulum which swings in one direction, but the rotation of the earth on its axis makes the pendulum appear to change its direction of swing.

Fracture (FRAK chur): An uneven break in a mineral that leaves rough surfaces.

Frost: Water vapor which condenses out of the air in the solid state rather than in the liquid state.

Fungi (FUHN jy): Plantlike organisms that obtain their food from organic material or other organisms.

Gabbro: Plutonic rock made up of dark-colored minerals and containing some feldspar.

Galaxy (GAL uhk see): A large system of stars.

Gastropods (GAS truh pahds): The group to which snails belong.

Gel: A moist, jellylike substance that slowly hardens, loses its water, and forms microscopic crystals.

Geochronologist (jee oh chron ALL oh gyst): A scientist mainly concerned with organizing geologic events into a time sequence.

Geode (JEE ohd): A cavity within a rock that is partly filled with crystals.

Geomorphologist (jee oh mah FALL oh gyst): The study of the earth's topographic features in order to explain the shape of the earth's crust.

Geosyncline (jee oh SIN klyn): A very large downward bend of the earth crust.

Geyserite (GY suh rite): A special variety of opal deposited on the edges of geysers and hot springs.

Glacial (GLAY shul) **cirque** (SORK): Curved, steep-walled excavation in a mountainside caused by a glacier eroding the mountain.

Glacial trough: A valley through which a glacier has passed.

Gneiss (nice): Rock that looks like granite except that it is layered.

Gondwanaland (Gond WA nuh land): Name given to an ancient continent believed by geologists to have included what is now the continents of Africa, South America, Australia, Antarctica and the subcontinent of India which were all joined together about 200 million years ago.

Granite: Plutonic rock made up chiefly of the light-colored minerals feldspar and quartz.

Gravity: The force that any two objects pull or exert on each other.

Guyots (GOO yos): Submerged volcanic mountains on the ocean floor which have flat tops.

Half-life: The length of time that it takes for half the original amount of a radioactive material to change into something else.

Hard water: Water containing dissolved calcium bicarbonate.

Helium: A colorless, very light gas which makes up only a very tiny amount of the atmosphere.

Hematite (HEM uh tyt): Magnetite which has united with oxygen from the air and has become the red oxide of iron.

Humus (HYOO muhs): Soil which is usually dark in color and contains decayed organic matter.

Hydrometer (hy DRAHM uh tuhr): An instrument for measuring the specific gravity of liquids.

Hygrometer (hi GROHM ee tur): An instrument which measures relative humidity.

Hygroscopic (hi gruh SKOP ik): Refers to substances that attract more than the usual amount of moisture.

Hypothesis (hy PAHTH uh suhs): A proposed solution or guess to solving a problem.

Igneous (IG nee uhs) **rock:** Rock that forms when molten rock cools.

Incinerator (in SIN er ay tuh): A furnace which burns trash, reducing it to an ash.

Independent variable: The variable that is manipulated by the experimenter.

Infrared (in frah RED): Invisible radiant energy beyond the red that our eyes can see.

Internal (IN tur nl) **mold:** A mold of the inside of an object, such as a shell.

Irregular galaxies: Galaxies which have no regularly recognized shape.

Isobars: Lines on weather maps joining points of equal air pressure.

Isotherm (EYE suh therm): Lines on a map which connect points that have the same air temperature.

Isotopes (EYE suh tohps): Atoms of the same element which contain different numbers of nuclear particles.

Jet streams: Strong winds which blow from west to east about 25 000 meters above the earth.

Joints: Cracks in rocks that become wider as a result of weathering.

Kame: A low long-shaped hill formed at the top of a ridge.

Kepler's Second Law of Planetary Motion: The speeds of planets change so that a line connecting the sun and a planet sweeps over equal areas during equal periods of time.

Kettle: Crater left in an outwash by a large block of ice that melted.

Kilocalorie: Amount of heat energy needed to raise the temperature of one liter of water one Celsius degree.

Lateral (LAT uh al) **moraine:** Ridges of unsorted materials formed along the sides of a glacier.

Laurasia (Lo RAY zha): Name given to an ancient continent believed by geologists to have included what is now the continents of North America and Eurasia which were joined together about 200 million years ago.

Lava (LAHV uh): Molten rock on the surface of the earth.

Lichens (LY kuhns): An organism composed both of algae and fungi.

Light year: The distance that light travels in one year, about 9½ trillion kilometers.

Lignite: A brown, low-grade coal which has little moisture.

Limey-sandstone: Sandstone that contains some calcium carbonate.

Limonite (LIME uhn ite): Hematite which has united with water and has become the yellow-brown oxide of iron.

Lithosphere (LITH uh sfear): The outermost solid part of the earth; the crust.

Load: Material carried by water waves.

Loam (LOOM): A soil which is 40% sand, 40% silt, and 20% clay.

Local group: Refers to about twenty galaxies in the cluster which contains our own galaxy.

Longshore drift: Waves which wash in and out along the shore and take with them a load of pebbles and sand.

Luster: The way light is reflected from the surface of a mineral.

Magma (MAG muh): Molten rock deep within the earth.

Magnetic (mag NET ik) **field:** The region of influence around a magnet.

Magnetite (MAG nuh tyt): Mineral which is a black oxide of iron and is attracted to a magnet.

Magnitude (MAG nuh tood): A measure of brightness.

Mantle (MANT uhl): The layer of the earth that is between the core and the crust.

Mantle (MANT uhl): A thin membrane located between the shells that comes to the edge of each shell of a clam.

Marble: Metamorphic rock resulting from the calcium carbonate in limestone recrystallizing into large crystals of calcite.

Maximum sustainable yield: The greatest amount of any harvest which can be made each year and still have the same amount available to harvest the following year.

Meanders (mee AN duhrs): A turn or a bend in the channel of a stream.

Meandering stream: A well-developed stream which has wide turns or bends.

Metallic luster: Describes a mineral that reflects light like a piece of polished metal.

Metamorphic (met uh MOR fik) **rock:** Rock changed from its original form.

Meteor (MEET ee uh): Fragments of rock which reach our atmosphere; "shooting stars."

Meteorite: A meteor which has landed on the earth.

Meteor shower: Nights when many meteors can be seen in one area of the sky.

Methane: A gas sometimes called marsh gas, a natural gas.

Microclimate: Weather condition in a very small place.

Milky Way Galaxy: The galaxy in which our solar system is located.

Mohorovicic (moh huh ROH vuh chich) **discontinuity** (dis kahnt uhn OO uht ee): The boundary of the earth that separates the solid crust from the mantle. Referred to as the Moho for short.

Mold: A reverse print of a shell or bone.

Mold: The impression of a substance made in modeling clay. Also a cavity left in a rock after the body of an organism has decayed.

Mollusks: Animals that live in water, most of which have shells.

Monitors: In meteorology, instruments which keep track of weather, measuring and reporting the data.

Moraine (muh RAYN): Ridges of unsorted materials left by a melting glacier.

Mudflows: Places where the soil has flowed almost like water.

Natural gas: General term for several gases produced during the chemical breakdown of plant or animal matter.

Natural levee (LEV ee): A natural bank that confines a stream to its channel.

Neap tide: The tide which occurs when the moon is in the first and third quarters; the tide of minimum range.

Nebulae (NEB yuh lee): Certain regions of our galaxy which contain vast amounts of dust of clouds of gas.

Neon: An inactive gas which makes up only a very tiny amount of the atmosphere.

Nitrogen: A colorless, fairly inactive gas which makes up about four-fifths of the atmosphere.

Nitrogen cycle (SI cle): The process by which nitrogen enters and leaves the soil.

Nodule (NAHJ ool): Bumps in the roots of plants where nitrogen-fixing bacteria live.

North sky pole: The point in the sky which is directly over the earth's North Pole.

Novas (NOH vuhs): Stars which increase in brightness by ten thousand times.

Nuclear (NOO klee uh) **fission** (FIZH uhn): The splitting apart of Uranium-235 atoms which releases large amounts of energy.

Nuclear fusion: The process of uniting nuclei of atoms together, releasing enormous amounts of energy.

Nuclear reactor (ree AK tuh): A furnace which releases nuclear energy.

Obsidian (ahb SID ee uhn): Lava which cooled quickly and hardened into a glassy material.

Oozes: Sediments on the ocean floor which are formed from the remains of living things.

Opal (OH pl): A semiprecious stone made from silicon dioxide.

Optical double stars: Double stars, one of which is much closer to us than the other.

Orbit (OR bit): The path that an object in the sky travels along.

Ores: Metallic minerals occurring in amounts large enough to be mined.

Outwash: Sand and gravel deposited by melted glacial water.

Overfishing: Catching more fish each year than can be replenished.

Oxbow lake: A lake which is formed when the meander of a stream is isolated from the main stream.

Ozone: A special form of oxygen gas in which each molecule has three atoms of oxygen instead of the usual two.

Paleontology (pay lee uhn TAHL uh jee): The study of prehistoric living things through the fossil record.

Parallax (PAR uh lacks): Describing what appears to be movement of an object but is really caused by an observer's changing position.

Partial lunar eclipse: Occurs when only part of the moon passes through the earth's umbra.

Pedalfer (PED al fir) **soils:** Soils containing relatively large amounts of aluminum and iron which are common in the eastern half of the United States.

Pedocal (PED uh cl) **soils:** Soils containing relatively large amounts of calcium which are common in the western half of the United States.

Pelecypods (pah LESS ee pahds): The group to which all clams, oysters, mussels and scallops belong.

Penumbra (puh NUHM brah): The outer, light part of a shadow.

Peridotite (puh RID uh tyt): Plutonic rock made up of dark-colored minerals but without feldspar.

Perigee (PEHR uh jee): The point where the moon is at its nearest distance to the earth.

Perihelion (per uh HEEL yuhn): The point in a planet's orbit that is nearest the sun.

Period: The time it takes water to rise to the crest of a wave, fall into the trough, and rise again.

Permanent hardness: Water that contains calcium sulfate and magnesium sulfate which cannot be driven out of solution by heating.

Petroleum (puh TROH lee uhm): Flammable liquid which seems to have been formed from marine sediments.

Phases (FAZES): The different shapes of the moon which are observed monthly from earth.

Phytoplankton (fy toh PLANGK tuhn): Microscopic marine plants which live at or near the ocean's surface.

Planetary nebulae: Nebulae which have small, hot stars located at their centers which give off ultraviolet light which our eyes cannot see.

Planetoids: Asteroids (minor planets).

Plutonic (ploo TAHN ik) **rock:** Rock that forms deep within the earth.

Porphyry (POR fuh ree): Rock with large crystals that are in a fine-grained mass.

Potholes: Round pits contained in the beds of many streams.

Precession (pree SESH uhn): The circular path taken by the top of the shaft of a gyroscope as the gyroscope spins.

Predator (PRED uht uhr): Animal that eats other animals.

Prime meridian: The 0° line of longitude.

Principle (PRIN suh pl) **of superposition** (SOO puh poe SIH shuhn): A basic idea of geologists that rock layers deposited first are covered by layers deposited later.

Principle of uniformity (yoo nuh FOR muh tee) **of process:** An idea that the earth's surface has been changed in the past in the same way it is being changed today.

Pumice (PUHM uhs): The hardened froth on solidified lava.

Pyrite (PY rite): Golden, metallic-looking mineral known as "fool's gold."

Quartzite: Metamorphic rock which results when quartz sandstone hardens.

Quasars (KWAY sahrs): Sources of radio energy from outer space, not stars.

Radiation fog: A fog that forms in cool air.

Radioactive (RAYD ee oh AK tiv) **decay:** Changing one element into another through a loss of charged particles within it.

Radioactivity: Giving off radiant energy in the form of particles or rays.

Radiometer (rayd ee OHM ee tuhr): An instrument for measuring radiant energy.

Recycling (ree SI kln): The process by which resources are recovered and used again.

Red shift: The shifting of spectral lines toward the red end of the spectrum.

Reduction (ruh DUK shuhn): The process of removing oxygen from a compound.

Relative humidity: A measure of the amount of water vapor in the air compared to the total amount of water vapor that the air can hold.

Relative time: Relating an event before and/or after other events.

Renewable (ree NEW ah ble): Types of energy which can be restored quickly by natural processes.

Rip current: Water which forms a fast flowing current.

Rock: Two or more minerals found together in nature.

Sandstone: A bedded sedimentary rock made up mostly of sand grains; it splits easily into thin layers.

Sandy soil: Soil which feels gritty and contains particles of sand which can be seen with the unaided eye.

Saturate (SACH uh rayt): Something filled with the maximum that it can absorb, such as sugar dissolved in water.

Saturated (SACH uh ray ted): Refers to air that holds all the moisture that it can.

Scale: The comparison or ratio between the distance on a map and the actual distance on the ground.

Schist: Metamorphic rock that shows some recrystallization and contains layers of dark minerals.

Scientific revolution: A dramatic change in an idea which is so different that scientists are forced to think about their subject in quite a different way.

Scoria (SKOHR ee uh): Lava which hardened into porous rock.

Sea arch: A section of a wall of a cave which remains after water has eroded through the back of the cave.

Sea caves: Cavities hollowed out of cliffs by the action of water waves.

Seamounts: Submerged volcanic mountains on the ocean floor.

Sediment (SED uh ment): Small bits of bare rock and minerals worn away from the earth's surface by wind and water.

Sedimentary (sed uh MENT uh ree) **rock:** A kind of rock made when a layer of sediment hardens.

Sediments (SED uh ments): Rock particles that are dropped from water (or from wind or ice).

Seismogram: The record recorded by a seismograph.

Seismograph (SYZ muh graf): An instrument which records vibration of the earth.

Selenite: Gypsum that is found as large, clear crystals.

Shadow zone: Regions that are 11 000 — 16 000 kilometers from an earthquake.

Shale: A rock formed largely by the hardening of clay; it is fine grained and splits easily into thin layers.

Shaley-sandstone: Sandstone that contains a little clay.

Shield volcano: A series of quiet lava flows from one volcano that gradually build up a mound that is much broader than it is high.

Shoreline of emergence (ee MUHR jens): Land at the shoreline which has been rising, or where the level of the sea has been falling.

Shoreline of submergence (sub MUHR jens): Land at the shoreline which has been falling, or where the level of the sea has been rising.

Sial rocks: Less dense continental rocks, rich in silicon and aluminum.

Silt soil: Soil which feels similar to flour or talcum powder, composed of particles larger than clay but smaller than sand.

Sima rocks: More dense ocean floor rocks, rich in silicon and magnesium.

Siphons: Tubelike organs in clams used for drawing in or forcing out liquid.

Sky equator: The imaginary line around the sky.

Slate: Rock formed from shale and containing microscopic crystals of mica.

Slickensides: Rocks along a fault which are polished smooth when the two blocks rub against each other.

Slope: The change in elevation per kilometer of a stream bed.

Smelting: The process of changing minerals to metals by using chemical and heat treatments.

Sodium chloride: The chemical name for table salt.

Soil compaction (com PAK shuhn): The degree to which a soil is packed.

Solar constant: Refers to sunlight hitting the earth and providing a relatively constant amount of energy to each square centimeter of surface.

Solar flare: A sudden, violent eruption of gas thrown out from the sun's surface.

Solar noon: The time when the sun casts the shortest shadow.

Solar prominence: Streams of gas rising from the sun's surface.

Sonar: Device which sends sound waves through water and registers the vibrations reflected back from the ocean floor or from an object in the water.

Sounding: A measurement of the depth of water.

Spatter cone volcano: A steep sided volcano with a thick, strong-walled cone.

Specific gravity: The weight of a substance compared with the weight of an equal volume of water.

Spectroscope (SPEK truh skohp): An instrument for studying the visible spectrum.

Spectroscopic (SPEK truh SKOP ik) **binaries:** Double stars, one of which moves at times toward us while the other is moving away from us.

Spectrum (SPEK truhm): Pattern of colors produced when light is separated into its colors.

Spiral galaxies: Galaxies that are shaped like spirals.

Spring tide: The tide which occurs when the moon is new or full; the tide of greatest range.

Stack: An arch which has widened by wave action until its top collapses leaving a column of supporting material.

Stalactite (stuh LAK tyt): Icicle-shaped deposit of calcite which grows and hangs down from the ceiling of a cave.

Stalagmite (stuh LAG myt): An upward-growing deposit of calcite on the floor of a cave caused by drops falling from the ceiling of the cave.

Stream profile: A graph which shows the slope of a stream bed.

Striations (STRY ay shuhns): Scratches found on the floor and walls of a valley caused by a glacier.

Subsoil: Soil which is hard, packed together, and difficult to plow.

Sunspots: Dark spots or groups of spots on the sun's surface.

Supernovas (soo puh NOH vuhs): Stars which suddenly increase in brilliance by over a hundred thousand times.

Supersaturate: Something filled with so much material that it cannot absorb any more.

Suspension (suh SPEN shuhn): A mixture of water and tiny particles which may not settle from the water for months or even years.

Swell: In calm seas, waves with rounded crests that tend to move in groups.

Syncline (SIN klyn): A downward bend in a rock layer.

Talus (TAY luhs): Heap of loose rock which has piled up at the bottom of a cliff.

Tar: A sticky, thick liquid obtained by the destructive distillation of wood.

Telescopic binaries: Double stars that revolve around a common center of mass.

Temporary hardness: Water that can have the dissolved calcium bicarbonate removed by heating.

Terminal (TUHRM uh nuhl) **moraine:** The ridge which marks the limit of a glacier's advance.

Theories (THEE uh rees): Principles based on a collection of facts which have held firm for a long time.

Theory (THEE uh ree): An idea which helps to explain an observation.

Theory of plate tectonics (tek TAHN iks): A theory that the earth's crust is divided into regions called plates.

Topographic (tahp uh GRAF ik) **maps:** Maps which accurately show the details of hills, valleys and other features of a particular region of the earth's surface.

Topsoil: The dark-colored top layer of the soil which contains humus.

Total lunar eclipse: Occurs when all of the moon passes through the earth's umbra.

Total solar eclipse: Occurs when the disk of the moon appears larger than the disk of the sun, and all direct light from the sun is blocked.

Travertine (TRAV uhr teen): A form of calcite deposited from hot spring water.

Trilobite (TRILL uh byte): Marine animal whose body has three lengthwise divisions, now extinct.

Trough: The lowest part of a water wave.

Tuff: Small particles which were shot out of volcanoes and were later blown away and became cemented together.

Ultraviolet: Invisible radiant energy beyond the violet end of the spectrum.

Umbra (UHM brah): The inner, dark part of a shadow.

Unconformity (uhn kahn FOR muh tee): A break between two sets of rock layers.

Universe (YOO nuh vuhrs): All the objects in space and all the space between the objects.

Upwellings: Deep ocean currents which rise to the surface.

Van Allen radiation (RAYD eee ay shuhn) **belts:** Doughnut-shaped regions around the earth which contain fast-moving charged particles.

Variable: A factor which can be changed in an experiment.

Variable stars: Stars that change in brightness.

Veins: Cracks in rocks filled with different material.

Vertical line: A line which points to the center of the earth.

Volatile (VAHL uh tyl): Able to vaporize and become a gas.

Volcanic (vahl CAN ik) **ash:** Pulverized rock produced by exploding volcanoes.

Volcanic breccia (BRECH ee uh): Cinders from a volcano that became cemented together.

Volcanic rock: Igneous rock that forms at or near the surface, such as when lava cools.

Vortex: A column of gas or liquid which flows with a whirling motion.

Warm front: The invisible boundary which is the beginning of a warm air mass.

Water of crystallization (KRIS tuh ly ZAT shuhn): The formula of a mineral or compound that is combined with molecules of water.

Water retention: The water holding power of the soil.

Wavelength: The distance between crests of two water waves.

Zooplankton (zoe PLANGK tuhn): Microscopic marine animals which live at or near the ocean's surface.

ACKNOWLEDGMENTS

Illustrators for this edition:

Lee Ames   Anthony d'Adamo   Holly Moylan
Frank Schwarz   Mel Erikson   Paul S. Weiner
Andre LaBlanc   Ray Burns

Photographs are credited below. (Photographs credited to A & B have been taken by Talbot D. Lovering, the Allyn and Bacon Staff Photographer.)

UNIT 1—CHAPTER 1

P. 1—Werner Stoy. **2** *both*—A & B. **3** *all*—A & B. **4** *all*—A & B. **5** *all*—A & B. **6** *all*—A & B. **7** *left*—A & B; *right*—Edwin L. Shay. **8** *across top*—Weston Kemp; *bottom*—B. M. Shaub. **9** *both*—A & B. **10** *left*—Edwin L. Shay; *right*—Weston Kemp. **11**—Diamond Crystal Salt Company. **12** *top*—National Park Service; *middle*—A & B; *bottom*—B. M. Shaub. **14** *all*—A & B. **15** *top*—A & B; *bottom*—Portland Cement Association. **16**—U. S. Steel. **19** *top*—A & B; *along side*—Edwin L. Shay. **20** *four from top and bottom left*—A & B; *bottom right*—John H. Gerard.

CHAPTER 2

P. 21—Werner Stoy. **22** *all*—A & B. **23**—B. M. Shaub. **24**—U. S. Geological Survey. **26** *all*—Edwin L. Shay. **28**—Robert E. Kilburn. **29** *top*—A & B; *bottom*—B. M. Shaub. **31**—Dick Hufnagle. **32**—Educational Expeditions International. **33** *top*—Werner Stoy; *middle and bottom*—Edwin L. Shay. **34** *all*—Edwin L. Shay. **37** *both*—A & B. **38** *all*—Edwin L. Shay. **39** *all*—Edwin L. Shay. **40**—William Schwarting. **41** *all*—Frank White. **42** *top*—Edwin L. Shay; *middle*—B. M. Shaub; *bottom*—W. E. Schomo.

CHAPTER 3

P. 43—Grant Heilman Photography. **44**—Ralph Grim. **49**—Weston Kemp. **50**—Harold R. Hungerford. **51** *top*—Weston Kemp; *bottom*—A. Devaney. **52**—Peter S. Howell. **53** *top left*—Grant Heilman; *top right*—A. L. Lang; *bottom*—Grant Heilman. **54**—A & B. **55** *both*—Department of Agronomy, Cornell University; *bottom*—Charles C. Ladd. **60**—Pfizer & Co., Inc. **61** *left*—Dr. William E. Colby; *second from left*—Roy W. Simonson; *right*—Howard W. Higbee. **62**—D. O. Thompson.

CHAPTER 4

P. 63—A & B. **66**—A & B. **68**—B. M. Shaub. **69**—Mildred Fenton. **70**—A & B.

CHAPTER 5

P. 74—B. M. Shaub. **76**—The Society of California Pioneers.

CHAPTER 6

P. 71—Van Cleve Photography. **82** *top*—A & B; *middle*—Bjorn Bolstad from Peter Arnold, Inc.; *bottom*—Yoram Kahana from Peter Arnold, Inc. **84**—Lillian Bolstad from Peter Arnold, Inc. **87** *top*—Bjorn Bolstad from Peter Arnold, Inc.; *middle and bottom*—U. S. Department of Energy. **88** *top*—U. S. Department of Energy; *bottom*—Pacific Gas and Electric Co. **89**—U. S. Department of Energy. **90** *top*—Inland Steel Co.; *bottom*—A & B. **91** *top*—Arizona Office of Tourism; *middle and bottom*—A & B. **93**—Courtesy of Caterpillar Tractor Company. **94**—Judith L. Aronson. **95**—A & B.

UNIT 2—CHAPTER 1

P. 97—Grant Heilman. **98–99**—A & B. **106**—A & B.

CHAPTER 2

P. 111—David Muench. **112** *top*—Mary Shaub; *bottom*—Mildred Fenton. **113** *top*—Robert E. Kilburn; *bottom*—Weston Kemp. **114** *top*—Mildred Fenton; *middle*—U. S. Geological Survey; *bottom*—Mildred Fenton. **115** *top*—Katherine Jenson; *bottom*—Josef Muench. **116** *both*—B. M. Shaub. **118**—Jerome Wyckoff. **119** *top*—Mildred Fenton; *bottom*—SHOSTAL, Mike Roberts. **120**—ALPHA PHOTOS, C. C. Maxwell. **121**—Peter S. Howell. **122**—A & B. **123** *left*—Peter S. Howell; *right*—Mildred Fenton. **124**—A & B. **126** *both*—ALPHA PHOTOS, Ed Gray. **128** *top*—FPG, Hallihan; *bottom*—Grant Heilman. **129** *both*—FPG, Arthur Griffin. **130** *both*—A & B. **131**—Edwin L. Shay. **132** Grant Heilman. **133**—Fairchild Aerial Survey. **135** *top*—A & B; *bottom*—Edwin L. Shay. **136**—Lowery Aerial Photography. **138** *left*—B. M. Shaub; *right*—Weston Kemp. **139**—Ben Glaha. **140** *top*—Edgar J. Boucher; *middle*—W. H. Bradley; *bottom*—Michael Ciampa. **141**—Monkmeyer, Rathbone **142** *top*—Mary S. Shaub; *bottom*—Weston Kemp.

CHAPTER 3

P. 143—Josef Muench; **145**—Robert E. Kilburn. **146** *all*—Peter S. Howell. **147** *all*—Mildred Fenton. **148** *top*—Barron Lambert, Jr.; *bottom*—Harold R. Hungerford. **149**—SHOSTAL, Ray Manley. **150** *both*—Frank White. **151** *top*—Peter S. Howell; *bottom*—John H. Gerard. **152**—Mary S. Shaub. **154**—Ray Atkeson. **155**—William C. Bradley. **156**—Mildred Fenton. **157** *both*—B. M. Shaub. **158**—SHOSTAL, Charles W. Miller. **159** *left*—Mildred Fenton; *right*—Robert E. Kilburn. **160**—B. M. Shaub. **161**—Mildred Fenton. **162**—John S. Shelton. **164**—Grant Heilman. **166** *top*—A & B; *middle*—National Park Service, W. S. Keller; *bottom*—Weston Kemp. **167** *top*—Edwin L. Shay; *bottom*—Peter S. Howell. **168**—Grant Heilman. **170**—TOM STACK & ASSOCIATES, Walter Staugaard. **172**—Mildred Fenton. **173**—"Glaciers and the Ice Age" From the Encyclopaedia Britannica Educational Corporation. **176** *top*—A & B; *next-to-top*—Weston Kemp; *next-to-bottom*—Mary S. Shaub; *bottom*—Harold R. Hungerford.

CHAPTER 4

P. 177 Katherine Jenson. **178** *left*—Grant Heilman; *middle*—Frank White; *right*—L. W. Brownell. **179**—Courtesy of the American Museum of National History. **180** *top right*—Edwin L. Shay; *top left*—Katherine Jenson; *middle*—Edwin L. Shay; *bottom left*—Frank White; *bottom right*—Courtesy of Field Museum of Natural History. **181** *top left*—ENCYCLOPAEDIA BRITANNICA FILMS, INC: "The Rise of the Dinosaurs"; *top right*—Mildred Fenton; *middle left*—University of Michigan, for Carroll Lane Fenton; *middle right*—American Museum of Natural History: William Schwarting; *bottom left*—Frank White; *bottom right*—Yale Peabody Museum: John Howard. **182**—Courtesy of the

American Museum of Natural History. **183** *left*—Frank White; *right*—American Museum of Natural History: William Schwarting. **184**—A & B. **186**—Weston Kemp; *bottom*—Frank White. **187** *top*—Mildred Fenton; *next-to-top and next-to-bottom*—Katherine Jenson; *bottom*—Carroll Lane Fenton. **188**—Smithsonian Institution. **189** *top*—Yale Peabody Museum; *bottom*—Jerry Greenberg. **190** *top*—TOM STACK & ASSOCIATES, Ron Church; *bottom left*—TOM STACK & ASSOCIATES, Keith Gillett; *bottom middle and bottom right*—Frank White. **191** *top left*—University of Michigan, for Carroll Lane Fenton; *top right and bottom*—Frank White. **192** *left*—U. S. National Museum; *right*—University of Michigan, for Carroll Lane Fenton; *middle*—Frank White; *bottom*—American Museum of National History: R. C. Murphy. **194** *all*—American Museum of Natural History: William Schwarting. **195** *both*—Educational Services, Inc., 1963 from *Bones*. **196**—Frank White. **197**—Courtesy of Field Museum of Natural History. **198**—Courtesy of the American Museum of Natural History. **199**—Frank White. **200** *all*—Frank White. **201**—Courtesy of the American Museum of Natural History. **202** *top*—Chicago Natural History Museum; *next-to-top and next-to-bottom*—Carroll Lane Fenton; *bottom*—American Museum of Natural History; William Schwarting.

CHAPTER 5

P. **203**—Fritz Goro—LIFE magazine, Time, Inc. **204**—A & B. **205**—A & B. **206**—WESKEMP: Alain Perceval. **210** *top*—Ira Gavrin; *bottom*—FPG, Arthur Griffin. **211**—Grant Heilman. **212** *top*—Grant Heilman; *middle*—W. C. Bradley; *bottom*—Courtesy of the American Museum of Natural History. **216**—U. S. Geological Survey.

CHAPTER 6

P. **217**—Taurus Photos. **218**—Peter S. Howell. **219**—The Bettmann Archive, Inc.

UNIT 3—CHAPTER 1

P. **225**—Bell System Science Series. **226–227**—Gerald S. Hawkins, author of STONEHENGE DECODED and BEYOND STONEHENGE. **230**—Norman Lickyer Observatory. **232** *top*—Courtesy of the American Museum of Natural History; *bottom*—Official U. S. Navy Photo. **238** *top*—Black Star, Emil Schulthess; *middle*—Mount Wilson and Palomar Observatories; *bottom*—Harvard Observatory.

CHAPTER 2

P. **239**—NASA. **240** *across top*—Hale Observatories; *bottom*—A & B. **241**—Historical Pictures Service. **242**—WESKEMP, Hood. **244** *top*—U. S. Naval Observatory Photograph; *bottom*—NASA Photo by the Lunar and Planetary Laboratory, University of Arizona. **246**—Dr. Harold D. Edgerton. **247** *left and right*—Yerkes Observatory. **249**—Lick Observatory.

CHAPTER 3

250—Sacramento Peak Observatory/Air Force Cambridge Research Laboratories. **251**—Lick Observatory. **254**—*along side*—Mount Wilson and Palomar Observatories. **255**—Yerkes Observatory Photograph. **256–257** *across top*—Professor John C. Duncan, Wellesley College Observatory. **258**—American Museum of Natural History, William Schwarting. **259**—Lick Observatory. **260** *left and right*—Canadian Government Travel Bureau. **261**—PSSC Physics, D. C. Heath & Co. **264**—Mount Wilson and Palomar Observatories. **265** *top right*—Hale Observatories; *top left and middle*—Mount Wilson and Palomar Observatories; *bottom*—G. W. Gartlein. **268** *top*—American Museum of Natural History, William

Schwarting; *middle*—Lick Observatory; *bottom left*—Mount Wilson and Palomar Observatories; *bottom right*—A & B.

CHAPTER 4

P. **269**—A & B. **271**—Yerkes Observatory. **276** *all*—Yerkes Observatory.

CHAPTER 5

P. **279**—NASA. **282** *both*—Robert E. Kilburn. **283** *both*—A & B. **286**—A & B.

CHAPTER 6

P. **287**—Hale Observatories. **289**—Tiara Observatory. **290**—High Altitude Observatory, National Center for Atmospheric Research. **292** *top three*—Lowell Observatory; *bottom three*—Yerkes Observatory. **296**—Charles Kulick. **298**—Yerkes Observatory. **299** *bottom*—American Museum, Hayden Planetarium. **300**—Hale Observatories. **302**—Yerkes Observatory.

CHAPTER 7

P. **303**—Courtesy of Arecibo Observatory. **304** *all*—Harvard College Observatory. **305**—Edward E. Jameson, Planetarium Director, Natick Public Schools. **306**—Eastman Kodak Co. **307** *top spectrum*—"Spectrum Chart," Courtesy of Welch Scientific Company; *second spectrum*—Bausch and Lomb; *third and fourth spectra*—"Spectrum Chart," Courtesy of Welsch Scientific Company; *middle*—A & B; *bottom spectrum*—"Spectrum Chart," Courtesy of Welsh Scientific Company. **308** *top*—Yerkes Observatory; *bottom*—Robert E. Kilburn. **309** *top*—Lowell Observatory; *bottom*—Hale Observatories. **310** *all*—Hale Observatories. **311** *all*—Yerkes Observatory Photographs. **313**—Hale Observatories. **314** *both*—Mount Wilson and Palomar Observatories. **315** *top left*—Mount Wilson and Palomar Observatories; *top right*—Hale Observatories; *bottom left*—California Institute of Technology; *bottom right*—Hale Observatories. **317** *top*—Hale Observatories; *bottom*—California Institute of Technology. **318** *all*—Hale Observatories. **319** *all*—Mount Wilson and Palomar Observatories. **320** *top*—California Institute of Technology; *middle*—Hale Observatories; *bottom*—Mount Wilson and Palomar Observatories.

UNIT 4—CHAPTER 1

P. **321**—NASA. **322–323**—Peter S. Howell. **326** *top*—Weston Kemp; *bottom*—FPG, Oxman. **327** *left*—Russ Kinne; *right*—Weston Kemp. **328** *both*—Peter S. Howell. **329** EPA-DOCUMERICA-Gene Daniels. **330** *top*—Jacques Jangoux from Peter Arnold, Inc.; *next-to-top*—Klaus D. Francke from Peter Arnold, Inc.; *next-to-bottom*—Aevar Johannesson; *bottom*—Mary Shaub. **332**—Grant Heilman. **334**—A & B. **335**—Walter Dawn. **337**—Mary Shaub. **338** *top*—Howard B. Bluestein; *bottom*—N. H. Fletcher. **341** *both*—A & B. **342**—Wide World Photos. **345**—Official U. S. Navy Photo. **346** *top*—Weston Kemp; *bottom*—NASA.

CHAPTER 2

P. **347**—Dr. E. R. Degginger. **348** *top*—A & B; *bottom*—Weston Kemp. **354**—U. S. Army Cold Regions Research and Engineering Laboratory. **355**—Longwood Gardens Photograph. **356**—Official U. S. Navy Photograph.

CHAPTER 3

P. **359**—NASA. **361**—Arthur Griffin. **362** *left*—Mary Shaub; *right*—Howard B. Bluestein. **363**—U. S. Department of Commerce, National Oceanic and Atmospheric Administration. **364**—U. S. Department of Agriculture. **365** *across top*—National Weather Service; *bottom*—Science Services. **367**—A & B. **368** *left*—B. M. Shaub; *right*—Dr. E. R. Degginger. **369** *top*—FPG, Ellis Sawyer; *bottom*—ALPHA PHOTOS, Elizabeth Hibbs. **372**—Russ Kinne. **373**—SHOSTAL, Bert Vogel.

CHAPTER 4

P. **377**—Grant Heilman. **380**—A & B. **381** *both*—Robert Walan. **383**—National Weather Service. **383** *left*—Peter S. Howell; *right*—Richard Weymouth Brooks. **385**—Willis Peterson. **388**—A & B. **390**—Grant Heilman. **391**—Jerome Wyckoff. **392** *top*—FPG, William Eymann; *bottom*—ALPHA PHOTOS, Ralph Mandol. **393**—Peter S. Howell. **394** *top*—Weston Kemp; *bottom*—Russ Kinne.

CHAPTER 5

P. **395**—A & B. **399**—CAMERA HAWAII, Werner Stoy. **400**—A & B.

CHAPTER 6

P. **406**—David Muench. **407** *top*—CAMERA HAWAII, Werner Stoy; *bottom*—Audrey N. Tomera. **410** *top*—Black Star, Archie Lieberman; *bottom*—FPG, Bob and Ira Spring.

CHAPTER 7

P. **413** *top left*—EPA-DOCUMERICA-Gene Daniels; *middle*—NASA; *top right*—Department of Commerce, National Oceanic and Atmospheric Administration; *bottom left*—The Bettmann Archive, Inc.; *bottom right*—The Bettmann Archive, Inc. **414** *top*—National Environmental Satellite Service, National Oceanic and Atmospheric Administration; *bottom*—NASA. **415** *top*—U. S. Department of Commerce, National Oceanic and Atmospheric Administration; *bottom*—NASA. **416** *along side from top to bottom*—Official U. S. Navy Photograph, by PHCS Joseph L. Edge, Jr.; Official U. S. Navy Photograph, by J01 Kirby Harrison; Mary Shaub; Official U. S. Navy Photograph, by PH22 T. M. Putnam; NASA; Official U. S. Navy Photograph, by W. M. Powers; *center of page*—U. S. Department of Commerce, National Oceanic and Atmospheric Administration.

UNIT 5—CHAPTER 1

P. **417**—Sea Library, Jim and Cathy Church. **418**—NASA. **425** *top*—Courtesy Dr. John Milliman/Woods Hole Oceanographic Institution; *bottom*—Lamont-Doherty Geological Observatory. **425**—Dave Bartruff, FPG.

CHAPTER 2

P. **430**—Robert E. Kilburn. **434**—Lamont-Doherty Geological Observatory. **439**—Western Ways Photo. **442**—Josef Muench, FPG.

CHAPTER 3

P. **443**—NASA. **444** *top*—TOM STACK AND ASSOCIATES, Tom Stack; *bottom*—A & B. **445**—A & B. **447** *left*—Editorial Photocolor Archives, Inc./Dan O'Neill; *right*—Dr. E. R. Degginger. **449**—Official U. S. Coast Guard Photograph.

CHAPTER 4

P. **451**—TOM STACK AND ASSOCIATES, Larry Moon. **455**—Robert Perron. **456**—ALPHA PHOTOS, Werner Stoy. **457** *top*—ALPHA PHOTOS, Hecht; *center*—Peter S. Howell; *bottom*—A & B. **458**—Laurence Lowry. **460**—Werner Wolff for BLACK STARR. **466** *both*—Peter S. Howell.

CHAPTER 5

P. **468**—Sea Library, Mick Church. **469**—Jerome Wyckoff. **473**—A & B. **475**—Walter Dawn.

CHAPTER 6

P. **477**—National Marine Fisheries Service, National Oceanic and Atmospheric Administration. **479** *top*—William A. Watkins/Woods Hole Oceanographic Institution; *bottom*—Sea Library, George Green. **480** *top and middle*—Courtesy of Peabody Museum of Salem; *bottom*—National Marine Fisheries Service, National Oceanic and Atmospheric Administration. **482** *top and middle*—Sea Library; *bottom*—U. S. Department of Commerce, National Oceanic and Atmospheric Administration. **484**—Sea Library, Mick Church.

CHAPTER 7

P. **486** *top*—EPA-DOCUMERICA-Gene Daniels; *bottom*—EPA-DOCUMERICA-Gary Miller. **487**—Laurence Lowry. **488**—Image courtesy General Electric Company, Beltsville Photographic Engineering Laboratory. **489** *top*—Gulf Oil Corporation; *middle*—Sea Library, Milton S. Love; *bottom*—National Audubon Society/F. Frances. **490**—A & B; *bottom*—Dr. E. R. Degginger.

CHAPTER 8

P. **491**—NASA. **492**—NASA. **493**—Gulf Oil Corporation. **494**—U. S. Department of Commerce, National Oceanic and Atmospheric Administration. **495** *top*—Sea Library, Ben Cropp; *bottom*—Sea Library, Al Giddings. **496**—U. S. Naval Oceanographic Office. **497** *top*—A & B; *middle*—Courtesy Miami Seaquarium; *bottom*—Woods Hole Oceanographic Institution. **498** *top*—Sea Library, Scripps Institute; *bottom*—U. S. Department of Commerce, National Oceanic and Atmospheric Administration.

CHAPTER 9

P. **499**—NASA. **500** *both*—Image courtesy General Electric Company, Beltsville Photographic Engineering Laboratory. **501** *both*—Image courtesy General Electric Company, Beltsville Photographic Engineering Laboratory. **502**—A & B. **503**—Image courtesy General Electric Company, Beltsville Photographic Engineering Laboratory. **504**—Image courtesy General Electric Company, Beltsville Photographic Engineering Laboratory. **505**—Image courtesy General Electric Company, Beltsville Photographic Engineering Laboratory. **507**—Image courtesy General Electric Company, Beltsville Photographic Engineering Laboratory. **508**—Image courtesy General Electric Company, Beltsville Photographic Engineering Laboratory. **509**—Image courtesy General Electric Company, Beltsville Photographic Engineering Laboratory. **510**—Image courtesy General Electric Company, Beltsville Photographic Engineering Laboratory. **511**—Image courtesy General Electric Company, Beltsville Photographic Engineering Laboratory.

INDEX

Recycling, **95**
Red iron ore. *See* Hematite
Red shift, **309**
Reduction, of compounds, **16**
Reflected energy, 415
Relative humidity, **332**–333, 382
Relative time, **217**, 218–221
Renewable energy, **82**, 87, 89
Revolving objects, throwing, 261
Rhyolite, 160
Ridges: mid-ocean, 435, 436, 511; submarine, 426–427
Rip current, **459**
Ripple marks, 150; in sedimentary rock, 421–422
Ripples, in water, 452
River valleys. *See* Valleys
Rocks, 21–40; abrasion of, 121; age of, 145–146, 427; bedding of, 24, 114; carbon in, 26; cavities in, 29; changes in, 36–37; collecting, 68; deposits in, 29; faults, 156; folds in, 152–154; formation of natural, 21, 25; geosynclines, 154–155; glaciation and, 166–176; layering of, 144–146; magnetic properties, 435; making artificial, 24; mineral forming, 21; movements of, 152–157; on ocean floor, 435; recrystallization in, 37; radioactive dating of, 146; sial, **440**; sima, **440**; slickensides, 157; specific gravity of, 69; studying continental drift through, 432–433; synclines and anticlines, 152–153; transported by streams, 121; unconformities, 155; veins in, 28–29; volcanic, 21, 32–34, 158–164
Rotation: of earth, 363, 366–367, 464; of sun, 265

Salt (sodium chloride), **11**, 483
Salts: in seawater, 445–446, 447; in solution, 445–446
Sand dunes. *See* Dunes
Sandstones, **26**; classifying, 27; identifying, 26, 38; strength of, 26
Sandy soil, **44**, 45; measuring soil air in, 47; penetration rate of water in, 50

Satellites, 499–500; monitoring by, 413; orbit of, 414; photography, 501, 507–512; weather, 413–416
Saturated solutions, **28, 445**
Saturation, in air, **332**, 334
Saturation Curve, 332–334
Saturn, 244, 290, 291, 297
Scale, of map, **102**
Scallops, 186
Schist, identifying, **39**, 153
Science: assumptions in, 47; beginnings of, 64; controlled experiments in, 64; laws of, 295; oversimplifications, 338; uncertainties in, 39; use of analogies in, 26, 248
Scientific revolution, **80, 429,** 442
Scoria, **34**
Sea anemone, 190
Sea arch, **457**
Sea caves, **457**
Seamounts, **425**, 510
Seasons: and changes in water, 387–389; explaining, 246–247
Seawater: analysis of, 446, 460–461; density of, 462; dissolved gases in, 446; dissolved materials in, 445–446, 447; dissolved minerals in, 483; elements in, 483; evaporation of, 444–445, 462; freezing of, 447–449; fresh water from, 483; in motion, 451–467; pressure in, 447–448; rain's effect on, 462; salts in, 444–446; sediments in, 421–422
Sedimentary rocks, **22**–31, 432; bedding in, 114; cements in, 25, 28; classification of, 27; colors of, 26, 149; crystals in, 28–29; correlation of, 218; formation of, 22, 25, 144; identifying, 26, 147; impure, 26–27; ripple marks in, 421–422; underwater, 421–427
Sediments, rock, 21, **130**; age of, 424; bedded, 24, 135, 144; cementing of, 25, 28, 144; in continental shelf, 421; evidence from, 424; in formation of deltas, 136; identifying, 22–23; making, 22–23; marine, 424–425;

oozes, 424; and relative time, 218–221; settling rate for, 423, 434; sorting, 24; sources of, 424; transportation by wind, 139; underwater, 421–422, 423; wave action on, 458–459
Seismograms, **76**–77; interpreting, 78–79
Seismograph, 76–77
Selenite, **12**
Shadow zone, **77**, 78
Shale, **26**, 38, 153; classifying, 27; extracting oil from, 88; identifying, 39, 69; strength of, 26
Shells: casts of, 182–183; clams, 185–186; gaining knowledge from, 185–192; molds of, 182
Shield volcano, **158**
Shorelines, 147–151; changing, 507; mapping the, 420
Shorelines of submergence, and emergence, **108**–109
Sial rocks, **440**
Siliceous oozes, **424**
Silicone dioxide, 25; deposits of, 29
Silt soil, **44**, 45; measuring soil air in, 47; penetration rate of water in, 50
Sima rocks, **440**
Sinkholes, 507
Siphons, **186**
Skeletons: assembling, 195; early amphibians, 196–197; inferences from, 194, 196–199; obtaining, 194–195; reconstructing ancient, 193–202
Sky: changes in latitude in, 237; classifying objects in, 235; color of, 327; equator, 231; making model of, 230–232; at night, 233–234; observing, 226–228, 230. *See also* Atmosphere
Sky equator, **231**
Slate: foliation of, 37, 39; formation of, **38**, 71, 153; identifying, 39
Slickensides, **157**
Slope: and rate of flow, 124; of stream bed, **127**
Slumping, 118
Smelting, **90**
Smog, **412**

water and, 450; and spectra, 306, 308; of sunlit surfaces, 378–379; and water density, 461

Temporary hardness, in water, **31**

Terminal moraines, **168**

Terraces, formation of, 456

Theories, **208;** in astronomy, **230, 292;** about earth, 208–209, 429, 439; of plate tectonics, 439–442

Thunderstorms, 362, 365

Tidal power, 88

Tidal range, 260, 466

Tides, **260,** 264; wave action affected by, 456

Time: absolute, 221–224; geologic, 217; relative, 218–221

Time zones, 241, 283, 285

Tin, 91, 483

Titanium, 483

Topographic maps, 98–110; interpreting, 108–109; making, 105; symbols, 101

Topsoil, **61**

Tornadoes, 364, **365**

Total lunar eclipse, **258**

Total solar eclipse, **258**

Tracks, studying, 178, 200

Travertine, **29**

Trees, creep and, 120

Trenches, submarine, 427, 436

Triangulation, **433**

Trilobites, 145, **192**

Tropical climate, **408**

Tropopause, **343,** 345

Troposphere, **343,** 345

True north, finding, 233–234

Tuff, **33**

Ultraviolet, **350,** 379

Umbra, **256**

Unconformity, **155**

Undersea pressure, 448

Uniformitarianism, 431

Universe, **287**–301; apparent motion in, 288–289; early concepts of, 290–293; sun-centered, 292–293, 295

Upwellings, **479**

Uranium, 91, 146, 483

Urban dust domes, 411–412

Valleys: age of, 127; erosion in, 313; glaciated, 169; hanging, 170; stream slopes and, 127; widening of, 131

Van Allen radiation belts, 267

Variable stars, **310,** 312

Variables, in experiments, **64, 204**

Veins: deposits in, 29; in rocks, **28–29**

Venus, motions of, 289, 290, 291, 292

Vermiculite, 14, 65

Vertical line, **234**

Volatile, **40**

Volcanic ash, **33**

Volcanic breccia, **33**

Volcanic mountains, 162; underwater, 425

Volcanic rock, 21, **32–34,** 158–164

Volcanoes: dormant, **159;** erosion of, 161; formation of islands, 425, 511; forms of, 158; location of, 80, 160; predicting activity of, 159, 161; temperature in, 75; underwater, 425

Vortex formation, **364,** 365

Warm front, **370,** 373

Waste disposal, **90,** 94–95, 485–487; radioactive, 490

Water: adhesion to glass, 51, 52; and angle of rest, 117; capillary, 52; combined, 52; conduction by, 386, 447; density changes in, 387, 448–449, 451; drainage, 52; drinking, 483; heat absorption in, 386; microclimates in, 386–389, 392; penetration rate in soil, 50; seasonal changes in, 387–389; in soil, 46, 50, 51, 65; soil retention of, 51, 65; solubility of minerals in, 28, 445; specific heat of, 450; volume, and rate of flow, 124; volume changes in, 387; weight of, 448. See also Oceans and Seawater

Water of crystallization, 12; in changing rocks, 37; replacing, 13

Water pressure: effect on dissolving gases, 446; effect on dissolving materials, 447; underwater, 444–446, 448; and underwater explorations, 448

Water vapor: in air, 324, 331–

338; condensation of, 331, 352, 443; dew, 335; fog, 335–336; frost, 335; rain, 338, 355, 372; 462; and saturation, 332–334; temperature and, 332–334

Wavelength, 350, 351, 352; absorption of, 379–380; of energy, 500–501; infrared, 500; of ocean waves, 452

Waves, 150; on beaches, 455; deposits, 459; Doppler effect and, 309; erosion by, 451, 456–457; materials carried by, 458; origin of, 453; period of the, 452; speed of, 454; swells, 454

Weather, **348;** affected by earth's rotation, 363, 366–367; in cities, 410–412; forecasting, 368, 374–375, 396–398, 413–416, 494; fronts, 370, 372–373, 375; human's effect on, 508; instruments, 396–397, 413; maps, 374–375, 398; pollution effect on, 355; satellites, 413–416; systems, 368–373; wind, 360–367; See also Climate

Weathering, 112–120; bedding, 114; chemical 112–113; creep, 119–120; exfoliation, 114; frost, 114; landslide, 118–119

Wegener, Alfred, 431, 432, 442

Whales, 479, 480

Whirlwinds, 364, 365

"Whitings", of lakes, 509

Wind: abrasion, 140; belts, 366; causes of, 360–367; convection cells, 360–362; cyclones, 370–371; earth's rotation and, 363, 366–367; erosion by, 139–142; formation of sand dunes by, 140–142; glacial, 361; sculpturing, 140; and surface ocean currents, 464; in thunderstorms, 362, 365; tornadoes, 364, 365; transportation of sediments by, 139; vortex formation, 364; waterspouts, 365; whirlwinds, 364

Zinc, 29; deposits of, 91; in plants, 53; in seawater, 483

Zooplankton, 478–479, 487